629.
132
6
SRI

7000490863

UWE BRISTOL
WITHDRAWN
LIBRARY SERVICES

AF795612

IET CONTROL ENGINEERING SERIES 74

Eigenstructure Control Algorithms

Other volumes in this series:

UWE Bristol
FET
1 9 JAN 2012
49927
Library Services

Eigenstructure Control Algorithms

Applications to aircraft/rotorcraft handling qualities design

S. Srinathkumar

The Institution of Engineering and Technology

Published by The Institution of Engineering and Technology, London, United Kingdom

The Institution of Engineering and Technology is registered as a Charity in England & Wales (no. 211014) and Scotland (no. SC038698).

© 2011 The Institution of Engineering and Technology

First published 2011

This publication is copyright under the Berne Convention and the Universal Copyright Convention. All rights reserved. Apart from any fair dealing for the purposes of research or private study, or criticism or review, as permitted under the Copyright, Designs and Patents Act 1988, this publication may be reproduced, stored or transmitted, in any form or by any means, only with the prior permission in writing of the publishers, or in the case of reprographic reproduction in accordance with the terms of licences issued by the Copyright Licensing Agency. Enquiries concerning reproduction outside those terms should be sent to the publisher at the undermentioned address:

The Institution of Engineering and Technology
Michael Faraday House
Six Hills Way, Stevenage
Herts, SG1 2AY, United Kingdom

www.theiet.org

While the author and publisher believe that the information and guidance given in this work are correct, all parties must rely upon their own skill and judgement when making use of them. Neither the author nor publisher assumes any liability to anyone for any loss or damage caused by any error or omission in the work, whether such an error or omission is the result of negligence or any other cause. Any and all such liability is disclaimed.

The moral rights of the author to be identified as author of this work have been asserted by him in accordance with the Copyright, Designs and Patents Act 1988.

British Library Cataloguing in Publication Data
A catalogue record for this product is available from the British Library

ISBN 978-1-84919-259-0 (hardback)
ISBN 978-1-84919-260-6 (PDF)

Typeset in India by MPS Ltd, a Macmillan Company
Printed in the UK by CPI Antony Rowe, Chippenham

To Usha, Shruti, Vamshi, Vedant and Kriti
for bringing much joy to my life

Contents

Preface

Eigenstructure control entails modification of both eigenvalues and eigenvectors of a dynamic system using feedback. Since the dynamic response of a system is fully characterised by its eigenstructure, its control provides a powerful tool to remove inherent dynamic performance deficiencies in a system by using feedback control. The main objective of the book is to develop algorithms, based on *eigenstructure control theory*, and demonstrate their use in the design of *practical* flight control systems.

Some of the key desirable attributes of any design technique to succeed in the evolution of *practical* control systems are: (i) the control problem formulation must give insight into the nature of the existing performance deficiencies of the system to be controlled, in order to evolve suitable control structures to alleviate these deficiencies, and (ii) the tunable design parameters in the controller should provide good transparency of the cause and effect relation between design parameter change and consequent system performance variation. This is especially required since the design is invariably iterative in nature and re-tuning of the design parameters is inevitable. Indeed eigenstructure assignment concept enjoys these attributes and in particular is tailor-made to address improvement of flight vehicle handling qualities, which forms an important flight control law design requirement.

In general, when a new attractive control theoretical property is discovered, many computational algorithms are proposed to illustrate enhancement of different facets of system performance using the property. Further the evolution of algorithm structure for solving the problem formulation is influenced by factors such as (i) to mathematically prove the existence conditions of the control theoretical property and (ii) to adapt numerically stable linear algebra algorithms to derive the solution. In this process, the design parameters available for performance optimisation very likely lose the earlier stated attribute of transparency required for evolving good designs. Thus, a further attribute of a *practical* control system design tool is to have a set of simple-to-use, computational algorithms that produce useful engineering solutions while, as far as possible, retaining the transparency attributes. The algorithms developed in this book are driven by such a philosophy. A suite of algorithms has been developed to form a core *design tool set* to handle the flight control design problem.

The present-day powerful computers, with their extraordinary computation capability, have made performance optimisation techniques an essential element in the control design process. Nevertheless, the design parameters that need to be optimised have to originate from a well-thought-out controller structure. In this regard, eigenstructure parameters, by virtue of their direct link to system response, are ideal candidates for design optimisation. In the flight control application chapters the role of optimisation is further highlighted.

Books dealing with flight control system design have to be interdisciplinary in nature encompassing both control theory and flight dynamics. In general, the current books fall into two categories, namely: (i) books primarily originating from university research that tend to be heavily biased towards theoretical development with simple tutorial flight control examples included for completeness and (ii) books originating from experimental flight research, industry experience, etc., that highlight wide range of flight test results. In this class of books, the design process adopted to achieve the final control law design is generally not highlighted in detail. However, infusion of a new theory/design process into an already well-established methodology in the aircraft industry would require significant amount of investment (time and financial). This entails learning the new control theoretical properties and adapting them to the specific control design requirements. As the theoretical framework becomes more sophisticated, the infusion effort becomes more involved and complicated.

One of the motivations for writing this book has been to address the above-stated difficulties. In this book, in addition to developing the theoretical concepts and associated algorithms, attempt is made to develop a basic reference flight control law design process, based on *eigenstructure control theory*, to meet the *analytical handling qualities* specifications for aircraft and rotorcraft. It is hoped that this approach will facilitate quick absorption and adaptation of the new concepts and algorithms into the practical design environment.

Design of modern fly-by-wire aircraft/rotorcraft flight control systems perhaps offers the control engineer the most exciting design challenges. Many multivariable control theories, such as linear quadratic optimal control, eigenstructure control, H_∞ loop shaping, non-linear dynamic inversion and model following, usually referred to as 'modern' control techniques, have been and continue to be developed to address this design problem. Despite the large number of analytical studies and some experimental flight test demonstration programs that have been undertaken, there is still some reluctance in the aircraft industry to infuse these techniques into their control design practices. The classical control design techniques still continue to be the preferred design approach. Another objective of this book is to examine how application of eigenstructure control technique can be structured to alleviate some of these justifiable apprehensions.

The book has been written in the format of a research monograph and thus assumes the reader to have a basic background in (i) state variable control system design, (ii) linear algebra/matrix theory, (iii) flight mechanics/flight control and (iv) flight vehicle handling qualities. Some of the excellent books on these basic topics have been listed in Bibliography.

The book, being interdisciplinary in nature, should be of interest to both control and aeronautical engineers. The book should be of use to graduate/doctoral students and flight control researchers in universities and flight control engineers in the aircraft/rotorcraft industry. In particular, control research engineers may find it interesting to adapt the philosophy of formulating other multivariable control design problems, not covered in this book, as an eigenstructure assignment problem. The topics covered in this book are also well suited to teach in a professional development short course for flight control engineers.

While the application chapters are focused on flight control, it is emphasised that the control algorithms and design principles developed in this book are equally applicable to any dynamic system, such as process control, that needs feedback control. A distillation plant control problem is illustrated in Chapter 5.

Finally the author will feel his efforts have been well worth it if the book helps flight control engineers to make a judicious/objective assessment of the applicability of *eigenstructure control* in their new aircraft/rotorcraft design projects.

Acknowledgements

The contents of this book have evolved over many phases, spanning over three decades. The concept of eigenstructure assignment using feedback was discovered during my PhD research at Oklahoma State University. I would like to thank Dr R.P. Rhoten, my thesis advisor, for many thought-provoking discussions during this period. I would like to express my heartfelt gratitude to Mr J.R. Elliott, of NASA Langley Research Center (LaRC), for his unstinting encouragement and support to explore eigenstructure control concept to design flight control systems, during my tenures at LaRC as a National Research Council research associate. During this period, it is a pleasure to record my collaboration with Mr W.M. Adams, Jr., and Mr M.R. Waszak of LaRC, on the design and testing of a flutter suppression controller for a wind tunnel model (Chapter 12).

The genesis of evolving detailed case studies on aircraft and rotorcraft control, using eigenstructure modification process, emerged during my association with aircraft control law design studies at National Aerospace Laboratories (NAL), India. During this phase of book development, I would like to thank Mr Shyam Chetty of NAL and Dr G.S. Deodhare of Aeronautical Development Agency (ADA), India, for many insightful and challenging discussions (Chapters 9 and 10).

To my wife Usha, my warmest appreciation for graciously accepting my *romance* with *eigenstructure control theory* all these years and for her infinite patience and understanding during the writing of this book.

Finally, I would like to express my appreciation to the Institution of Engineering and Technology (IET), for undertaking publication of the book. My sincere thanks are due to Dr Nigel Hollingworth, Senior Commissioning Editor (*Control Series*), for his enthusiastic support of this book project. I would also like to record my appreciation to the IET *Control Series* Editors, Dr Lihua Xie and Dr S. Jagannathan, for recommending the book for publication. Thanks to Ms Lisa Reading, Commissioning Editor – Books, for her valuable support.

Chapter 1
Introduction

1.1 Multivariable system synthesis

Once a physical system has been modelled using appropriate mathematical structure, the synthesis problem is to modify the system dynamic characteristics to a more desirable form using feedback. If the modelling structure is a linear time-invariant state variable formulation, many synthesis procedures, such as quadratic optimal control, pole placement, H-infinity synthesis, non-linear dynamic inversion, model following control, etc. are available for system performance optimisation. One of the key desirable qualities of any synthesis procedure is the direct mapping of the design parameters available for modification to those of the system response parameters. While this mapping may not be always possible due to non-linear relationships, etc., it would be desirable to have at least a good visibility of the cause and effect relation between design and response parameters. For example, while the optimal control formulation is well suited for trajectory optimisation problems, difficulty is encountered in selection of the design parameters if the performance specifications are formulated in the frequency domain (as is the case in aircraft/helicopter handling qualities (HQ) design). These criteria are commonly defined as desirable locations of closed-loop system poles (eigenvalues in state variable formulation), and pole placement techniques become appropriate synthesis procedures. It is well known that if a system is controllable, arbitrary pole assignment for the closed-loop system is possible using state variable feedback [1]. Using this key property, many algorithms based on pole placement concepts have been proposed. However, none have proved to be practical design procedures. The common difficulty is not the fault of the algorithms but is really due to the inherent inability of pole locations to actually characterise the state variable responses. Further, for multi-input systems, the state feedback control law assigning a set of poles is not unique. That is, an infinite set of control laws will yield the same pole locations but can lead to radically different output responses. The reason becomes clear when one recognises that the eigenvectors play an important role in shaping the dynamic response of the system. This key concept of rationally utilising the freedom available in the feedback structure to satisfy eigenvalue as well as eigenvector specifications leads to the formulation and discovery of the eigenstructure assignment property [2,3].

1.2 Eigenstructure assignment formulations

The freedom in selection of both eigenvalue and eigenvector (eigenstructure) using feedback is now well known. There are alternative formulations available in the literature to synthesise the eigenvector for a selected eigenvalue. These can be categorised as (i) direct eigenvector elements selection [2], (ii) selection from achievable eigenvector subspaces [3] and (iii) selection by parametric representation [4]. The direct eigenvector element selection [2] enjoys the closest relation between the free design parameters and the dynamic response of the closed-loop system, a desirable property alluded to earlier. The selection of free parameters proposed in References 3 and 4 is farther removed from this direct mapping. A simple numerical example will now be presented to establish the fundamental constraint on the achievable eigenspace using state variable feedback and also briefly explain the eigenstructure assignment formulations used in References 2 and 3.

Consider the linear time-invariant multi-input controllable system

$$\dot{\mathbf{x}} = \mathrm{A}\mathbf{x} + \mathrm{B}\mathbf{u} \tag{1.1}$$

and a state feedback law

$$\mathbf{u} = \mathrm{K}\,\mathbf{x} \tag{1.2}$$

where $\mathbf{x} \in \mathrm{R}^{\mathrm{n}}$ and $\mathbf{u} \in \mathrm{R}^{\mathrm{m}}$. Then the closed-loop system matrix satisfies the relation

$$[\mathrm{A}+\mathrm{BK}]\mathrm{x} = \lambda \mathrm{x} \tag{1.3}$$

where λ and x are the desired closed-loop eigenvalue and eigenvector, respectively.

Example 1.1: Let

$$\mathrm{A} = \left[\begin{array}{ccc} 1 & 2 & 3 \\ 4 & 5 & 6 \\ \hline 1 & 1 & -1 \end{array}\right], \qquad \mathrm{B} = \left[\begin{array}{cc} 1 & 0 \\ 0 & 1 \\ \hline 0 & 0 \end{array}\right] \tag{1.4}$$

The control distribution matrix B in (1.4) is intentionally shown in a special canonical form to reveal the fundamental eigenspace assignment constraint property. This canonical form is always possible by a co-ordinate transformation, provided B is a full rank matrix.

For this example, (1.3) takes the form

$$\begin{bmatrix} 1+k_1 & 2+k_2 & 3+k_3 \\ 4+k_4 & 5+k_5 & 6+k_6 \\ 1 & 1 & -1 \end{bmatrix} \left[\begin{array}{c} z_1 \\ z_2 \\ \hline w \end{array}\right] = \lambda \left[\begin{array}{c} z_1 \\ z_2 \\ \hline w \end{array}\right], \qquad \mathrm{K} = \begin{bmatrix} k_1 & k_2 & k_3 \\ k_4 & k_5 & k_6 \end{bmatrix} \tag{1.5}$$

where the eigenvector x is shown in partitioned form as $\mathrm{x} = [z_1 \;\; z_2 \,|\, w]^{\mathrm{T}}$

Note that the third equation in (1.5) is independent of the feedback gain matrix elements. This leads to the following closed-loop eigenstructure assignment constraint relationship that is *solely* dependent on the system matrices A and B:

$$[1 \quad 1 \mid -(1+\lambda)]\begin{bmatrix} z_1 \\ z_2 \\ \hline w \end{bmatrix} = 0 \tag{1.6}$$

Equation (1.6) represents one equation in three unknowns where two elements z_1 and z_2 can be arbitrarily chosen and w can be solved using

$$w = \frac{z_1 + z_2}{-(1+\lambda)} \tag{1.7}$$

Equation (1.6) forms the basis of the algorithm development in Reference 2. An alternate way of interpreting (1.3) is given by

$$Sv = w \tag{1.8}$$

where $S = (\lambda I - A)^{-1}B$, $S \in R^{n \times m}$ and $w = Kv$. This implies that the closed-loop eigenvector lies in an m-dimensional subspace spanned by S, where m is the number of inputs. The vector w is computed for a desired eigenvector shape v_d using the least squares fit

$$w = (S^T S)^{-1} S^T v_d \tag{1.9}$$

This is a modified, widely used version of the original formulation given in Reference 3.

Returning to the example problem, if it is desired to have a decoupled eigenvector of the form $v = [0 \quad 0 \quad 1]^T$, wherein the third state variable response exhibits only the selected mode λ, then from (1.7) it is immediately clear that *only* for $\lambda = -1$, such an eigenvector shape can be assigned with z_1 and z_2 set to zero and w selected as unity. On the other hand, using the formulation (1.8), it is not possible to identify the unique mode $\lambda = -1$, which yields the desired decoupled eigenvector structure. The desired eigenvector $v_d = [0 \quad 0 \quad 1]^T$ for eigenvalue $\lambda = -1$, however, can be synthesised using (1.9), where

$$S = \begin{bmatrix} -1 & 0.5 \\ 1 & -0.5 \\ -0.3333 & 0 \end{bmatrix} \quad \text{and} \quad w = \begin{bmatrix} -3 \\ -6 \end{bmatrix} \tag{1.10}$$

Since the desired eigenvector lies in the subspace of S, the solution is exact. It is also apparent that (1.10) is computationally more involved compared with (1.7) and the design parameter vector w only indirectly synthesises the eigenvector as against the direct synthesis in (1.7).

This example reveals that the formulation in Reference 2 fully identifies both the eigenvalue and eigenvector that achieve the desired decoupled eigenstructure.

Similar additional characterisations will be further highlighted in Chapter 2. Thus, the direct eigenvector elements selection algorithm has some distinct advantages in terms of (i) complete characterisation of the achievable solution space, (ii) design visibility (especially in an interactive/iterative design process), (iii) computational simplicity, (iv) an algorithmic process that guarantees computing the feedback solution if a given eigenstructure specification indeed has a solution and (v) providing visibility to the modal robustness of the solution as the eigenstructure is being progressively synthesised. These desirable attributes of the synthesis procedure will be further highlighted in the subsequent chapters.

1.3 Algorithm development

The first part of the book focuses on development of a set of algorithms for design of control systems using the eigenstructure assignment formulation. The philosophy of algorithm development is to emphasise the advantages of direct eigenvector selection process using state feedback (Chapters 2 and 3) and output feedback (Chapter 4). The problem of selection of eigenstructure for improving system robustness to model uncertainty is addressed in Chapter 5. Finally it is shown that the concept of eigenstructure assignment can be extended to the design of modal canonical observers (Chapter 6) and model following controllers (Chapter 7).

A large number of application papers based on eigenstructure control, primarily for flight control law design, have been published over the past three decades that is indicative of the level of interest and maturity of the theory. Books by Liu and Patton [5] and Magni [6] present two other perspectives to the development of eigenstructure control theory. The book by Liu and Patton also gives an exhaustive bibliography on the first two decades of development (1975–98).

1.4 Flight control system design

As indicated in Preface, even though the analytical application papers of 'modern' multivariable control techniques to the flight control problem have been exhaustive, its infusion into the aircraft industry design practices has not been successful. Serious efforts are being made to understand and bridge this gap. The book edited by Magni *et al.* [7] presents an analytical study of application of modern control design methods. Two realistic benchmark aircraft flight control law design problems, RCAM (Research Civil Aircraft Model) and HIRM (High Incidence Research Model), have been formulated along with design objectives typically used by the industry. Control law designs for these problems, using 12 different multivariable methods, including eigenstructure control, have been analysed for their strengths and shortcomings. The outcome of the study has been mutually beneficial both to the research community and the industry. The book edited by Tishler [8] establishes the true complexity of flight control law design using actual aircraft and rotorcraft control system design case studies presented by experienced control system designers. Extensive flight test results are included to highlight the design verification process. The book edited by Pratt [9] provides a comprehensive review of the industry's practices in analysis and design of flight control systems. The importance of ground and flight-testing in the development of modern fly-by-wire aircraft flight control systems has been highlighted. The NATO report [10] is an excellent document, which examines the true problems associated with aircraft flight control law design and what modern/advanced design techniques really have to offer. A detailed study of flight test incidents exposing design deficiencies and 'lessons learned' from them have been documented. The study recommends a set of *best design practices* for the complete control law development cycle, starting from aircraft modelling to real-time piloted simulation and finally developmental flight-testing.

The success of classical control system design for multivariable control problems relies on the ability to convert the multi-loop design into a set of single-loop designs that can be optimised one at a time. There is a possibility in this methodology that the desired characteristics achieved in a previous loop closure may be lost in the subsequent loop design. This may lead to several trial/error design iterations. By contrast, in the multivariable design techniques, all loops are closed simultaneously to meet the design objectives. This is especially a desirable design feature when the system has heavy cross axes coupling. A general consensus that emerges at present is that flight control systems for conventional aircraft with small number of control effectors can be designed using classical control techniques. By virtue of the extensive experience gained in the various aircraft projects to date, the single-loop-at-a-time design process is fairly well understood and standardised. However, it is expected that the situation may change in the future, when more ambitious aircraft projects are contemplated with many control effectors, both aerodynamic and propulsive, and demanding the aircraft to operate at very high angle of attack, resulting in heavy cross axes coupling. The NASA F-18 HARV aircraft flight test program [11] is a typical example. Further in case of rotorcraft, the inter-axes coupling between the pitch, roll and yaw axes is substantial and design of full authority control laws for rotorcraft, using classical control techniques, will pose significant problems. Thus, it is reasonable to expect that in such future projects, multivariable control design techniques will play an important role in the design process. The above review clearly indicates that for multivariable design techniques to succeed in an industry environment, it is imperative that the control design process be structured to address practical implementation issues that have been highlighted in References 7–10.

1.5 Flight vehicle handling qualities design

The development of aircraft flight control systems poses some unique and difficult design challenges. The pilot flies the aircraft as a closed-loop system with *his dynamics* being part of the loop. He uses the aircraft attitudes as visual cues and accelerations as motion cues for manoeuvring the aircraft. If the pilot's dynamics were to be modelled as a variable gain and phase lag system, the overall pilot-in-the-loop stability of the system is substantially influenced by the pilot's dynamic response. If the pilot tries to control an aircraft that does not respond to his expectations, there is a very good likelihood the pilot-in-the-loop closed-loop system will become unstable. Such instabilities are usually referred to as Pilot Involved Oscillations (PIO). Thus, the primary goal of flight control system design is to provide a satisfactory pilot–vehicle interface that is usually referred to as *good* handling qualities (HQ). Unfortunately these HQ requirements are difficult to quantify due to the subjective nature of the pilot's perception to accomplish a given flying task. Substantial HQ research has been conducted over the decades to convert these pilot's subjective observations into key templates of control law parameter metrics that can be used by flight control designers to tune their systems. Extensive flight test experiments and in-flight simulation techniques [12] have been used to generate these critical *analytical* HQ metrics. A set of these HQ performance design guidelines for military aircraft is provided in MIL-HDBK-1797 [13] and for military rotorcraft in ADS-33E-PRF [14].

The pilot's flying tasks can be broadly classified into (i) large-amplitude gross acquisition manoeuvres wherein the pilot demands high accelerations from the aircraft and (ii) small-amplitude precision flying tasks such as formation flying, aerial refuelling and hover (station keeping) mode in a rotorcraft that demand intense pilot workload. It is in these intense workload tasks that the pilot can go out of synchronisation with aircraft response triggering a PIO incidence. The aircraft/rotorcraft application chapters (Chapters 9–11) address the small-amplitude precision flight control problem that can be analysed/designed using linear perturbation models of the aircraft at reference equilibrium conditions over the flight envelope and using linear control design techniques. It should be emphasised that *linear* control system design, based on *linear perturbation models*, forms a *non-trivial* part of the total flight control law design effort.

Figure 1.1 gives a broad schematic of the pilot-in-the-loop closed-loop system. The HQ design specifications contribute to the tuning of both the feedback

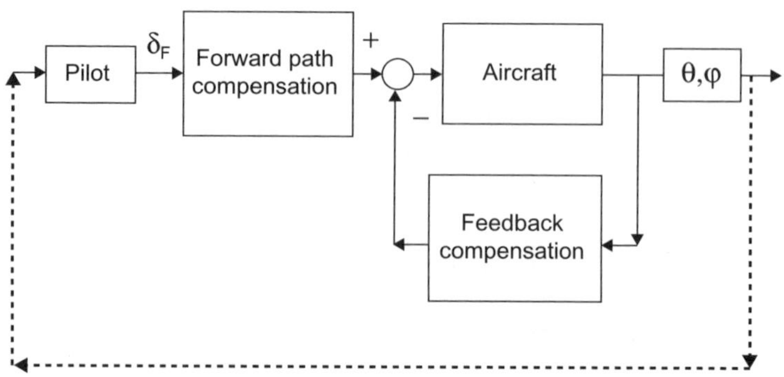

Figure 1.1 Schematic of pilot-in-the-loop flight control law

and forward path controllers. The feedback path controller design objectives are to provide adequate inner loop stability margins, good transient response, reduced inter-axes coupling and good disturbance rejection characteristics. The forward path controller primarily addresses the pilot-in-the-loop stability characteristics (loop shown dotted in Figure 1.1). The HQ specifications are provided as desirable frequency response characteristics templates for aircraft attitude/pilot control, transfer functions $\frac{\theta(s)}{\delta_F(s)}$ (pitch) and $\frac{\varphi(s)}{\delta_F(s)}$ (roll) with the pilot modelled as a pure gain element [15,16].

1.6 Flight control law design process

The second part of the book (Chapters 8–11) addresses the above-mentioned HQ design issues for small-amplitude precision flying tasks. Eigenstructure assignment algorithms developed in Chapters 2–7 will be used to derive these control laws. In these chapters, it is implicitly assumed that the reader is familiar with aspects of aircraft HQ, flight mechanics and flight control principles. The book by Hodgkinson [17] provides a lucid coverage of the development of *analytical* HQ design criteria for aircraft. The books by McLean [18] and Stevens and Lewis [19] give an excellent introduction to both flight mechanics analysis and basic flight control design. The reader is also encouraged to read books referred to earlier [7–10], to get a comprehensive view of the flight control system design challenges and additional guidelines for good control system design practices.

As pointed out earlier, the application of eigenstructure control technique is uniquely suited to improve small-amplitude HQ. The ability to map the HQ requirements as equivalent eigenvalue and eigenvector specifications makes this design process quite attractive.

Chapter 8 briefly reviews the relevant HQ metrics and flight control design guidelines, usually used by the industry, as applicable to the flight control applications of Chapters 9–11. Chapter 9 discusses in detail the development of lateral–directional HQ design process of a military aircraft. It is demonstrated how the typical classical control law structure consisting of a roll damper and stability axis yaw damper can be expanded to formulate an eigenstructure control problem using dynamic feedback, with the attendant enhancement in the design freedom. Chapter 10 addresses the control of the pitch axis of an aircraft. Both single- and multi-input designs are detailed, and the advantages of multi-input control are highlighted. The utility of model following control algorithms is also demonstrated. Chapter 11 deals with the complex rotorcraft handling problem. The decoupling of the pitch–roll–yaw axes is achieved by formulating a dynamic feedback structure, based on functional observers developed in Chapter 6. The functional observer estimates a benchmark state feedback controller. This results in a dynamic feedback compensator, using only inertial rate sensors as feedback variables. The forward path attitude command system design is based on the tunable explicit model following control algorithm developed in Chapter 7. It is shown that the combined feedback/forward path compensators result in a low dynamic order rotorcraft controller. Chapter 12, not related to HQ design, discusses the application of eigenstructure assignment techniques to control of aeroelastic modes of a flight vehicle. A wind tunnel model flutter control problem is addressed. In this example it is demonstrated how a fundamental understanding of the flutter mechanism from a control point of view leads to a very simple controller structure. The experimental verification and validation of the controller performance is also included.

Finally in the flight control application chapters, some of the aircraft state variable model parameters and flight condition data are retained in FPS units. This has been done to maintain the authenticity of the original source of information. To facilitate ready conversion of these data to SI units, a unit conversion table is included in Appendix E.

References

1. WONHAM, W.M.: 'On the pole assignment in multi-input controllable linear systems', *IEEE Trans. Automat. Control*, 1967, **AC-12**, (6), pp. 660–665
2. SRINATHKUMAR, S., and RHOTEN, R.P.: 'Eigenvalue/eigenvector assignment for multivariable systems', *Electron Lett.*, 1975, **11**, (6), pp. 124–125
3. MOORE, B.C.: 'On the flexibility offered by state feedback in multivariable systems beyond closed loop eigenvalue assignment', *IEEE Trans. Automat. Control*, 1976, **AC-21**, (5), pp. 689–692
4. FAHMY, M.M., and O'REILLY, J.: 'On eigenstructure assignment in linear multivariable systems', *IEEE Trans. Automat. Control*, 1982, **27**, (3), pp. 690–693
5. LIU, G.P., and PATTON, R.J.: *Eigenstructure Assignment for Control System Design* (John Wiley & Sons, Chichester, 1988)
6. MAGNI, J.-F.: *Robust Modal Control with a Toolbox for Use with MATLAB* (Kluwer Academic/Plenum Publishers, New York, 2002)
7. MAGNI, J.-F., BENNANI, S., and TERLOUW, J. (eds.): *Robust Flight Control – A Design Challenge*. Lecture Notes in Control and Information Sciences, vol. 224 (Springer-Verlag, Berlin, 1997)
8. TISHLER, M.B. (ed.): *Advances in Aircraft Flight Control* (Taylor & Francis, London, 1996)
9. PRATT, R. (ed.): *Flight Control Systems: Practical Issues in Design and Implementation*. IEE Control Engineering Series, no. 57 (IET, England, 2000)
10. ANONYMOUS: 'Flight Control Design – Best Practices', December 2000, NATO RTO Technical Report 29
11. BOWERS, A.H., PAHLE, J.W., WILSON, R.J., FLICK, B.C., and ROOD, R.D.: 'An Overview of the NASA F-18 High Alpha Research Vehicle', 1996, NASA TM-4772
12. SHAFER, M.F.: 'In-Flight Simulation Studies at the NASA Dryden Flight Research Facility', 1992, NASA TM 4396

13. ANONYMOUS: 'Flying Qualities of Piloted Aircraft', 1997, MIL-HDBK-1797
14. ANONYMOUS: 'Aeronautical Design Standard – Performance Specification: Handling Qualities Requirements for Military Rotorcraft', March 2000, ADS-33E-PRF
15. GIBSON, J.C.: 'Handling qualities for unstable aircraft', *Proceedings of the ICAS Congress*, paper ICAS-86-5.3.4, 1986, London, pp. 433–445
16. MUIR, E.: 'HIRM design challenge problem description' In MAGNI, J.-F., BENNANI, S., and TERLOUW, J. (eds.): *Robust Flight Control – A Design Challenge*. Lecture Notes in Control and Information Sciences, vol. 224 (Springer-Verlag, Berlin, 1997)
17. HODGKINSON, J.: *Aircraft Handling Qualities* (AIAA Education Series, AIAA, Washington, 1999)
18. MCLEAN, D.: *Automatic Flight Control Systems* (Prentice-Hall, London, 1990)
19. STEVENS, B.L., and LEWIS, F.L.: *Aircraft Control and Simulation*, 2nd edn (John Wiley & Sons, New Jersey, 2003)

Chapter 2
Eigenstructure assignment characterisation

2.1 Definitions

Consider an $n \times n$ real matrix A. The matrix has n-eigenvalues (λ_i), n-right eigenvectors (x_i) and n-left eigenvectors (y_i) such that

$$Ax_i = \lambda_i x_i, \qquad A^T y_i = \lambda_i y_i \tag{2.1}$$

Superscript 'T' in (2.1) indicates the transpose operator. Throughout the book, the right eigenvector (x_i) will be referred to as 'eigenvector'. Since x_i is a vector, it is customary to normalise it to unit length.

For a real matrix, complex eigenvalues in (2.1) appear in conjugate pairs. For numerical computations, it is convenient to use real arithmetic for complex eigenvalue/eigenvector representation. The following transformation is used for this purpose.

Consider the complex eigenvalue/eigenvector relation

$$A(\alpha + j\beta) = (a + j\beta)(u + jv) \tag{2.2}$$

where $\alpha + j\beta$ is the complex eigenvalue of A and $u + jv$ is the associated eigenvector (with $j = \sqrt{-1}$). Equation (2.2) can be solved for the real and imaginary parts as

$$A[u \mid v] = [u \mid v]\begin{pmatrix} \alpha & \beta \\ -\beta & \alpha \end{pmatrix} \tag{2.3}$$

where u and v constitute the real eigenvector pair and the 2×2 real eigenvalue matrix, $\begin{pmatrix} \alpha & \beta \\ -\beta & \alpha \end{pmatrix}$, represents the complex eigenvalue in quasi-diagonal form. The eigenvector pair is normalised to unit length such that

$$\sqrt{u^T u + v^T v} = 1 \tag{2.4a}$$

The real eigenvectors are normalised to unit length such that

$$x^T x = 1 \tag{2.4b}$$

2.2 Introduction

The need to control simultaneously the modes (eigenvalues) and the associated mode shapes (eigenvectors) to achieve acceptable dynamic response of the output variables was established in Chapter 1. Unfortunately the process of coupling the individual modes to the output variables through the entries of the eigenvector is non-linear as the following relationship shows.

Consider the linear time-invariant multi-input controllable system

$$\begin{aligned} \dot{\mathbf{x}} &= \mathrm{A}\mathbf{x} + \mathrm{B}\mathbf{u} \\ \mathbf{y} &= \mathrm{C}\mathbf{x} \end{aligned} \tag{2.5}$$

where $\mathbf{x}$ is an n-state vector, $\mathbf{u}$ is an m-control vector ($m \leq n$, $m > 1$), and A and B are matrices of compatible dimensions. Assume B is full rank.

The system response to an initial condition $\mathbf{x0}$ of the state vector, in modal canonical form, is given by

$$\mathbf{x}(t) = \sum_{i=1}^{n} e^{\lambda_i t}\, x_i y_i^T \mathbf{x0} \tag{2.6}$$

Let $\sigma_i = y_i^T \mathbf{x0}$, then (2.6) can be written as

$$\mathbf{x}(t) = \sum_{i=1}^{n} \sigma_i e^{\lambda_i t} x_i \tag{2.7}$$

Clearly from (2.6) the state variable responses are not directly controlled by the eigenvectors alone but are also influenced by the left eigenvector y_i. However, it may be required to identify certain closed-loop response structures, which need specific mode decoupling from specific response variables. For example, it is often the case that higher order systems may be considered as a coupling of lower order systems, each with its own set of acceptable performance specifications. In such a case, the eigenvectors should be selected such that the eigenvalues appropriate to one set of response variables do not unduly influence the other responses. Similarly it may be desirable to ensure that systems with both real and complex eigenvalues have eigenvectors selected so that minimal oscillatory response will arise in those responses associated with real eigenvalues. For example, in (2.7), if the kth entry of an eigenvector (x_i) corresponding to a mode (λ_i) is zero, then the kth state variable will not contain the mode λ_i, irrespective of the structure of the left eigenvector (y_i).

The time response of a selected output ($\mathbf{y_i}$) of the system (2.5) to a selected control input ($\mathbf{u_j}$) is given by

$$\mathbf{y_i}(t) = \sum_{k=1}^{n} (c_i x_k)(y_k^T b_j) \times \int_0^t e^{\lambda_k \tau}\, \mathbf{u_j}(t-\tau)\, d\tau \tag{2.8}$$

For a step command to input $\mathbf{u}_j$, (2.8) reduces to

$$\mathbf{y}_i(t) = \sum_{k=1}^{n} \frac{R_o R_i}{\lambda_k} e^{\lambda_k (t)}, \quad R_o = (c_i x_k), \quad R_i = (y_k^T b_j) \tag{2.9}$$

The magnitudes of the modal residues R_o and R_i determine the system response due to mode λ_k. Finally the transfer function can be expanded in partial fraction form as

$$\frac{\mathbf{y}_i(s)}{\mathbf{u}_j(s)} = \sum_{k=1}^{n} \frac{R_o R_i}{s - \lambda_k} \tag{2.10}$$

The modal residues provide a vital link between the system response and the right and left eigenvectors. This facilitates the determination of eigenvector shapes to achieve desired response characteristics. As indicated in Chapter 1, for multi-input systems, eigenvectors can be modified without disturbing pole locations due to the non-uniqueness of the modal control process. Equally clearly, there is not sufficient freedom to arbitrarily select the eigenvectors, except in the pathological case of an n-state, n-input system. The eigenstructure synthesis formulation to be detailed in the following sections will identify the exact freedom available for eigenvector selection.

2.3 State feedback design

The eigenvalue/eigenvector selection problem using state feedback can be formulated as follows [1, 2].

Given the system (2.5), find a state feedback law

$$\mathbf{u} = \mathbf{Kx} \tag{2.11}$$

such that the closed-loop system matrix satisfies

$$[A + BK] x_i = \lambda_i x_i \quad (i = 1\text{–}n) \tag{2.12}$$

where λ_i and x_i are the desired *distinct* closed-loop eigenvalues and eigenvectors, and the notation "i = 1–n," in (2.12) implies the sequence i = 1, 2, ..., n. In order to identify the freedom in selection in (2.12), partition (2.12) as

$$\left\{ \left[\begin{array}{c|c} A_{11} & A_{12} \\ \hline A_{21} & A_{22} \end{array} \right] + \left[\begin{array}{c} B_1 \\ \hline B_2 \end{array} \right] [K_1 \quad K_2] \right\} \left[\begin{array}{c} z_i \\ \hline w_i \end{array} \right] = \lambda_i \left[\begin{array}{c} z_i \\ \hline w_i \end{array} \right], \quad i = 1\text{–}n \tag{2.13}$$

where A_{11}, B_1 and K_1 are m × m matrices and other matrices are compatibly dimensioned. Also assume that B_1 is non-singular (if necessary by at most reordering the state variables). z_i is an m-vector with $x_i^T = [z_i^T \mid w_i^T]$. Completing the multiplications of the partitioned quantities in (2.13) yields

$$[A_{11} + B_1K_1]z_i + [A_{12} + B_1K_2]w_i = \lambda_i z_i \tag{2.14}$$

$$[A_{21} + B_2K_1]z_i + [A_{22} + B_2K_2]w_i = \lambda_i w_i \tag{2.15}$$

From (2.14), it follows that

$$K_1z_i + K_2w_i = B_1^{-1}[\lambda_i z_i - A_{11}z_i - A_{12}w_i] \tag{2.16}$$

Substituting $K_1z_i + K_2w_i$ from (2.16) into (2.15) yields the following fundamental closed-loop eigenvector constraining relationship:

$$[\lambda_i I_{n-m} - F]w_i = [G + \lambda_i H]z_i, \quad i = 1\text{–}n \tag{2.17}$$

where I_{n-m} is an (n – m)th-order identity matrix and F, G and H are matrices defined by

$$F = A_{22} - HA_{12}, \quad G = A_{21} - HA_{11}, \quad H = B_2B_1^{-1} \tag{2.18}$$

For complex conjugate pair eigenvalue $\lambda = \alpha \pm j\beta$ assignment in quasi-diagonal form, the eigenvector constraint relationship takes the form

$$\left[\begin{array}{c|c} \alpha I_{n-m} - F & -\beta I_{n-m} \\ \hline \beta I_{n-m} & \alpha I_{n-m} - F \end{array}\right]\left[\begin{array}{c} w_1 \\ \hline w_2 \end{array}\right] = \left[\begin{array}{c|c} G + \alpha H & -\beta H \\ \hline \beta H & G + \alpha H \end{array}\right]\left[\begin{array}{c} z_1 \\ \hline z_2 \end{array}\right] \tag{2.19}$$

where

$$[x_1 \mid x_2] = \left[\begin{array}{c|c} z_1 & z_2 \\ \hline w_1 & w_2 \end{array}\right] \tag{2.20}$$

Equation (2.20) implies two real eigenvectors are simultaneously synthesised.

Now (2.17) constitutes a set of (n – m) equations in the n-elements of the eigenvector. Thus, if λ_i is not an eigenvalue of F, the m-components of the eigenvector (z_i), at most (m − 1) components, can be arbitrarily chosen with one component utilised for eigenvector scaling (scaled to unit length). The remaining (n – m) components of the eigenvector (w_i) are determined using (2.17) as

$$w_i = C_iz_i; \quad C_i = [\lambda_i I_{n-m} - F]^{-1}[G + \lambda_i H], \quad i = 1\text{–}n \tag{2.21}$$

where C_i is defined as the modal coupling matrix corresponding to λ_i. Modal coupling matrix for complex eigenvalue pair is derived using (2.19). The above analysis shows that using state variable feedback, all n eigenvalues and n*(m – 1) eigenvector entries can be arbitrarily assigned. Thus, (2.17) and (2.19) provide a mapping of the non-unique parameters of K into those of the modal parameters, which can be meaningfully used for response shaping. For a feedback solution to exist, for a given eigenvalue and eigenvector assignment specification, the resulting eigenvector (modal) matrix

$$X = [x_1 \quad x_2 \quad \dots \quad x_n] \tag{2.22}$$

must be non-singular. The above analysis can now be stated as follows.

Theorem 2.1: *For the system (2.5),* n*-eigenvalues can be arbitrarily assigned. In addition,* n*-eigenvectors can be partially assigned with* (m − 1) *entries in each eigenvector arbitrarily chosen.*

From the above result, it is worth noting that the m*n arbitrary gain elements of the state feedback matrix K have been meaningfully mapped to the assignment of n-eigenvalues and n*(m – 1) eigenvector parameter selection totalling to m*n parameters.

It is also important to note that (2.17) shows the complete relationship between the desired eigenvalues and its associated eigenvectors. This relationship not only illustrates the eigenvector forms that can be attained using state feedback, but also points out immediately that certain preconceived objectives may be impossible to achieve. Later examples will illustrate this aspect. Hence, the design approach is complete in the sense that when it fails to meet certain specifications, it does so by showing that no other state feedback law could satisfy them. Thus, the analysis provides a complete 'spectral characterisation' of all the closed-loop eigenvector realisations given the triple (A, B, λ_i). With a non-singular modal matrix X and the eigenvalue matrix $\Lambda = \mathrm{Diag}(\lambda_1 \lambda_2 \dots \lambda_n)$, chosen subject to the constraint of (2.17), the closed-loop system matrix $\hat{A} = A + BK$ is uniquely determined by

$$\hat{A} = \left[\begin{array}{c|c} \hat{A}_{11} & \hat{A}_{12} \\ \hline \hat{A}_{21} & \hat{A}_{22} \end{array}\right] = X \Lambda X^{-1} \tag{2.23}$$

The inverse of eigenvector matrix X is related to the left eigenvectors as $[X^{-1}]^T = [y_1 \quad y_2 \quad \dots \quad y_n]$. The norm of the left eigenvectors $(y_i, i = 1:n)$ is indicative of the numerical ill-conditioning of the eigenvector matrix X for inversion. This property will be used in Chapter 5 to derive system robustness metrics.

The required feedback matrix $K = [K_1 \mid K_2]$ that synthesises the closed-loop system matrix $\hat{A}$ is computed using the relations

$$K_1 = B_1^{-1}[\hat{A}_{11} - A_{11}], \quad K_2 = B_1^{-1}[\hat{A}_{12} - A_{12}] \tag{2.24}$$

The eigenstructure synthesis procedure can be summarised as follows.

Algorithm 2.1:
Step 1. Compute matrices F, G and H. (See (2.18))
Step 2. For i = 1–n:

Specify desired eigenvalues λ_i, $\{\lambda_i \not\subset \lambda(F)\}$, and z_i-vector.
Solve (2.21) and synthesise the eigenvector x_i.

Step 3. Construct eigenvector matrix X.

If X is non-singular go to step 4.
Else change z_i specifications and go to step 2.

Step 4. Construct state feedback gain matrix K. (See (2.24))

Algorithm 2.1 is the simplest way of implementing the state feedback eigenstructure assignment procedure and can be used to compute engineering solutions.

However, an algorithm in Chapter 3 will be developed that will sequentially synthesise the desired eigensystem while ensuring the inverse of the modal matrix X. The numerical connotation of the non-singularity of X (with eigenvectors scaled to unit length) on the robustness of the design to system parameter variations will be discussed in Chapter 5. Finally it is to be noted that for the trivial case of m = n, the entire eigenstructure can be assigned.

The concept of eigenstructure modification process will now be illustrated using simple numerical examples.

2.4 Examples

Example 2.1: This tutorial example is used to describe the eigenvector synthesis process. Let the system (2.1) be given by

$$\begin{bmatrix} \dot{\mathbf{x}}_1 \\ \dot{\mathbf{x}}_2 \\ \dot{\mathbf{x}}_3 \end{bmatrix} = \begin{bmatrix} 1 & 0 & 1 \\ 0 & 1 & 1 \\ 3 & 4 & 1 \end{bmatrix} \begin{bmatrix} \mathbf{x}_1 \\ \mathbf{x}_2 \\ \mathbf{x}_3 \end{bmatrix} + \begin{bmatrix} 1 & 0 \\ 0 & 1 \\ 1 & 1 \end{bmatrix} \begin{bmatrix} \mathbf{u}_1 \\ \mathbf{u}_2 \end{bmatrix} \tag{2.25}$$

and the control law $\mathbf{u} = \mathbf{Kx}$ to be selected so that

(i) Eigenvalues are at $\lambda_1 = -1$, $\lambda_2 = -2$ and $\lambda_3 = -3$.
(ii) The responses must be decoupled such that $\mathbf{x}_1$ should exhibit only e^{-2t} transients, $\mathbf{x}_2$ only e^{-3t} transient and $\mathbf{x}_3$ only e^{-t} transient.

The initial condition response for the closed-loop system in terms of modal matrix X can be expressed as

$$\mathbf{x}(t) = X e^{\Lambda t} X^{-1} \mathbf{x}(0) \tag{2.26}$$

or in terms of the individual components as

$$\begin{bmatrix} \mathbf{x}_1(t) \\ \mathbf{x}_2(t) \\ \mathbf{x}_3(t) \end{bmatrix} = \begin{bmatrix} \mathbf{x}_{11} & \mathbf{x}_{12} & \mathbf{x}_{13} \\ \mathbf{x}_{21} & \mathbf{x}_{22} & \mathbf{x}_{23} \\ \mathbf{x}_{31} & \mathbf{x}_{32} & \mathbf{x}_{33} \end{bmatrix} \begin{bmatrix} \alpha_1 e^{\lambda_1 t} \\ \alpha_2 e^{\lambda_2 t} \\ \alpha_3 e^{\lambda_3 t} \end{bmatrix} \tag{2.27}$$

where $\alpha = X^{-1}\mathbf{x}(0)$ and $\alpha = [\alpha_1 \quad \alpha_2 \quad \alpha_3]^T$. The design specifications set forth above will be satisfied if X is of the form

$$X = \begin{bmatrix} 0 & 1 & 0 \\ 0 & 0 & 1 \\ 1 & 0 & 0 \end{bmatrix} \tag{2.28}$$

For the system (2.25), the matrices of the eigenvector constraint relationship (2.18) are

$$F = -1; \qquad G = [2 \quad 3]; \qquad H = [1 \quad 1]$$

and from (2.17) we have

$$(\lambda + 1)w = [2 + \lambda \quad 3 + \lambda]\begin{bmatrix} z_1 \\ z_2 \end{bmatrix} \tag{2.29}$$

where λ, z_1 and z_2 are the arbitrary modal parameters that can be chosen and w is computed from (2.21) to complete the eigenvector synthesis.

For $\lambda_1 = -1$, choosing z_1 and z_2 as zero results in w to be arbitrarily chosen since the eigenvalue to be assigned coincides with matrix F. Let $w = 1$. This results in the eigenvector $x_1 = [0 \quad 0 \quad 1]^T$. It is to be noted that if the eigenvalue to be assigned coincides with that of matrix F, the eigenvector has a unique structure wherein z is a null matrix and w is the eigenvector of the F matrix. This coincident eigenvalue case will be further discussed in the next chapter in the context of synthesising mode decoupled eigenvector structures. Continuing the synthesis, for $\lambda_2 = -2$, choosing $z_2 = 1$ and $z_1 = 0$ results in $w = 0$ resulting in the eigenvector $x_2 = [1 \quad 0 \quad 0]^T$. By inspection we can see that the two eigenvectors synthesised are linearly independent. For higher order systems an algorithmic process (to be described in Chapter 3) is required. Finally for $\lambda_1 = -3$, choosing $z_1 = 0$ and $z_2 = 1$ results in $w = 0$. The resultant eigenvector $x_3 = [0 \quad 1 \quad 0]^T$. The eigenvector structures synthesised for x_2 and x_3 are also special cases wherein the z-vector is in the null space of matrix $(G + \lambda H)$. This structure will also be discussed in Chapter 3 for synthesising mode-decoupled eigenstructure. Thus, for the eigenvalue matrix $\Lambda = \text{Diag}(-1, -2, -3)$, we have the eigenvector matrix and its inverse as

$$X = \begin{bmatrix} 0 & 1 & 0 \\ 0 & 0 & 0 \\ 1 & 0 & 0 \end{bmatrix}, \qquad X^{-1} = \begin{bmatrix} 0 & 0 & 1 \\ 1 & 0 & 0 \\ 0 & 1 & 0 \end{bmatrix} \tag{2.30}$$

The closed-loop system matrix from (2.23) is

$$\hat{A} = \begin{bmatrix} -2 & 0 & 0 \\ 0 & -3 & 0 \\ 0 & 0 & -1 \end{bmatrix} \tag{2.31}$$

The state feedback gain matrix, which assigns the above eigenstructure, can now be computed using (2.24) as

$$K = \begin{bmatrix} -3 & 0 & -1 \\ 0 & -4 & -1 \end{bmatrix} \tag{2.32}$$

It is seen that from (2.31) the desired full mode decoupling has been achieved.

In summary, the proposed synthesis procedure provides transparency on the type of mode coupling that may accrue for a given pole assignment. The 'modal coupling matrices' C_i of (2.21) play a key role in characterising the achievable eigenvector shapes for a desired eigenvalue specification. In practical control system design, eigenvalue specifications are seldom required to be a specific value but are required to lie in a region in the complex plane. Thus, it is possible to search for the best possible eigenvector shape in this assignable region through the characterisation provided by the modal coupling matrices.

Example 2.2: Another simple tutorial example will now be presented to establish the need for an algorithmic process to accomplish the eigenstructure synthesis. It will illustrate that certain procedural idiosyncrasies, as simple as selecting the sequence of synthesising the eigenvectors, may induce problems that could be resolved by inspection for low-order systems, but do require a well-defined algorithm for more complex systems.

Let the system be defined as

$$\begin{bmatrix} \dot{x}_1 \\ \dot{x}_2 \\ \dot{x}_3 \end{bmatrix} = \begin{bmatrix} 1 & 2 & 3 \\ 4 & 5 & 6 \\ -1 & -1 & 0 \end{bmatrix} \begin{bmatrix} x_1 \\ x_2 \\ x_3 \end{bmatrix} + \begin{bmatrix} 1 & 0 \\ 0 & 1 \\ 0 & 0 \end{bmatrix} \begin{bmatrix} u_1 \\ u_2 \end{bmatrix} \tag{2.33}$$

It is required to assign eigenvalues at $\lambda_1 = -1$, $\lambda_2 = -2$ and $\lambda_3 = -1$, Further it is also required that the modal matrix have the structure

$$X = \begin{bmatrix} 1 & -1 & -1 \\ 0 & 1 & 1 \\ w_1 & w_2 & w_3 \end{bmatrix} \tag{2.34}$$

The w-elements are computed based on pole specifications. For this example from (2.18) the respective matrices are $F = 0$, $G = [-1 \quad -1]$ and $H = [0 \quad 0]$. From (2.17)

$$w = \frac{-(z_1 + z_2)}{\lambda} \tag{2.35}$$

It is seen that the eigenvectors for λ_2 and λ_3 are identical with w_2 and w_3 being zero. Hence, the modal matrix is singular and the desired assignment is not possible. For this simple example, examining the constraint equation (2.35) would suggest appropriate perturbation of λ and/or z_1 and z_2 are required to ensure a non-singular modal matrix. For a higher order problem, a well-defined algorithm will be needed. Such an algorithm will be developed in Chapter 3, and this problem will be re-examined as an illustration of its utility.

2.5 Summary

In this chapter, the eigenstructure synthesis using state variable feedback has been analysed. It is shown that the solution scheme is simple and just involves solving a set of linear equations to synthesise each eigenvector for a specified eigenvalue. The modal characterisation elegantly maps the n*m arbitrary feedback gain parameters to the arbitrary selection of n-eigenvalues and n*(m – 1) eigenvector elements (total of n*m parameters). Since the system dynamic response is fully characterised by its eigenstructure, the direct control of the eigenstructure as discussed in this chapter provides a useful basis for feedback control design. The special mode decoupling eigenvector structures assignable, as illustrated in Example 2.1, demonstrates the unique advantage of the direct eigenvector element selection formulation alluded to in Chapter 1.

References

1. SRINATHKUMAR, S.: ‘Spectral characterization of multi-input dynamic systems’, PhD thesis, Oklahoma, Oklahoma State University, May 1976
2. SRINATHKUMAR, S.: ‘Modal Control Theory and Application to Aircraft Lateral Handling Qualities Design’, 1978, NASA TP 1234

Chapter 3
Eigenstructure synthesis algorithm

3.1 Introduction

The development of the eigenstructure synthesis algorithm in this chapter is primarily based on Reference 1. As discussed in Chapter 2, the central question in the eigenstructure assignment is the guaranteed generation of a non-singular modal matrix satisfying the eigenvector constraint relation (2.21) given by

$$w_i = C_i z_i; \qquad C_i = [\lambda_i I_{n-m} - F]^{-1}[G + \lambda_i H], \quad i = 1\text{–}n \tag{3.1a}$$

where the vector z_i is the designer-specified part of the eigenvector and w_i is the computed part of the eigenvector, thus forming the eigenvector as $x_i^T = [z_i^T \mid w_i^T]$. This eigenvector constraint can also be written as

$$R_i x_i = 0; \qquad R_i \in R^{(n-m)\times n} \tag{3.1b}$$

where $R_i = [G + \lambda_i H \mid F - \lambda_i I_{n-m}]$ and $x_i^T = [z_i^T \mid w_i^T]$.

Examination of (3.1b) reveals that eigenvector x_i is orthogonal to the (n − m)-dimensional subspace defined by R_i. In other words, the eigenvector spans an m-dimensional subspace. Thus, the synthesis problem reduces to selecting a non-singular set of n-eigenvectors with one vector included from an eigen-subspace associated with each eigenvalue. System controllability assures the existence of such a set, and for the multi-input case, it is an infinite set. However, this is an inefficient way to synthesise, since the designer loses direct control of the selection of the arbitrary elements of the eigenvector (z_i). Also any algorithm would be computationally intractable since it would involve pairing n vectors from a set of n*m vectors until a non-singular set results.

Alternatively, if the z-vectors are chosen arbitrarily, then it is important to keep track of the linear independence of the eigenvectors as they are sequentially generated. It would appear that an easy way to accomplish this would be to construct the projector of the subspace spanning the eigenvectors already synthesised in the sequence as

$$P^{(k-1)} = N[N^T N]^{-1} N^T \tag{3.2}$$

where $N = [x_1 x_2 \ldots x_{k-1}]$ are the first (k − 1) linearly independent eigenvectors, and selecting x_k such that

$$P^{(k-1)} x_k \neq x_k \tag{3.3}$$

Unfortunately this procedure is potentially susceptible to the generation of a singular set of eigenvectors since it is quite likely that a combination of eigenvalues and eigenvectors specification may not have a solution. This is exactly what occurred in Example 2.2. However, it is possible to detect the occurrence of such a situation, and by slightly relaxing the eigenvalue (λ) and/or the corresponding design vector (z), it is possible to synthesise a non-singular modal matrix. The degree of perturbation, which determines the ill-conditioning of the modal matrix to inversion, has implications of the robustness of the solution to system parameter perturbation. This aspect will be discussed in detail in Chapter 5. An algorithm will now be presented that incorporates the condition (3.3) without having to explicitly compute the projection matrices $P^{(k)}$ and also allow maximal flexibility in the selection of the design vector (z). It is assumed that the eigenvalues to be assigned are distinct. The algorithm will be described for assigning real eigenvalues, and procedures to assign complex pair of eigenvalues will be indicated.

3.2 Eigenstructure synthesis

An algorithm is developed below that forms a *constructive proof* for Theorem 2.1. The algorithm establishes the conditions required for a solution to exist for a specified eigenstructure (λ_i, z_i: i = 1–n).

Algorithm 3.1: The following notations will be used in the algorithm presentation:

1. e_r is an n-vector with rth entry equal to unity and all other entries zero.
2. $x_k^T = [z_k^T \mid w_k^T]$ is the kth eigenvector, and z_k is a designer-specified m-vector.
3. $x_k^{(k)} = Q^{(k-1)} x_k$, $Q^{(0)} = I_n$ and $Q^{(i)}, i \neq 0$ is defined in (6).
4. $x_{rk}^{(k)} = \sigma_k$ is the rth entry of $x_k^{(k)}$.
5. $M^{(k)}$ is an (n × n) deflation matrix of the form

$$M^{(k)} = \left[\begin{array}{c|c|c} I_{r-1} & m_a^{(k)} & 0 \\ \hline 0 & 1 & 0 \\ \hline 0 & m_b^{(k)} & I_{n-r} \end{array}\right] \tag{3.4}$$

where the vectors $m_a^{(k)}$ and $m_b^{(k)}$ are computed by a deflation technique as given in Algorithm 3.2 for a real eigenvector. Similar procedure can be written for real eigenvector pair corresponding to complex eigenvalues. Further,

$$M^{(k)}x_k^{(k)} = \sigma_k e_r, \quad r \in \{\Delta^{(k)}\} \tag{3.5}$$

where $\Delta^{(k)}$ is a subset of integers $\{1, 2, \ldots, n\}$ containing the pivots not already used in the construction of the matrices $M^{(1)}, M^{(2)}, \ldots, M^{(k-1)}$ and $\Delta^{(1)}$ is the complete set $\{1, 2, \ldots, n\}$. Note $M^{(k)}$ can be constructed if and only if $\sigma_k \neq 0$.

6. $$Q^{(k-1)} = M^{(k-1)}M^{(k-2)} \cdots M^{(1)} \tag{3.6}$$

Algorithm 3.1 proceeds as follows:

Step 1. For $k = 1 - n$, do steps 2–5.
Step 2. For $\lambda = \lambda_k$ and $\lambda_k \not\subset \lambda(F)$
(for $\lambda_k \subset \lambda(F)$; see section 3.4, case 2)

Compute

$$C_k = [\lambda_k I_{n-m} - F]^{-1}[G + \lambda_k H] \tag{3.7}$$

Step 3. For some $r \in \{\Delta^{(k)}\}$

(a) Compute

$$[g_r^{(k)}]^T = f_r^{(k-1)} + h_r^{(k-1)} C_k \tag{3.8}$$

where $[f_r^{(k-1)} \mid h_r^{(k-1)}]$ is the rth row of $Q^{(k-1)}$ and $g_r^{(k)}$ is an m-vector.

(b) Compute

$$\sigma_k = [g_r^{(k)}]^T z_k \tag{3.9}$$

where z_k is the arbitrarily specified design vector.

(i) If $\sigma_k \neq 0$, compute w_k (3.1a) and $M^{(k)}$ (3.4) and go to step 1.
(ii) If $\sigma_k = 0$, select another $r \in \{\Delta^{(k)}\}$ and return to step 3(a).
(iii) If $\sigma_k = 0$, for all $r \in \{\Delta^{(k)}\}$, go to step 4.

Step 4. For some $r \in \{\Delta^{(k)}\}$

(a) If $g_r^{(k)} \neq 0$ (No Solution)
Perturb $z_k = z_k + \delta z_k$ to make $\sigma_k \neq 0$.
Compute w_k and $M^{(k)}$ and go to step 1.

(b) If $g_r^{(k)} = 0$, select another $r \in \{\Delta^{(k)}\}$ and repeat step 4(a).

(c) If $g_r^{(k)} = 0$, all $r \in \{\Delta^{(k)}\}$ (No Solution)

Perturb $\lambda_k = \lambda_k + \delta\lambda_k$ and go to step 2.

Step 5. Compute the feedback gain matrix K (2.24).

In order to clearly see that the kth linearly independent eigenvector x_k can be synthesised provided $\sigma_k \neq 0$, assume without loss of generality that the first $(k-1)$ eigenvectors are generated with the pivots $r = \{1, 2, \ldots, k-1\}$. Then these vectors are transformed into the canonical form under $Q^{(k-1)}$ as

$$Q^{(k-1)}[x_1 : x_2 : \cdots x_{k-1}] = \frac{\mathrm{Diag}(\sigma_1, \sigma_2, \cdots, \sigma_{k-1})}{0} \tag{3.10}$$

with

$$\sigma_i \neq 0; \quad i = 1-(k-1)$$

and the projector spanning the subspace of these transformed eigenvectors has the simple form

$$P^{(k-1)} = \left[\begin{array}{c|c} I_{k-1} & 0 \\ \hline 0 & 0 \end{array}\right] \tag{3.11}$$

Now choosing the kth eigenvector so that its transformed eigenvector $x_k^{(k)} = Q^{(k-1)}x_k$ causes $\sigma_k \neq 0$; $r \in \{k, k+1, \ldots, n\}$, ensures the linear independence of x_k since (3.3) is clearly satisfied.

Remarks:

1. Algorithm 3.1 can be directly extended to assign complex pair eigenvalues in quasi-diagonal form by noting that two real eigenvectors are synthesised in one iteration. Further to ensure linear independence between these vectors, the test condition $\sigma_k \neq 0$ in step 3(b)(i) gets modified to ensure the non-singularity of a 2×2 matrix Σ_k. This matrix is constructed by selecting two rows of the transformation $Q^{(k-1)}$ and constructing the transformation similar to (3.5) corresponding to the real eigenvector pair $[x_k \mid x_{k+1}]$ associated with the complex pair eigenvalues.
2. The iterative procedure in step 3(b)(ii) attempts to meet exact eigenvalue/eigenvector specification by searching for an available pivot index for which $\sigma_k \neq 0$. In step 4(a) exact eigenvalue specification is met while perturbing the eigenvector specification to meet the linear independence of the eigenvectors. The test in step 4(c) indicates that perturbation of both eigenvalue and eigenvector specification is warranted to meet the eigenvector linear independence requirement.
3. Since the matrices $M^{(k)}$ have only one non-trivial column, co-ordinate transformations in (3.6) reduce to simple vector multiplications. Further the inverse of the modal matrix (X) required for computing the feedback gain matrix K is easily evaluated by noting that $Q^{(n)}X$ has the general form

$$Q^{(n)}X = \mathrm{Diag}[\sigma_1, \sigma_2, \ldots, \sigma_n]L \tag{3.12}$$

where L is an elementary column permutation matrix dependent on the sequence of generating the pivot indices $r \in \{\Delta^{(k)}\}$ in steps 3 and 4. The scalars σ_i in (3.12) are the non-zero elements of (3.5). The eigenvector matrix inverse (X^{-1}) is readily computed from (3.12). It is also worth noting that the determinant of X is related to σ_i as

$$|\text{Det}[X]| = \left|\prod_{i=1}^{n} \sigma_i\right| \tag{3.13}$$

This is due to $|\text{Det } Q^{(n)}|$ and $|\text{Det } L|$ being unity. The symbol $|x|$ indicates absolute value. Thus, the numbers σ_i provide a good measure of the linear independence between the normalised eigenvectors x_i. If complex eigenvalues are assigned, the equivalent σ in (3.13) is given by the determinant of the 2×2 matrix Σ_k defined in Remark 1 above. The role of the determinant as a system robustness metric will be explored in Chapter 5.

4. A noteworthy feature of Algorithm 3.1 is that the eigenvectors do not explicitly undergo any change in the sequence of transformations (3.6). This keeps the mode-coupling characteristics transparent during synthesis, a very desirable feature for an interactive computer-aided design.

Algorithm 3.2:

```
function [M] = deflate (x, k, n)
% x is an n-vector
% k is a pivot from the set Δ = {1, 2, ..., n} such that x(k) ≠ 0
M = I_n;      % Initialise
    for i = 1:n
            If i ≠ k
            M(i, k) = −(x(i)/x(k));
            end
    end
return
```

3.3 Example

Example 3.1: Example 2.2 will now be used to highlight the features of Algorithm 3.1. Applying the algorithm step by step to the system (2.33) yields the following synthesis sequence.

Mode 1: $\lambda_1 = -1,\ z = [1 \quad 0]^T$

$$\Delta^{(1)} = \{1,2,3\};\ \text{choose } r = 3. \text{ Then } C_1 = [1 \quad 1];\ g_1^{(1)} = [1 \quad 1]^T;\ \sigma_1 = 0.7071.$$

Since $\sigma_1 \neq 0$, the eigenvector is synthesised as $x_1 = [0.7071 \quad 0 \quad 0.7071]^T$. The transformation $Q^{(1)}$ becomes

$$Q^{(1)} = \begin{bmatrix} 1 & 0 & -1 \\ 0 & 1 & 0 \\ 0 & 0 & 1 \end{bmatrix}$$

Mode 2: $\lambda_2 = -2,\ z = [-1 \quad 1]^T$

$\Delta^{(2)} = \{1, 2\}$; choose r = 2. Then $C_2 = [0.5 \quad 0.5]$; $g_2^{(2)} = [0 \quad 1]^T$; $\sigma_2 = 0.7071$.

Since σ_2 is not zero, the eigenvector is synthesised as $x_2 = [-0.7071 \quad 0.7071 \quad 0]^T$. The transformation $Q^{(2)}$ becomes

$$Q^{(2)} = \begin{bmatrix} 1 & 1 & -1 \\ 0 & 1 & 0 \\ 0 & 0 & 1 \end{bmatrix}$$

Mode 3: $\lambda_3 = -1,\ z = [-1 \quad 1]^T$

$\Delta^{(3)} = \{1\}$. Thus, r = 1. Then $C_3 = [1 \quad 1]$; $g_3^{(3)} = [0 \quad 0]^T$; $\sigma_3 = 0$.

Since $\sigma_3 = 0$, the third eigenvector cannot be synthesised. Further since $g_3^{(3)}$ is a null vector, the eigenvalue specification cannot be met. This is obvious since the basis vector spanning the two-dimensional subspace corresponding to the eigenvalue $\lambda = -1$ is already spanned by the first two mode assignments. Thus, a second eigenvector corresponding to $\lambda = -1$ cannot be synthesised as revealed by the vector $g_3^{(3)}$. This implies λ_3 must be perturbed. Let $\lambda_3 = (-1 - \varepsilon)$, $\varepsilon > 0$. Then

$$C_3 = \left[\frac{1}{1+\varepsilon} \quad \frac{1}{1+\varepsilon}\right]; \quad g_3^{(3)} = \left[\frac{-\varepsilon}{1+\varepsilon} \quad \frac{-\varepsilon}{1+\varepsilon}\right]^T; \quad \text{and again } \sigma_3 = 0$$

This implies the third eigenvector still cannot be synthesised. In this case since $g_3^{(3)}$ is not the null vector, the eigenvalue specification can be met but the eigenvector specification requires a perturbation. Let $z_3^T = [-1 \quad 1+\delta]$, $\delta \neq 0$. Then

$$\sigma_3 = \frac{-\varepsilon\delta}{1+\varepsilon} \neq 0$$

and the eigenvector can be synthesised as

$$x_3 = \left[-1 \quad 1+\delta \quad \frac{\delta}{1+\varepsilon}\right]$$

It should be noted that the eigenvector x_3 is not scaled to unit length. The closed-loop eigenvalue and eigenvector matrices are

$$\Lambda = \mathrm{Diag}[-1, \quad -2, \quad -1-\varepsilon] \quad \text{and} \quad X = \begin{bmatrix} 0.7071 & -0.7071 & -1 \\ 0 & 0.7071 & 1+\delta \\ 0.7071 & 0 & \dfrac{\delta}{1+\varepsilon} \end{bmatrix}$$

Further

$$|\mathrm{Det}[X]| = |\sigma_1\sigma_2\sigma_3| = 0.5\left|\frac{\varepsilon\delta}{1+\varepsilon}\right|$$

since $\sigma_1 = \sigma_2 = 0.7071$. Thus, in this case setting a tolerance on the value of σ_3 would directly control the numerical ill-conditioning of X and consequently influence the choice of the perturbations ε and δ.

Continuing with the example, choosing $\varepsilon = 0.1$ and $\delta = 0.5$, we have $\Delta^{(3)} = \{1\}$. Thus, $r = 1$. Then

$$C_3 = [0.9091 \quad 0.9091]; \quad g_3^{(3)} = [0.0909 \quad 0.0909]^T; \quad \sigma_3 = 0.0244.$$

Since σ_3 is not zero, the eigenvector is synthesised as

$$x_3 = [-0.5379 \quad 0.8068 \quad 0.2445]^T$$

The transformation $Q^{(3)}$ becomes

$$Q^{(3)} = \begin{bmatrix} 1 & 1 & -1 \\ -33 & -32 & 33 \\ -10 & -10 & 11 \end{bmatrix}$$

The closed-loop eigenvalue and eigenvector matrices are

$$\Lambda = \text{Diag}[-1, \quad -2, \quad -1.1] \quad \text{and} \quad X = \begin{bmatrix} 0.7071 & -0.7071 & -0.5379 \\ 0 & 0.7071 & 0.8068 \\ 0.7071 & 0 & 0.2445 \end{bmatrix}$$

The numerical condition of the eigenvector matrix is

(i) $|\text{Det}[X]| = |\sigma_1\sigma_2\sigma_3| = 0.0122$.
(ii) Condition number $\kappa_2(X) = 161.7462$.

The state feedback gain matrix is given by

$$K = \begin{bmatrix} -32.8 & -31.8 & 27.8 \\ 25.7 & 22.7 & -35.7 \end{bmatrix}$$

It is also interesting to note that if the sequence of assignments of the modes were changed to $\lambda_1 = -1$, $\lambda_2 = -1$ and $\lambda_3 = -2$, then two eigenvectors corresponding to $\lambda = -1$ could be synthesised as

$$[x_1 \mid x_2] = \begin{bmatrix} 0.7071 & -0.7071 \\ 0 & 0.7071 \\ 0.7071 & 0 \end{bmatrix}$$

This is possible since the eigenvectors span an m-dimensional ($m = 2$) subspace.

3.4 Special eigenvector structures

In aircraft/helicopter control system design, it is generally required to reduce the inter-axes coupling using feedback. This design requirement results in choosing individual eigenvector shapes of the aircraft modes to dominantly exhibit selected response variables. This mode decoupling design approach also has additional benefit in that the modal robustness of the feedback system improves. The following eigenvector decoupling structures find use in such design efforts (used already for illustration in Example 2.1):

Case 1. Selection of design vector z
For the state space system (2.5), if $(n - m) < m$, then for a specified eigenvalue λ, if the z-vector can be chosen to be in the null space of the matrix $[G + \lambda H]$ then (2.17) has a solution if and only if w-vector is the null vector. This results in the decoupled eigenvector $x^T = [z^T \mid 0]$.

Case 2. Selection of eigenvalue λ
If the closed-loop eigenvalue (λ) is chosen to coincide with an eigenvalue of the matrix F, then (2.17) has a solution if and only if z-vector is a null vector. The w-vector satisfies the relation, $Fw = \lambda w$. That is, w-vector is an eigenvector of the matrix F corresponding to λ. This results in the decoupled system eigenvector $x^T = [0 \mid w^T]$. Note that the algorithm outlined in section 3.2 (step 2) required that the specified eigenvalue should not coincide with that of the matrix F. Algorithm 3.1 can easily be extended to handle the case of the assigned mode coinciding with the eigenvalues of matrix F in light of the discussions in this section.

3.5 Assignment of repeated eigenvalues

In all the discussions so far it has been assumed that the eigenvalues to be assigned are distinct. Multiple eigenvalue assignment by feedback is generally of academic interest since, as already emphasised earlier, exact pole assignment is seldom required in practical control system design. However, from (3.1b) it is seen that the eigenvectors span an m-dimensional subspace and thus theoretically m-repeated roots could be assigned to the closed-loop system. In Example 3.1 it was also illustrated that two eigenvectors corresponding to $\lambda = -1$ could be assigned if they were the first two vectors to be synthesised.

3.6 Summary

In this chapter an algorithm has been detailed for eigenstructure assignment. The algorithm sequentially constructs the specified eigenstructure while allowing maximal flexibility in the specification of eigenvalues/eigenvectors and assures the generation of an n-dimensional eigenspace required for the solution. The numerical condition of the eigenstructure being synthesised, which is closely related to the robustness of the resulting solution, is transparent to the designer, through the σ-parameters, as the eigenvectors are sequentially synthesised. The simplicity of the algorithm makes it well suited for computer implementation as an iterative/interactive design tool for multivariable control system synthesis.

Reference

1. SRINATHKUMAR, S.: 'Spectral characterization of multi-input dynamic systems', PhD thesis, Oklahoma, Oklahoma State University, May 1976

Chapter 4
Eigenstructure assignment by output feedback

4.1 Introduction

The eigenstructure assignment using state variable feedback was analysed in detail in Chapters 1 and 2. It is interesting to examine the extent of eigenstructure assignment that is possible if only output feedback is employed (number of outputs less than the number of states). At the outset, the pole placement problem with output feedback has received much attention by researchers over the years. In the state feedback case, all system poles can be arbitrarily assigned if the system is controllable. Unfortunately such a definitive statement in case of output feedback continues to be elusive. However, in Reference 1 it is shown that, given a controllable–observable system, consisting of n-states, m-controls and r-outputs, min(n, m + r − 1) eigenvalues can 'almost' always be arbitrarily assigned using output feedback, where n, m and r are the number of states, inputs and outputs, respectively. Further, as a direct consequence, it is shown that by augmenting the system with a dynamic compensator of order $q = (n - m - r + 1)$, all the augmented system poles can be arbitrarily assigned. Extensions in Reference 2 show that for a class of systems where $r > \mu$ and $m \geq \nu$, arbitrary eigenvalues can 'almost' always be assigned, if $n < (m + r + \nu - 1)$. Here ν and μ are the controllability and observability indices, respectively. In Reference 3 also it is shown that min(n, m + r − 1) arbitrary eigenvalues can be assigned for 'almost' all pairs of (B, C) matrices and a counterexample is also provided to emphasise that for some systems this upper bound of pole assignment is not achievable. In Reference 4 it is shown that if $m^*r \geq n$, the system is pole assignable provided the gain matrix is allowed to be complex. Focus of current research has been mainly to characterise the pole assignment by output feedback for the case $m^*r \geq n$ [5,6]. Eigenstructure assignment using output feedback has also been characterised in References 7 and 8. Finally a survey paper on the complexity of the output feedback problem is given in Reference 9.

In this chapter, eigenstructure assignment possibility using output feedback, which is consistent with the concepts developed in Chapters 2 and 3, is now presented in detail. An algorithm to assign $(m + r - 1)$ arbitrary eigenvalues and $(r - 1)$ eigenvectors (partially) is also described and its features illustrated with tutorial examples. Finally the structure of a dynamic compensator for the case $n > (m + r - 1)$ is also illustrated. The algorithm development is primarily based on Reference 8.

4.2 Problem formulation

Consider a linear, time-invariant, multivariable controllable and observable system:

$$\dot{\mathbf{x}} = \mathbf{A}\mathbf{x} + \mathbf{B}\mathbf{u}$$
$$\mathbf{y} = \mathbf{C}\mathbf{x} \tag{4.1}$$

where **x** is an n-vector, **u** is an m-vector, **y** is an r-vector, m, r < n, and B and C are full rank, and further assume m, r > 1 (multi-input/multi-output). For notational simplicity the system (4.1) will be referred to by the triple {C, A, B}. The problem is to find a control law of the form

$$\mathbf{u} = \mathbf{K}\mathbf{y} \tag{4.2}$$

to assign arbitrary eigenvalues to the closed-loop system matrix (A + BKC).

4.2.1 Assignment of max(m, r) eigenvalues

The closed-loop system matrix (A + BKC) after applying feedback (4.2) satisfies the relation

$$[A + BKC]x_i = \lambda_i x_i \tag{4.3}$$

where λ_i and x_i are the desired eigenvalues and eigenvectors. In order to identify the freedom in selection in (4.3), partition (4.3) as

$$\left\{\left[\begin{array}{c|c} A_{11} & A_{12} \\ \hline A_{21} & A_{22} \end{array}\right] + \left[\begin{array}{c} B_1 \\ \hline B_2 \end{array}\right] KC\right\}\left[\begin{array}{c} z_i \\ \hline w_i \end{array}\right] = \lambda_i \left[\begin{array}{c} z_i \\ \hline w_i \end{array}\right], \quad i = 1\text{–}n \tag{4.4}$$

where A_{11} and B_1 are m × m matrices and other matrices are compatibly dimensioned. Also assume that B_1 is non-singular (if necessary by at most reordering the state variables). z_i is an m-vector with $x_i^T = [z_i^T : w_i^T]$. Completing the multiplications of partitioned matrices and matrix algebra manipulations similar to (2.13)–(2.16), the following constraint equations result for real eigenvalue assignment:

$$[\lambda_i I_{n-m} - F]w_i = [G + \lambda_i H]z_i \tag{4.5}$$

$$[A_1 + B_1 KC]x_i = \lambda_i z_i \tag{4.6}$$

where the matrices F, G and H are as defined in (2.18) and $A_1 = [A_{11} \mid A_{12}]$. Derivation of constraint relations to complex eigenvalue pair assignment is similar to those developed in Chapter 2. Now (4.5) constitutes a set of (n − m) equations in the n-elements of the eigenvector. Thus, if λ_i is not an eigenvalue of F, of the m-components of the eigenvector (z_i), at most (m − 1) components can be arbitrarily chosen with one component utilised for eigenvector scaling (scaled to unit length). The remaining (n − m) components of the eigenvector (w_i) are determined using (4.5). Examination of (4.6) (which is an additional constraint compared with the state feedback case) reveals that at most r-eigenvalues and r-eigenvectors satisfying (4.5) can be assigned to the system (4.1) by the feedback matrix

$$K = B_1^{-1}[Z_r\Lambda_r - A_1X_r][CX_r]^{-1} \tag{4.7}$$

where Λ_r is the diagonal matrix of eigenvalues, $Z_r = [z_1 \mid z_2 \mid \cdots \mid z_r]$ and $X_r = [x_1 \mid x_2 \mid \ldots \mid x_r]$. The solution to (4.7) exists provided the selection of r-eigenvalues and eigenvectors results in the non-singularity of the matrix $[CX_r]$. Extension of Algorithm 3.1 is straightforward to guarantee $[CX_r]$ in (4.7) is non-singular. The procedure involves checking the linear independence of the eigenvectors in a modified co-ordinate axis system as follows.

Assume that the C matrix is of the form $C = [C_a \mid C_b]$ with C_a being non-singular. Since C is full rank, this can be achieved with at most a reordering of the state variables. Now apply a co-ordinate transformation

$$\{C, A, B\} \Rightarrow \{CT_0, T_0^{-1}AT_0, T_0^{-1}B\} \tag{4.8}$$

where

$$T_0 = \left[\begin{array}{c|c} C_a^{-1} & -C_a^{-1}C_b \\ \hline 0 & I_{(n-r)} \end{array}\right], \quad CT_0 = [I_r \mid 0] \quad \text{and} \quad X_0 = T_0^{-1}X_r \tag{4.9}$$

The sequential synthesis of Algorithm 3.1 applied to the transformed system (4.9) is equivalent to assigning the first r-eigenvalues and r-eigenvectors $X_0^T = [Z_0^T \mid W_0^T]$ with Z_0 arbitrarily chosen.

Notice that if the analysis is carried out on the dual system of (4.1), represented by the triple $\{B^T, \ A^T, \ C^T\}$, it can be shown that m-eigenvalues and m-left eigenvectors (with r − 1 elements in each eigenvector arbitrary) can be arbitrarily assigned to the system. The above analysis can now be summarised as follows.

Theorem 4.1: *The system (4.1) can be assigned* max(r, m) *eigenvalues 'arbitrarily close' to the desired set. In addition,* max(r, m) *right eigenvector elements or left eigenvector elements (by duality) can be partially assigned with* min(m − 1, r − 1) *entries in each eigenvector or left eigenvector arbitrarily chosen.*

Theorem 4.1 indicates that only r-eigenvalues and r-right eigenvectors (partial) can be assigned to the system. The connotation 'arbitrarily close' in Theorem 4.1 indicates that slight perturbation in eigenvalue specification may be needed in the following situations: (i) assigned eigenvalues coincide with the eigenvalues of matrix F (2.17) and (ii) an exact combination of eigenvalue/eigenvector specifications may not result in a non-singular matrix CX_r in (4.7).

Notice that the m*r arbitrary gain elements of the feedback matrix K have been successfully mapped to r-eigenvalues selection and r*(m − 1) eigenvector parameter selection totalling m*r parameters. However, by sacrificing some freedom in the selection of eigenvectors, it is possible to extend the number of eigenvalues to be assigned to (m + r − 1) as discussed in the next section.

The simple Algorithm 2.1 is directly applicable to the eigenstructure assignment stated in Theorem 4.1. In this case instead of checking X is non-singular in Algorithm 2.1, the non-singularity of the matrix CX_r need to be ensured.

4.2.2 *Assignment of (m + r − 1) eigenvalues*

The basic approach to the development of this algorithm is to construct the output feedback gain matrix as $K = (K_1 + K_2)$. The gain matrix K_1 assigns $p = (r - 1)$ eigenvalues and associated partial eigenvectors. The gain matrix K_2 assigns additional m-eigenvalues while protecting the p-eigenvalues/eigenvectors already assigned. The construction procedure of the algorithm also leads to a set of sufficient conditions for assigning (m + r − 1) eigenvalues. For the case where $(m + r - 1) > n$, judicious elimination and/or a linear combination of inputs or outputs could be done to ensure $n = (m + r - 1)$ and thus all system poles can be assigned.

Algorithm 4.1: The algorithm proceeds as follows:

Step 1. Assign p-eigenvalues and p-linearly independent eigenvectors

$$X_p = [x_1 \,\vdots\, x_2 \,\vdots\, \cdots \,\vdots\, x_p] \tag{4.10}$$

Let K_1 be a non-unique feedback matrix corresponding to this assignment given by

$$K_1 = B_1^{-1}[Z_p\Lambda_p - A_1X_p][CX_p]^+ \tag{4.11}$$

where the superscript '+' indicates the Moore–Penrose inverse.

Let the closed-loop matrix be

$$\tilde{A} = A + BK_1C \tag{4.12}$$

Extension of the algorithm of Chapter 3 is straightforward to guarantee that CX_p in (4.11) is full rank. The choices of pivotal indices need to be restricted to those corresponding to the first (r − 1) outputs in the co-ordinate axes $\{\bar{C}, \bar{A}, \bar{B}\}$ defined in (4.13). In this transformation, the output matrix is in a canonical form defined in (4.16). It should be emphasised that the algorithm checks the linear independence of the transformed eigenvectors (4.17). This is equivalent to the assignment of first p-eigenvalue/eigenvectors of the state feedback case discussed in Algorithm 3.1.

Step 2. In order to accomplish the second-stage assignment of m-eigenvalues, it is required to transform the closed-loop system of the first-stage assignment $\{C, \tilde{A}, B\}$ into a special canonical form. Assume that the C matrix is of the form $C = [C_a \mid C_b]$, with C_a being non-singular. Since C is full rank, this can be achieved with at most a reordering of the state variables. Now apply a co-ordinate transformation

$$\{C, \tilde{A}, B\} \Rightarrow \{CT_0, T_0^{-1}\tilde{A}T_0, T_0^{-1}B\} = \{\bar{C}, \bar{A}, \bar{B}\} \tag{4.13}$$

where

$$T_0 = \left[\begin{array}{c:c} C_a^{-1} & -C_a^{-1}(C_b - C_c) \\ \hdashline 0 & I_{(n-r)} \end{array}\right] \tag{4.14}$$

with

$$C_c = \left[\begin{array}{c} 0_{p\times(n-p)} \\ \hdashline c \end{array}\right], \quad c = [1 \;\; 1 \;\; \cdots \;\; 1] \tag{4.15}$$

The output matrix C in the transformed co-ordinates will be of the form

$$\bar{C} = \left[\begin{array}{c:c} I_p & 0 \\ \hdashline 0 & c \end{array}\right] = [\bar{C}_1 \;\vdots\; \bar{C}_2] \tag{4.16}$$

The eigenvectors X_p assigned in step 1 also undergo a co-ordinate transformation as

$$M = \left[\begin{array}{c} M_1 \\ \hdashline M_2 \end{array}\right] = T_0^{-1} X_p \tag{4.17}$$

where M_1 is a $p \times p$ matrix.

Step 3. If M_1 is non-singular, go to step 4.

Else go to step 1, perturb the eigenvector specifications to ensure M_1 is non-singular.

Step 4. Apply a co-ordinate transformation

$$\{\bar{C}, \bar{A}, \bar{B}\} \Rightarrow \{\bar{C}T_1, T_1^{-1}\bar{A}T_1, T_1^{-1}\bar{B}\} = \{\hat{C}, \hat{A}, \hat{B}\} \tag{4.18}$$

where

$$T_1 = \left[\begin{array}{c:c} M_1 & 0 \\ \hdashline M_2 & I_{(n-p)} \end{array}\right] \tag{4.19}$$

The transformed system has the form

$$\dot{\hat{\mathbf{x}}} = \left[\begin{array}{c:c} \Lambda_p & \hat{A}_{12} \\ \hdashline 0 & \hat{A}_{22} \end{array}\right] \hat{\mathbf{x}} + \left[\begin{array}{c} \hat{B}_1 \\ \hdashline \hat{B}_2 \end{array}\right] \mathbf{u}$$
$$\mathbf{y} = [\hat{C}_1 \;\vdots\; \bar{C}_2]\hat{\mathbf{x}} \tag{4.20}$$

where Λ_p is the diagonal matrix of p-eigenvalues assigned in step 1. It is to be noted that matrix $\bar{C}_2$ is invariant under this transformation.

Step 5. In order to assign additional m-eigenvalues to the transformed system (4.20), while protecting the p-eigenvalues assignment of step 1, the second feedback gain matrix K_2 is restricted to be of unity rank, of the form $K_2 = qs^T$, with q being an m-vector and s being an r-vector, chosen such that

$$s^T[\hat{C}_1 \mid \bar{C}_2] = [0 \mid c] \tag{4.21}$$

Now q must be chosen such that

$$[\hat{A}_{22} + \hat{B}_2 qc] \tag{4.22}$$

is assigned m-eigenvalues. Since $(\hat{A}_{22}, \hat{B}_2)$ is controllable, the following result can be stated.

Theorem 4.2: *The single output subsystem* $\{c, \hat{A}_{22}, \hat{B}_2\}$ *can be assigned* m-*eigenvalues 'arbitrarily close' to the desired set if and only if: (a)* M_1 *in (4.15) is non-singular, (b)* $(c, \hat{A}_{22})$ *is observable and (c)* $\hat{B}_2$ *is full rank.*

This result follows directly from Theorem 4.1. Further, conditions (a)–(c) restrict the admissible eigenstructure in step 1. To identify these parametric restrictions, express the submatrix $\hat{A}_{22}$ in terms of the corresponding partitioned matrices in (4.13) and (4.20) as

$$\hat{A}_{22} = [\bar{A}_{22} + \bar{B}_2 K_1 \bar{C}_2] + M_0[\bar{A}_{12} + \bar{B}_1 K_1 \bar{C}_2] \tag{4.23}$$

where $M_0 = -M_2 M_1^{-1}$ and let $K_1 \bar{C}_2 = \kappa c$ with $\kappa \in R^m$. Now (4.23) can be visualised as a closed-loop system matrix derived by applying feedback to a dynamical system

$$\begin{aligned} \dot{\xi} &= \bar{A}_{22}^T \xi + \bar{A}_{12}^T \mu_1 + c^T \mu_2 \\ \eta &= \bar{B}_2^T \xi + \bar{B}_1^T \mu_1 \end{aligned} \tag{4.24}$$

with $\xi \in R^{n-p}$, $\eta \in R^m$, $\mu_1 \in R^p$, $\mu_2 \in R^1$; and feedback laws, $\mu_1 = M_0^T \xi$, $\mu_2 = \kappa^T \eta$. From (4.22) it can be shown that condition (b) of Theorem 4.2, which is equivalent to

$$\{c, [\bar{A}_{22} + M_0 \bar{A}_{12}]\} \text{ is observable} \tag{4.25}$$

and condition (c) of Theorem 4.2, which is equivalent to

$$\hat{B}_2 = \bar{B}_2 + M_0 \bar{B}_1 \text{ is full rank} \tag{4.26}$$

Equations (4.25) and (4.26) clearly indicate the restrictions on selection of eigenvectors in step 1 (matrix M). This leads to the following result.

Theorem 4.3: *The system* {C, A, B} *can be assigned* (m + r − 1) *eigenvalues 'arbitrarily close' to the desired set, if the first* p-*eigenvalues and eigenvectors in step 1 are chosen such that*

I. M_1 *is non-singular;*

II. $\hat{B}_2 = \bar{B}_2 + M_0\bar{B}_1$ *is full rank;*

III. $\{[\bar{A}_{22}^T + \bar{A}_{12}^T M_0^T], c^T\}$ *is controllable.*

The connotation 'arbitrarily close' in Theorems 4.2 and 4.3 indicates that slight perturbation in eigenvalue specification may be needed in the following situations: (i) assigned eigenvalues coincide with eigenvalues of matrix F, (ii) an exact combination of eigenvalue/eigenvector specifications in step 1 may not result in a non-singular M_1 and (iii) coincident eigenvalue of matrix F situation similar to (i) exists for the single output subsystem assignment of Theorem 4.2.

Condition III derived from (4.24) implies that the feedback from input μ_1 must be such that the controllability of the feedback system is preserved with respect to input μ_2. Conditions II and III yield non-linear algebraic constraints on the eigenvector parameters to be assigned in step 1 and thus, in general, can only be used as test conditions for each assignment in step 1.

Step 6. If conditions of Theorem 4.3 are satisfied, compute the feedback gain matrix q of (4.22) for the single output subsystem $\{c, \hat{A}_{22}, \hat{B}_2\}$ to assign m-eigenvalues using the results of Theorem 4.1.

Step 7. The composite output feedback matrix is

$$K = K_1 + K_2 \tag{4.27}$$

where K_1 is computed from (4.11) and $K_2 = qs^T$, with s defined in (4.21) and q computed in step 6. The output feedback matrix K assigns (m + r − 1) eigenvalues to the system.

From the above analysis the following final result can be stated.

Theorem 4.4: *The system (4.1) can be assigned* (m + r − 1) *eigenvalues 'arbitrarily close' to the desired set. In addition,* (r − 1) *eigenvectors can be partially assigned with* (m − 1) *entries in each eigenvector arbitrarily chosen.*

From the above result, it is worth noting that the m*r arbitrary gain elements of the output feedback matrix K have been meaningfully mapped to (m + r − 1) eigenvalues selection and (r − 1)*(m − 1) eigenvector parameter selection totalling to m*r parameters.

4.3 Eigenstructure assignment for systems with proper outputs

Consider the state space system

$$\begin{aligned} \dot{\mathbf{x}} &= \mathbf{A}\mathbf{x} + \mathbf{B}\mathbf{u} \\ \mathbf{y} &= \mathbf{C}\mathbf{x} + \mathbf{D}\mathbf{u} \end{aligned} \tag{4.28}$$

and a control law $\mathbf{u} = \mathrm{K}\mathbf{y} + \mathbf{u}_{ref}$. Substituting the control law in (4.28) yields

$$\mathbf{u} = \mathrm{PKC}\mathbf{x} + \mathrm{P}\mathbf{u}_{ref} \tag{4.29a}$$

where $\mathrm{P} = [\mathrm{I_m} - \mathrm{KD}]^{-1}$. If G is the output feedback matrix with $\mathrm{D} = 0$, we have the following relationships between K and G:

$$\mathrm{K} = \mathrm{G}[\mathrm{I_r} + \mathrm{DG}]^{-1} \tag{4.29b}$$

$$\mathrm{G} = [\mathrm{I_m} - \mathrm{KD}]^{-1}\mathrm{K} \tag{4.29c}$$

Thus, a feedback law $\mathbf{u} = \mathrm{G}\mathbf{y}$, computed using the algorithms outlined for $\mathrm{D} = 0$ in section 4.2, are readily converted to those systems where $\mathrm{D} \neq 0$ using the relation in (4.29b).

The closed-loop system takes the form

$$\begin{aligned} \dot{\mathbf{x}} &= (\mathrm{A} + \mathrm{BPKC})\mathbf{x} + \mathrm{BP}\mathbf{u}_{ref} \\ \mathbf{y} &= (\mathrm{C} + \mathrm{DPKC})\mathbf{x} + \mathrm{DP}\mathbf{u}_{ref} \end{aligned} \tag{4.30}$$

It is important to note that the output feedback law of proper systems ($\mathrm{D} \neq 0$) introduces a control-mixing matrix P and also modifies the output matrix.

4.4 Eigenstructure assignment with dynamic output feedback

Theorems 4.1–4.3 established conditions under which min(n, m + r – 1) eigenvalues can be assigned. However, for systems with $n > m + r - 1$, all poles cannot be assigned. Kimura [1] in addition to establishing this upper bound on the number of eigenvalues that can be assigned by static gain output feedback also introduced the design of a dynamic compensator to assign all eigenvalues of the system for the case $n > m + r - 1$. The following analysis is based on Reference 1.

The problem of pole assignment for the system (4.1) using a dynamic compensator can be reduced to the pole assignment of an augmented system

$$\begin{aligned} \begin{bmatrix} \dot{\mathbf{x}} \\ \dot{\mathbf{x}}_a \end{bmatrix} &= \begin{bmatrix} \mathrm{A} & 0 \\ 0 & 0 \end{bmatrix} \begin{bmatrix} \mathbf{x} \\ \mathbf{x}_a \end{bmatrix} + \begin{bmatrix} \mathrm{B} & 0 \\ 0 & \mathrm{I_q} \end{bmatrix} \begin{bmatrix} \mathbf{u} \\ \mathbf{u}_a \end{bmatrix} \\ \begin{bmatrix} \mathbf{y} \\ \mathbf{y}_a \end{bmatrix} &= \begin{bmatrix} \mathrm{C} & 0 \\ 0 & \mathrm{I_q} \end{bmatrix} \begin{bmatrix} \mathbf{x} \\ \mathbf{x}_a \end{bmatrix} \end{aligned} \tag{4.31}$$

where q denotes the order of the dynamic compensator. It is clearly seen that the augmented system (4.31) is pole assignable with distinct eigenvalues provided that

$$\mathrm{q} = \mathrm{n} - \mathrm{m} - \mathrm{r} + 1 \tag{4.32}$$

Applying Algorithm 4.1 to the (n + q)-order augmented system (4.31) results in the output feedback control law

$$\begin{bmatrix} \mathbf{u} \\ \mathbf{u}_a \end{bmatrix} = \begin{bmatrix} K_1 & K_2 \\ K_3 & K_4 \end{bmatrix} \begin{bmatrix} \mathbf{y} \\ \mathbf{y}_a \end{bmatrix} \tag{4.33}$$

This results in the following composite static and dynamic feedback law:

$$\mathbf{u} = K_1\mathbf{y} + K_2(sI_q - K_4)^{-1}K_3\mathbf{y} \tag{4.34}$$

The closed-loop augmented system with the control law (4.34) will have the specified eigenvalues. However, one major limitation of this design procedure is that the stability of the dynamic compensator in (4.34), determined by the eigenvalues of K_4, is not guaranteed. Thus, to ensure compensator stability, either a trial and error method or a constrained optimisation procedure has to be adopted. An alternative dynamic controller design based on functional observers will be discussed in Chapter 6, which guarantees the stability of both the closed-loop system and the compensator. Finally an alternative bound for the minimum order (q) of the dynamic compensator discussed above, based on controllability and observability indices, is also given in Reference 10.

4.5 Examples

Example 4.1: Let the system (4.1) be defined by the matrices

$$A = \begin{bmatrix} -1 & -1 & 1 \\ 1 & -1 & -1 \\ -1 & -1 & 0 \end{bmatrix}, \quad B = \begin{bmatrix} 1 & 0 \\ 0 & 1 \\ 1 & -1 \end{bmatrix}, \quad C = \begin{bmatrix} 1 & 0 & 0 \\ 0 & 1 & 1 \end{bmatrix} \tag{4.35}$$

This tutorial example has been carefully constructed to highlight the conditions under which a preconceived eigenstructure may not be assignable. It will also be used to walk through the important steps of Algorithm 4.1. Since in this example $n = m + r - 1$, all eigenvalues can be arbitrarily assigned to the system and one eigenvector can be assigned with one element arbitrarily chosen.

Case 1. Let the eigenstructure to be assigned be $\lambda_1 = -1$, with its associated partial eigenvector $z_1 = [1 \quad 0.5]^T$. Let $\lambda_2 = -1 \pm j1$.

The eigenvector constraint matrices are $F = [-2]$, $G = [1 \quad -1]$ and $H = [1 \quad -1]$.

Step 1. Assigning $\lambda_1 = -1$ yields the eigenvector $X_p = [0.8944 \quad 0.4472 \quad 0]^T$.

The gain matrix is $K_1 = \begin{bmatrix} 0.4 & 0.2 \\ -0.8 & -0.4 \end{bmatrix}$.

Step 2. Since the C matrix is already in the desired canonical form, $T_0 = I_3$. $M = X_p$, with M_1 being non-zero and $M_0^T = [-0.5 \quad 0]$.

Step 3. From Theorem 4.3 (condition III)

$$[\bar{A}_{22}^T + \bar{A}_{12}^T M_0^T] = \begin{bmatrix} -1 & -1 \\ -1 & 0 \end{bmatrix} + \begin{bmatrix} -1 \\ 1 \end{bmatrix} [-0.5 \quad 0] = \begin{bmatrix} -0.5 & -1.0 \\ -1.5 & 0 \end{bmatrix} \quad \text{and} \quad c^T = \begin{bmatrix} 1 \\ 1 \end{bmatrix}$$

and $\left\{ \begin{bmatrix} -0.5 & -1.0 \\ -1.5 & 0 \end{bmatrix}, \begin{bmatrix} 1 \\ 1 \end{bmatrix} \right\}$ is not controllable. Hence, subsystem pole assignment is not possible. Thus, the specified eigenstructure assignment is not possible.

Case 2. Let the eigenstructure to be assigned be $\lambda_1 = -1$, with its associated partial eigenvector $z_1 = [1 \quad 1]^T$ and $\lambda_2 = -1 \pm j1$.

Step 1. Assigning $\lambda_1 = -1$ yields the eigenvector $X_p = [0.7071 \quad 0.7071 \quad 0]^T$.
The gain matrix is $K_1 = \begin{bmatrix} 0.5 & 0.5 \\ -0.5 & -0.5 \end{bmatrix}$.

Step 2. Since the C matrix is already in the desired canonical form, $T_0 = I_3$.
$M = X_p$, with M_1 being non-zero and $M_0^T = [-1 \quad 0]$.

Step 3. From Theorem 4.3 (condition II)

$$\hat{B}_2 = \bar{B}_2 + M_0 \bar{B}_1 = \begin{bmatrix} 0 & 1 \\ 1 & -1 \end{bmatrix} + \begin{bmatrix} -1 \\ 0 \end{bmatrix} [1 \quad 0] = \begin{bmatrix} -1 & 1 \\ 1 & -1 \end{bmatrix} \text{ is not full rank.}$$

Hence, subsystem pole assignment is not possible. Thus, the specified eigenstructure assignment is not possible.

Case 3. Let the eigenstructure to be assigned be $\lambda_1 = -1$, with its associated partial eigenvector $z_1 = [1 \quad 0]^T$ and $\lambda_2 = -1 \pm j1$.

Step 1. Assigning $\lambda_1 = -1$ yields the eigenvector $X_p = [1 \quad 0 \quad 0]^T$.
The gain matrix is $K_1 = \begin{bmatrix} 0 & 0 \\ -1 & 0 \end{bmatrix}$.

Step 2. Since the C matrix is already in the desired canonical form, $T_0 = I_3$.
$M = X_p$, with M_1 being non-zero and $M_0^T = [0 \quad 0]$.

Step 3. All conditions of Theorem 4.3 are satisfied.
Assigning $\lambda_2 = -1 \pm j1$ to the subsystem yields

$$s^T = [0 \quad 1], \qquad q = \begin{bmatrix} -1 \\ 3 \end{bmatrix}, \qquad K_2 = q s^T = \begin{bmatrix} 0 & -1 \\ 0 & 3 \end{bmatrix}$$

Step 4. The composite output feedback, which assigns the desired eigenstructure, is

$$K = K_1 + K_2 = \begin{bmatrix} 0 & -1 \\ -1 & 3 \end{bmatrix}$$

Case 4. Let the eigenstructure to be assigned be $\lambda_1 = -2$, with its associated partial eigenvector $z_1 = [1 \quad -1]^T$ and $\lambda_2 = -1 \pm j1$.

Step 1. Since $\lambda_1 = -2$ coincides with the eigenvalue of F matrix, the eigenvector has a special form as $X_p = [0 \quad 0 \quad 1]^T$ as discussed in Chapter 2. Further $M_1 = 0$, and condition I of Theorem 4.3 is violated. Thus, the desired eigenstructure cannot be assigned.

In summary, Example 4.1 illustrates that assignment of an exact combination of eigenvalue and eigenvector may not always be possible. In Cases 1–3, eigenvector modification was adequate to assign $\lambda_1 = -1$. In Case 4, an eigenvalue perturbation of $\lambda_1 = -2$ will be required.

The counterexample proposed in Reference 3 will now be used to show conclusively (using Theorem 4.3) that $(m + r - 1)$ eigenvalues cannot be assigned to the system.

Example 4.2: Let the system matrices be given by

$$A = \begin{bmatrix} 0 & 0 & 0 & 0 \\ 0 & 0 & 0 & 0 \\ 1 & 0 & 0 & 0 \\ 0 & 1 & 0 & 0 \end{bmatrix}, \quad B = \begin{bmatrix} 1 & 0 \\ 0 & 1 \\ 0 & 0 \\ 0 & 0 \end{bmatrix}, \quad C = \begin{bmatrix} 0 & 0 & 1 & 0 \\ 0 & 0 & 0 & 1 \end{bmatrix} \tag{4.36}$$

The state variables in Reference 3 have been permuted to obtain the matrices in (4.36) to ensure that the matrix B_1 (4.4) is non-singular. The eigenvector constraint matrices are

$$F = \begin{bmatrix} 0 & 0 \\ 0 & 0 \end{bmatrix}, \quad G = \begin{bmatrix} 1 & 0 \\ 0 & 1 \end{bmatrix}, \quad H = \begin{bmatrix} 0 & 0 \\ 0 & 0 \end{bmatrix} \tag{4.37}$$

Now for assigning an eigenvalue λ, the eigenvector constraint relationship (4.5) is given by

$$\begin{bmatrix} \lambda & 0 \\ 0 & \lambda \end{bmatrix} \begin{bmatrix} x_3 \\ x_4 \end{bmatrix} = \begin{bmatrix} 1 & 0 \\ 0 & 1 \end{bmatrix} \begin{bmatrix} x_1 \\ x_2 \end{bmatrix} \tag{4.38}$$

Choosing $x_1 = 1$ (arbitrary selection of one element), the eigenvector using (4.38) in literal form is

$$x^T = \begin{bmatrix} 1 & x_2 & \dfrac{1}{\lambda} & \dfrac{x_2}{\lambda} \end{bmatrix} \tag{4.39}$$

The construction of the transformation in step 2 of Algorithm 4.1 to assign $(m + r - 1)$ eigenvalues requires that the matrix C_a be non-singular. For this purpose permute the states with the co-ordinate transformation

$$T_s = \begin{bmatrix} 0 & 0 & 1 & 0 \\ 0 & 0 & 0 & 1 \\ 1 & 0 & 0 & 0 \\ 0 & 1 & 0 & 0 \end{bmatrix} \tag{4.40}$$

The transformed eigenvector takes the form

$$x_s = T_s x \quad \text{and} \quad x_s^T = \begin{bmatrix} \frac{1}{\lambda} & \frac{x_2}{\lambda} & 1 & x_2 \end{bmatrix} \tag{4.41}$$

The transformation T_0 (4.14) for the permuted system is

$$T_0^{-1} = \begin{bmatrix} 1 & 0 & 0 & 0 \\ 0 & 1 & -1 & -1 \\ 0 & 0 & 1 & 0 \\ 0 & 0 & 0 & 1 \end{bmatrix} \tag{4.42}$$

The eigenvector also undergoes a transformation as (4.17)

$$M = T_0^{-1} x_s \quad \text{and} \quad M^T = \begin{bmatrix} \frac{1}{\lambda} & \left(\frac{x_2}{\lambda} - 1 - x_2\right) & 1 & x_2 \end{bmatrix} \tag{4.43}$$

The subsystem of Theorem 4.3 is

$$\bar{A}_{22}^T = \begin{bmatrix} 0 & 0 & 0 \\ 0 & 0 & 0 \\ 1 & 0 & 0 \end{bmatrix}, \qquad \bar{A}_{12}^T = \begin{bmatrix} 0 \\ 1 \\ 0 \end{bmatrix}, \qquad c = \begin{bmatrix} 1 \\ 1 \\ 1 \end{bmatrix} \tag{4.44}$$

and

$$M_0^T = [(\lambda x_2 + \lambda - x_2) - \lambda - \lambda x_2] \tag{4.45}$$

Applying the controllability check (condition III of Theorem 4.3) yields the controllability matrix

$$C_0 = \begin{bmatrix} 1 & 0 & 0 \\ 1 & -x_2 & 0 \\ 1 & 1 & 0 \end{bmatrix} \tag{4.46}$$

It is seen that C_0 has rank < 3 for any admissible eigenstructure assignments in (4.38) and thus the system cannot be assigned three eigenvalues $(m + r - 1)$. The algorithm proposed in Reference 3 does not provide an explicit condition when $(m + r - 1)$ eigenvalues cannot be assigned. To cover these pathological cases, the authors concluded that for ‘almost’ all (B, C) pairs $(m + r - 1)$ eigenvalues assignment is possible. Theorem 4.3 however gives explicit conditions when a solution indeed exists. However, from Theorem 4.1, two eigenvalues can be assigned to this system.

4.6 Summary

In this chapter eigenstructure assignment using output feedback has been characterised. The upper bound on the number of eigenvalues that can be assigned still remains as min(n, m + r − 1) as proved in Reference 1. Current research is still addressing the generic solvability of the pole placement problem when $m*r \geq n$. Using the concepts of Chapter 2, an eigenstructure assignment algorithm has been developed to assign min(n, m + r − 1) arbitrary eigenvalues and in addition assign (m − 1) entries in each of the (r − 1) eigenvectors arbitrarily. Sufficient conditions for the existence of the solution are also derived. In situations where $n > m + r - 1$, it is possible to explore the possibility of designing a dynamic compensator to assign all poles of the system. This procedure however does not guarantee the stability of the resulting compensator. An alternative low-order dynamic compensator design based on functional observers will be discussed in Chapter 6.

References

1. KIMURA, H.: 'Pole assignment by gain output feedback', *IEEE Trans. Automat. Control*, 1975, **AC-20**, (4), pp. 509–516
2. KIMURA, H.: 'A further result on the problem of pole assignment by output feedback', *IEEE Trans. Automat. Control*, 1977, **AC-22**, (3), pp. 458–463
3. DAVISON, E.J., and WANG, S.H.: 'On pole assignment in linear multivariable systems using output feedback', *IEEE Trans. Automat. Control*, 1975, **AC-20**, (4), pp. 516–518
4. HERMANN, R., and MARTIN, C.F.: 'Application of algebraic geometry to systems theory – part I', *IEEE Trans. Automat. Control*, 1977, **AC-22**, (1), pp. 19–25
5. EREMENKO, A., and GABRIELOV, A.: 'Pole placement by static output feedback for generic linear systems', *SIAM J. Control Optim.*, 2002, **41**, (1), pp. 303–312
6. FU, M.: 'Pole placement via static output feedback is NP hard', *IEEE Trans. Automat. Control*, 2004, **49**, (5), pp. 855–857
7. SRINATHKUMAR, S.: 'Eigenvalue/eigenvector assignment using output feedback', *IEEE Trans. Automat. Control*, 1978, **AC-23**, (1), pp. 79–81
8. SRINATHKUMAR, S.: 'Eigenvalue/Eigenvector Assignment Using Output Feedback', 1978, NASA TP 1118
9. KIMURA, H.: 'Pole assignment by output feedback: a long-standing open problem', *Proceedings of 33rd Conference on Decision and Control*, 1994, Lake Buena Vista, Florida, pp. 2101–2105
10. BRASCH, F.M., and PEARSON, J.B.: 'Pole placement using dynamic compensator', *IEEE Trans. Automat. Control*, 1970, **AC-15**, pp. 34–43

Chapter 5
Robust eigenstructure assignment

5.1 Introduction

The design of feedback systems based on a 'mathematical model' of the plant should be insensitive to uncertainty associated with the parameters of the model. This is required since the system model parameters, in most cases, are only a best estimate of the real plant. The concepts of gain and phase margins evolved to guarantee closed-loop stability in the presence of specified uncertainty of the plant gain/phase characteristics. Thus, the design of 'robust' feedback systems has been extensively studied in the control literature. In the multivariable control setting, a wide range of robustness metrics has been postulated to characterise system behaviour under plant ignorance. The nature of the feedback design technique also influences the selection of appropriate robustness criteria, and this is also true for eigenstructure assignment.

During the characterisation of eigenstructure assignment using state and output feedback, it was emphasised that for the solution to exist, the eigenvector matrix (X) must be non-singular. In this chapter this non-singularity of X will be numerically quantified and its relationship to the robustness of the resulting solution will be explained. These aspects were also briefly alluded to during the development of Algorithm 3.1.

5.2 Robustness metrics

The classical linear algebra textbooks [1,2] have established that the sensitivity of eigenvalues of a square matrix to perturbations in its parameters is directly related to the ill-conditioning of the matrix of eigenvectors (modal matrix) to inversion. In Reference 1, a vector of robustness measures has been derived that indicates the conditioning of the individual eigenvalues. In Reference 2 a scalar robustness metric that indicates the overall condition of the modal matrix is defined. Consider the state space system defined in (2.5). Following the sensitivity analysis in Reference 1, the closed-loop system matrix under state feedback ($\hat{A} = A + BK$) satisfies the relation

$$\hat{A}x_i = \lambda_i x_i, \quad i = 1\text{–}n \tag{5.1}$$

The first-order differential changes in (5.1) can be written as

$$d\hat{A}\, x_i + \hat{A}\, dx_i = \lambda_i\, dx_i + d\lambda_i\, x_i \tag{5.2}$$

For real eigenvalues let y_i be the left eigenvector satisfying

$$\hat{A}^T y_i = \lambda_i y_i, \quad i = 1\text{–}n \tag{5.3}$$

Premultiply (5.2) by y_i^T and using (5.3), the real eigenvalue perturbation can be expressed as

$$d\lambda_i = \frac{y_i^T\, d\hat{A}\, x_i}{y_i^T x_i} \tag{5.4}$$

The corresponding eigenvector change is given by

$$dx_i = \sum_{j=1}^{n-1} d_{ij} x_j \tag{5.5}$$

with

$$d_{ij} = 0\ (i = j) \quad \text{and} \quad d_{ij} = \frac{y_j^T\, d\hat{A}\, x_i}{(\lambda_i - \lambda_j)(y_j^T x_i)} \quad (i \neq j) \tag{5.6}$$

From (5.4) and (5.5) the following bounds for eigenvalue and eigenvector results:

$$|d\lambda_i| \leq \frac{\|d\hat{A}\| \|x_i\| \|y_i\|}{|y_i^T x_i|} = c_i \|d\hat{A}\|, \quad \text{with bounds } 1 \leq c_i \leq \infty \tag{5.7}$$

where c_i is defined as the 'condition number' of λ_i. The symbol $\|x\|$ indicates the two-norm of a matrix or a vector and $|x|$ indicates the absolute value. If the eigenvectors are normalised such that $y_i^T x_i = x_i^T x_i = 1$, then $c_i = \|y_i\| \geq 1$. The corresponding bound on the eigenvector perturbation is given by

$$\|dx_i\| \leq \|d\hat{A}\| \sum_{j=1, j \neq i}^{n} \frac{c_i}{|\lambda_i - \lambda_j|} \tag{5.8}$$

The mode condition numbers c_i ($i = 1$–n) constitute a *vector* of robustness measures. Thus, the norm of the left eigenvector, represented by c_i, indicates the sensitivity of the corresponding eigenvalue to model uncertainty.

The analysis in Reference 2 defines the condition number for the eigenvector matrix X as

$$\kappa_2(X) = \|X\| \|X\|^{-1}, \quad \text{with bounds } 1 \leq \kappa_2(X) \leq \infty \tag{5.9}$$

The condition number $\kappa_2(X)$ can be used as a scalar robustness measure. The condition number is also the ratio of maximum to minimum singular values of the eigenvector matrix X.

Another candidate scalar robustness measure is the determinant of X. It is well known that, in general, the determinant of a matrix is a poor indicator of its condition to inversion. However, the modal matrix determinant is a well-defined condition number to estimate the ill-conditioning of X since its columns are normalised to unit length. The following analysis establishes the bounds of determinant (X).

Assume X is non-singular. Let $Q = X^T X$, and let $\Delta = \mathrm{Det}(X)$. Since Q is positive definite, it follows that [3]

$$\Delta^2 = \mathrm{Det}(Q) \le \prod_{i=1}^{n} q_{ii} \tag{5.10}$$

Since $q_{ii} = x_i^T x_i = 1$, $|\Delta| \le 1$. The upper bound $\Delta = 1$ is attained when $X^T X = I_n$, implying that X is orthonormal. Thus, $\kappa_2(X)$ and Δ simultaneously reach their lower and upper bounds, respectively, when X is perfectly conditioned. Further

$$|\Delta| = \prod_{i=1}^{n} \sigma_i, \quad \sigma_1 \ge \sigma_2 \ge \cdots \sigma_n > 0 \tag{5.11}$$

where σ_i are the singular values of X with $\kappa_2(X) = \sigma_1/\sigma_n$. When X is singular, $\sigma_n = 0$ and $\Delta = 0$ (lower bound). Thus

$$0 \le |\Delta| \le 1 \tag{5.12}$$

There is also an interesting geometric interpretation of $|\Delta|$. The determinant can be viewed as the volume of a parallelepiped with the eigenvectors as the sides. The volume is a maximum when the parallelepiped is a cube with the eigenvectors being orthonormal. This corresponds to X being perfectly conditioned with $|\Delta| = 1$. The fact that the determinant represents a volume is established by proving that the volume satisfies the characteristic properties of a determinant [4, p. 229].

Finally if complex eigenvalues are represented in quasi-diagonal form as defined in (2.3) and real eigenvector pair is normalised as in (2.4a), the bounds get modified as (i) $2 \le c_i \le \infty$ (for complex eigenvalue) and (ii) $0 \le |\Delta| \le 1/2^p$, where p is the number of complex pair eigenvalues. If the system has a combination of real and complex eigenvalues, the modal matrix condition number is bounded by $\sqrt{2} \le \kappa_2(X) \le \infty$.

5.3 Robust eigenstructure characterisation

Having defined some of the modal robustness measures, it is natural to explore the possibility of selecting eigenstructure of the closed-loop system to improve robustness. A very comprehensive characterisation of the properties of robust eigenstructure has been reported in Reference 5. Properties relating the numerical conditioning of X with (i) eigenvalue selection, (ii) impact on feedback gain magnitudes, (iii) closed-loop stability bounds, etc. have been identified. These properties have been found to be of importance from engineering design considerations [6].

Consider the fundamental eigenstructure constraint relation for real eigenvalues (3.1b)

$$R_i\begin{bmatrix} z_i \\ \hline w_i \end{bmatrix} = 0, \quad \text{where } R_i = [G + \lambda_i H \;\vdots\; F - \lambda_i I_{n-m}] \tag{5.13}$$

Let

$$S_i = \text{Null}\{R_i\}, \quad S_i \in R^{n\times m} \tag{5.14}$$

where Null{·} implies the null space with orthonormal basis vectors. It is seen that the eigenvector x_i lies in the subspace spanned by S_i. Define the *eigenspace* matrix, which is a concatenation of the subspaces of all eigenvalues, as follows:

$$S = [S_1 S_2 \cdots S_n], \quad S \in R^{n\times n^*m} \tag{5.15}$$

For complex eigenvalue (λ_c) assignment, using (2.19) and (2.20), the eigenvector constraint takes the form

$$R_c\begin{bmatrix} x_r \\ \hline x_c \end{bmatrix} = 0, \quad R_c \in R^{2(n-m)\times 2n} \tag{5.16}$$

where

$$R_c = \left[\begin{array}{c|c|c|c} G + \alpha H & F - \alpha I_{n-m} & -\beta H & \beta I_{n-m} \\ \hline \beta H & -\beta I_{n-m} & G + \alpha H & F - \alpha I_{n-m} \end{array}\right] \tag{5.17}$$

The corresponding eigenspace submatrix $\hat{S}_c = \text{Null}(R_c)$ takes the form

$$\hat{S}_c = \begin{bmatrix} S_a \\ \hline S_b \end{bmatrix}, \qquad \hat{S}_c \in R^{2^*n\times 2^*m}, \quad (S_a, S_b) \in R^{n\times 2^*m} \tag{5.18}$$

To construct the complete eigenspace matrix along with the real eigenvalue submatrices as in (5.15), the submatrix corresponding to the complex eigenvalue pair in (5.18) is concatenated as

$$S_c = [S_a \;\vdots\; S_b] \tag{5.19}$$

The matrix S_c is the representation of the complex eigenspace matrix corresponding to $\lambda_c = \alpha \pm j\beta$ using real arithmetic. The linear combination of the basis vectors of S_a and S_b determines the eigenvectors x_r and x_c in (5.16).

The results, derived in Reference 5, which are of importance to the algorithm development to follow, are stated below:

Result 1 [5, Theorem 8]. A lower bound for the condition of the modal matrix (X) is given by

$$\min \kappa_2(X) \geq \frac{\kappa_2(S)}{\sqrt{n}} \tag{5.20}$$

This important property relates the numerical condition of X and S. If S is ill-conditioned, based on the selection of the eigenvalues, the attainable condition of X is also poor. Thus, the property clearly establishes the link between the selection of eigenvalues (S matrix) and the attainable lower bound on the condition number of X. It should also be noted that the lower bound of $\kappa_2(S) = 1$.

Result 2 [5, Theorem 5]. A bound on the attainable feedback gain norm, for a selected eigenvalue and eigenvector set, is given by

$$\|K\|_2 \le \frac{\|A\|_2 + \max_j\{|\lambda_j|\}\kappa_2(X)}{\sigma_m(B)} \tag{5.21}$$

where $\sigma_m(B)$ is the smallest singular value of the state space system matrix B (2.5). This is a useful property to estimate the gain magnitudes for a given choice of eigenvalues. Equation (5.21) also indicates the adverse effect of B matrix ill-condition on the resulting feedback gains. This property can also be used to estimate an upper bound on the norm of feedback gain matrix K for a given selection of eigenvalues, if $\kappa_2(X)$ is replaced by its estimate from (5.20).

5.4 Robust eigenstructure assignment

Algorithm development for robust eigenstructure assignment brings into focus the role of optimisation into the synthesis process. In general, a direct non-linear constrained optimisation procedure can be set up to simultaneously select eigenvalues (λ) from a defined region in the complex plane (Ω) and the free parameters of the eigenvectors (z), to minimise any of the robustness metrics postulated. This optimisation is adequate to produce reasonable engineering solutions. The issue of the solution converging to local minimum is usually resolved by generating multiple solutions with different initial guesses. The other issue of concern in such an optimisation process is the time taken for the solution to converge. This could pose a problem for high-order system optimisation. However, time of computation may not be an issue with present-day high-speed computing. Nevertheless, efforts have been made to devise algorithms with rapid convergence properties and also use standard, well-proven, linear algebra computation codes to produce accurate solutions. The algorithm described in Reference 5, popularly known as the KNV algorithm, addresses this problem. The KNV algorithm has been studied for its algorithmic features, and extensions to the algorithm by defining a well-posed optimisation problem also have been reported [7]. Experience of using the KNV algorithm on flight control problems indicates that while the solutions do converge to low $\kappa_2(X)$, better solutions are invariably obtained by the direct non-linear optimisation algorithms. The direct optimisation solutions of course do take longer time to converge compared with the KNV algorithm. The KNV algorithm constructs a 'robust' modal matrix for a predefined set of eigenvalues. Equation (5.20) shows a way to select an eigenvalue set to improve the condition of the S matrix, resulting in enhanced modal robustness. However, this property has not been explicitly used in the algorithm development in Reference 5. In Reference 6 the relation between $\kappa_2(S)$ and $\kappa_2(X)$ has been exploited to set up a two-step optimisation algorithm to optimise both eigenvalues and eigenvectors. This is an alternative optimisation procedure compared with the direct non-linear optimisation of simultaneously selecting eigenvalues and eigenvectors to minimise $\kappa_2(X)$. The two-step algorithm is also found to be less time-consuming. In summary, the following constrained optimisation procedures can be used for robust eigenstructure assignment.

Algorithm 5.1:

$$\min_{\lambda_i \in \Omega, \|x_i\|_2 = 1} \{\kappa_2(X) \text{ or } -|\det(X)|\} \quad \text{subject to } R_i x_i = 0 \tag{5.22}$$

Algorithm 5.2:
Step 1. Robust eigenvalue assignment

$$\min_{\lambda_i \in \Omega} \{\kappa_2(S)\} \quad \text{subject to } R_i x_i = 0 \tag{5.23}$$

Step 2. Robust eigenvector assignment

$$\min_{\|x_i\|_2 = 1, \lambda = \lambda_{opt} \text{ (from Step 1)}} \{\kappa_2(X) \text{ or } -|\det(X)|\} \quad \text{subject to } R_i x_i = 0 \tag{5.24}$$

5.5 Examples

Example 5.1: Let the system be defined by the matrices

$$A = \begin{bmatrix} -1 & -1 & 1 \\ 1 & -1 & -1 \\ -1 & -1 & 0 \end{bmatrix}, \qquad B = \begin{bmatrix} 1 & 0 \\ 0 & 1 \\ 1 & -1 \end{bmatrix} \tag{5.25}$$

This tutorial example has been constructed to illustrate the role of $\kappa_2(S)$ in the solution process. For the eigenvalue set $\Lambda = \{-1, -1, -2\}$, the closed-loop system $A + BK = \text{Diag}\{-1, -1, -2\}$, for a gain matrix

$$K = \begin{bmatrix} 0 & 1 & -1 \\ -1 & 0 & 1 \end{bmatrix} \tag{5.26}$$

The modal matrix is I_3 and has ideal condition $\kappa_2(X) = 1$; the eigenspace matrix

$$S = \left[\begin{array}{cc|cc|cc} \lambda = -1 & & \lambda = -1 & & \lambda = -2 & \\ \hline 1 & 0 & 1 & 0 & -0.7071 & 0 \\ 0 & 1 & 0 & 1 & -0.7071 & 0 \\ 0 & 0 & 0 & 0 & 0 & 1 \end{array}\right] \tag{5.27}$$

has a condition number $\kappa_2(S) = \sqrt{3}$ ($\sqrt{n}$, n = 3). From (5.20), for $\kappa_2(X) = 1$, the upper bound of $\kappa_2(S) = \sqrt{3}$. Algorithms 5.1 and 5.2 with different initial parameter guess also yielded other perfectly conditioned modal matrix solutions. Table 5.1 summarises the results.

Table 5.1 Summary of optimal solutions

Method	λ_{opt}*	$\kappa_2(S)$†	**K**‡
None	−1.0 −1.0 −2.0	$\sqrt{3}$	$\begin{bmatrix} 0 & 1 & -1 \\ -1 & 0 & 1 \end{bmatrix}$
Algorithm 5.2§	−0.7075 −1.3911 −2.4432	$\sqrt{2}$	$\begin{bmatrix} -0.6738 & 0.7477 & -1.4214 \\ -1.2523 & 0.1506 & 0.5971 \end{bmatrix}$
Algorithm 5.1‖	−1.0889 −1.4918 −5.9474	1.52	$\begin{bmatrix} -0.0998 & 0.9964 & -1.0962 \\ -1.0036 & -2.1679 & 3.1643 \end{bmatrix}$

*Admissible region (Ω): $-6.0 \le \lambda \le -0.5$.
†$\kappa_2(X)$ and c_i ($i = 1 - 3$) are unity in all cases.
‡Gains rounded to four places.
§Step 2 – minimise $\kappa_2(X)$.
‖Minimise $\kappa_2(X)$.

From Table 5.1 it is seen that $\kappa_2(X) = 1$ solution is attained for $\kappa_2(S)$ in the range $1 \le \kappa_2(S) \le \sqrt{n}$. The next example will show that it is not necessarily true that if $\kappa_2(S)$ is in the range $1 \le \kappa_2(S) \le \sqrt{n}$, $\kappa_2(X) = 1$ solution will be attained.

Example 5.2: The system considered is a distillation plant [5]. The state space system matrices are

$$A = \begin{bmatrix} -3.149 & 1.597 & 0 & 1.547 & 0 \\ 0.9807 & -2.132 & 1.306 & 0 & 0 \\ 0 & 0.0628 & -0.1094 & 0 & 0 \\ 2.632 & 0.0355 & 0 & -4.257 & 1.855 \\ 0 & 0.0023 & 0 & 0.1636 & -0.1625 \end{bmatrix}$$

$$B^T = \begin{bmatrix} 0.0838 & 0.0638 & 0 & 0.1004 & 0.0063 \\ -0.1396 & 0 & 0 & -0.2060 & -0.0128 \end{bmatrix} \tag{5.28}$$

The state variables in Reference 5 have been permuted as (x_3, x_2, x_1, x_4, x_5) in (5.28).

This example highlights the importance of selection of eigenvalues to achieve robust feedback solutions. In Reference 5 it was noted that the choice of the assignable eigenvalue set $\Lambda = \{-0.2, -0.5, -1, -1 \pm j1\}$ results in poor conditioning of the eigenspace matrix (S) and hence one cannot expect the condition of the modal matrix (X) to be any better. However, in Reference 5 no method was suggested to improve the situation. It is here that both Algorithms 5.1 and 5.2 can be effectively used to determine a set of eigenvalues, which improves the condition of S and consequently the modal matrix condition. Tables 5.2 and 5.3 summarise the results.

Table 5.2 Modal condition numbers

Case no.	Method	$\kappa_2(S)$	$\kappa_2(X)$	$\lVert K \rVert_2$
1	Open loop	2.3190	4.0688	–
2	Min $\kappa_2(X)$	46.7881	39.3740	297.7036
3	Algorithm 5.1	1.8738	1.5568	20.8800
4	Algorithm 5.2	1.6597	1.5323	75.5917
5	Algorithm 5.2	1.6499	1.3693	22.6800

Table 5.3 Mode condition numbers

Case no.	Method		Modes (λ_i*) and condition numbers (c_i)				
1	Open loop	λ_i	−0.0773	−0.0142	−0.8953	−2.8408	−5.9822
		c_i	1.5125	1.9002	1.6217	1.2020	1.0639
2	Min $\kappa_2(X)$	λ_i	−0.2	−0.5	−1.0	−1 ± j1	
		c_i	3.13	8.8814	6.8928	31.9517	
3	Algorithm 5.1	λ_i	−0.10	−0.1673	−5.809	−1.678 ± j0.705	
		c_i	1.0073	1.0229	1.022	2.028	
4	Algorithm 5.2	λ_i	−0.1001	−0.1559	−5.6075	−6.323 ± j1.7269	
		c_i	1.0096	1.0382	1.0025	2.0482	
5	Algorithm 5.2	λ_i	−0.1016	−0.1185	−0.3847	−2.9375	−5.9269
		c_i	1.01	1.0466	1.0314	1.0388	1.0154

*Admissible region (Ω):
1. Real roots: $-10.0 \le \lambda \le -0.1$.
2. Complex roots: $-10.0 \le \text{Real}(\lambda) \le -0.1$ and $0.1 \le |\text{Imag.}(\lambda)| \le 10.0$.

Table 5.4 Selected state feedback solutions

Case no.	Gain matrix (K)
2	$\begin{bmatrix} -197.0904 & 91.3653 & -91.8028 & 165.8610 & -42.6087 \\ -36.4371 & 24.1686 & -48.3730 & 22.6629 & 0.3990 \end{bmatrix}$
5	$\begin{bmatrix} 10.5529 & -7.4923 & -15.0855 & 2.5679 & -4.9754 \\ 6.0979 & -3.0622 & -7.5034 & -2.7779 & 1.0258 \end{bmatrix}$

From the results in Tables 5.2 and 5.3, the following observations can be made. The basic plant itself has reasonably good modal condition. The choice of eigenvalue assignment in Case 2 (used in Reference 5) has substantially worsened the robustness characteristics. The problem is further compounded by high feedback gains as indicated by the norm of the gain matrix in Table 5.2 and actual gains in Table 5.4. The mode condition number corresponding to the mode $-1 \pm j1$ is very high ($c = 31.3597$), giving a clear indication of the need to modify this mode. In Cases 3 and 4 an attempt is made to find an optimal eigenvalue set with three real roots and one complex conjugate pair root. In both cases, there is a significant improvement in $\kappa_2(S)$ and $\kappa_2(X)$. Both algorithms have assigned overly damped complex conjugate root to improve mode condition as well as overall modal conditioning. The optimal solutions indicate that assigning a complex root with critical damping to the plant seems unnatural (Case 2). This observation is also substantiated when one notes that the basic plant has all real roots. In Case 5, the system is specified to have only a real eigenvalue set. Solution obtained by Algorithm 5.2 is satisfactory with low gain norm (Tables 5.3 and 5.4). Finally it should be emphasised that in Cases 3–5 even though $\kappa_2(S)$ lies in the interval $1 \leq \kappa_2(S) \leq \sqrt{5}$ ($\sqrt{n}$, $n = 5$), it does not imply that the ideal condition $\kappa_2(X) = 1$ is attained.

5.6 Summary

In Chapter 2 it was pointed out that for a specified eigenstructure assignment to be realised by state feedback, the modal matrix (X) must be non-singular. In this chapter numerical characterisation of this non-singularity has been provided and its relation to the sensitivity (robustness) of the feedback solution to model parameter uncertainties has been established. A vector of sensitivity measures called mode condition numbers (c_i) and scalar sensitivity measures such as $\kappa_2(X)$ and $|\det(X)|$ can be effectively utilised to assess the robustness of resulting feedback design. It is worth recalling that the eigenstructure synthesis algorithm of Chapter 3 provides the designer visibility of the 'robustness' of the solution as the eigenvectors are sequentially synthesised through the parameters σ_k (3.9). The condition number of the eigenspace matrix (S) characterised in Reference 5 plays a pivotal role in the selection of eigenvalues to improve robustness of the modal matrix (X). The ability to optimally select an eigenvalue set from an admissible region in the complex plane to improve system robustness constitutes a complete characterisation of the robust eigenstructure assignment problem.

References

1. FADDEEV, D.K., and FADDEEVA, V.N.: *Computational Methods of Linear Algebra* (W.H. Freeman and Co., San Francisco, 1963)
2. WILKINSON, J.H.: *The Algebraic Eigenvalue Problem* (Oxford University Press, Oxford, 1965)
3. BECKENBACH, E.F., and BELLMAN, R.E.: *Inequalities* (Springer-Verlag, New York, 1965)
4. LANG, S.: *Introduction to Linear Algebra*, 2nd edn (Springer-Verlag, New York, 1986)
5. KAUTSKY, J., NICHOLS, N.K., and VAN DOOREN, P.: 'Robust pole assignment in linear state feedback', *Int. J. Control*, 1985, **41**, (5), pp. 1129–1155
6. SRINATHKUMAR, S.: 'Robust eigenvalue/eigenvector selection in linear state feedback systems', *Proceedings of the 27th Conference on Decision and Control*, 7–9 December 1988, Austin, Texas, pp. 1303–1307
7. CHU, E.K.: 'Optimization and pole assignment in control system design', *Int. J. Appl. Math. Comput. Sci.*, 2001, **11**, (5), pp. 1035–1053

Chapter 6
Modal canonical observers

6.1 Introduction

The state feedback control synthesis techniques naturally carry over to the so-called '*dual*' problem, namely the design of observers or state estimators. The problem of state estimation arises from the fact that in practical systems only a subset of states/linear combination of states (outputs) is available for measurement. Thus, some form of estimating the inaccessible states becomes mandatory. This is especially true if a state variable feedback law has to be implemented using only available outputs. It is well known that for an n-state, m-input, r-output controllable, observable state space system, an nth-order filter with arbitrary eigenvalues, called a 'full-order' observer, can be constructed to asymptotically estimate all the states [1]. Luenberger [2] established that an $(n - r)$th-order filter with arbitrary eigenvalues is adequate to estimate all the states. This observer is called a 'reduced-order' observer. There are two variants of observers depending on the system signals available for state reconstruction. The unknown input observer (UIO) utilises only the output signals for state reconstruction, and the known input observer (KIO) utilises both the inputs and outputs of the system. The classical 'full-order' and Luenberger's 'reduced-order' observers fall under the category of known input state observers. Finally, to implement a state feedback control law of the form $u = Kx$, estimation of a linear combination of the states is all that is required. Thus, it may be possible to construct an observer with an order less than the 'reduced order'. This observer is called a 'functional' observer and can be either KIO or UIO. Finally the design algorithms must cater to systems that contain a mix of strictly proper and proper outputs. Each of the above-listed categories of observers has been addressed individually using different problem formulations in the literature (with the strictly proper outputs case receiving the most attention). Note that the observer is a dynamic system defined by matrices associated with its dynamics and input/output connectivity. This state space structure of the observer is itself flexible, and many such structures have been studied leading to different design algorithms. Some of the results and additional references on the design of known input functional observers can be found in References 3–7 and on UIOs in References 8–12.

An important application of observers is found in the fault detection and isolation (FDI) methodology [13,14]. The use of *analytical* (*software*) rather than physical (*hardware*) redundancy in dynamic system elements such as sensors, actuators, etc. has been seriously considered for this purpose. The primary motivation for incorporating such *software* redundancy is cost. For example, to provide a single failure operational reliability in sensor systems, it may be sufficient to use two identical hardware sensors and use *analytical* redundancy to perform fault isolation task in the event of a mismatch in the duplex hardware. The analytical redundancy scheme employs a mathematical model of the dynamics of the system to generate a third sensor signal, which is used as an arbitration signal to isolate the faulty hardware sensor. State estimation forms a key element in this type of FDI design. Typically deterministic full/reduced-order observers or Kalman filters in the stochastic case (when noise estimates are available) are used for state estimation. However, these observers require information of both the input and output of the system for state reconstruction. There are cases when some of the inputs to the system are not measurable. A typical example is the gust input influencing the dynamics of an aircraft. In such situations the UIO is employed.

In this chapter a unified approach to the design of observers will be presented that covers all categories of observers discussed above. Algorithms for the design of observers will be derived using the eigenstructure assignment concept developed in Chapter 2. The general analysis and algorithm development will be detailed for the design of UIO with mixed (strictly proper and proper) outputs, and modifications of the algorithm to other special cases will be identified.

6.2 Problem formulation

Consider the following linear time-invariant system:

$$\begin{aligned} &\dot{\mathbf{x}} = \mathrm{A}\mathbf{x} + \mathrm{B}\mathbf{u} \\ &\mathbf{y}_1 = \mathrm{C}_1\mathbf{x} + \mathrm{D}_1\mathbf{u}, \qquad \mathbf{y}_2 = \mathrm{C}_2\mathbf{x} + \mathrm{D}_2\mathbf{u} \\ &\mathbf{y} = \mathrm{C}\mathbf{x} + \mathrm{D}\mathbf{u}, \qquad \mathbf{y}^{\mathrm{T}} = [\mathbf{y}_1^{\mathrm{T}} \mid \mathbf{y}_2^{\mathrm{T}}] \end{aligned} \tag{6.1}$$

where $\mathbf{x} \in R^n$, $\mathbf{u} \in R^s$ and $\mathbf{y}_1 \in R^{rp}$ (proper outputs), $\mathbf{y}_2 \in R^{rs}$ (strictly proper outputs) and $\mathbf{y} \in R^r$ (r = rp + rs). A, B, C, D, C_1, C_2, D_1 and D_2 are matrices of compatible dimensions. Matrix $D_1 \in R^{rp \times s}$. Matrix $D_2 \in R^{rs \times s}$ is a zero matrix. Superscript 'T' indicates matrix/vector transpose. It is assumed that B and C are full rank and the pair (A, C) is observable.

Given a linear functional of the state vector

$$f = K\mathbf{x} \tag{6.2}$$

and a linear transformation of the states

$$z = T\mathbf{x} \tag{6.3}$$

where $f \in R^m$ and $z \in R^p$. It is assumed that $m \le (n - r)$. The problem is to design a pth-order observer with the general structure

$$\dot{\delta} = F\delta + G\mathbf{u} + H\mathbf{y}$$
$$\theta = M\delta + R\mathbf{u} + N\mathbf{y} \tag{6.4}$$

where $\delta \in R^p, \theta \in R^m$ and F, G, H, M, N and R are matrices of compatible dimensions. It is required that (i) the observation vector δ asymptotically approach z and consequently θ approach f, (ii) the matrix F has arbitrary eigenvalues and (iii) the observer order (p) is minimal.

The method to derive conditions for the observer states δ in (6.4) asymptotically approaches the state transformation z = T**x** in (6.3), and θ approach f = K**x** in (6.2) is well established. However, it will be derived here for completeness.

Let e = (δ − z) be the error between the linear transformation and its estimate. Then the error dynamics is given by

$$\dot{e} = F\delta + G\mathbf{u} + H\mathbf{y} - T(A\mathbf{x} + B\mathbf{u}) \tag{6.5}$$

Adding and subtracting Fz, (6.5) can be written as

$$\dot{e} = Fe + (FT - TA + HC)\mathbf{x} + (G - TB + HD)\mathbf{u} \tag{6.6}$$

For $\dot{e}$ to approach zero asymptotically, it is clearly seen from (6.6) that eigenvalues of F should be on the left half complex plane (stable roots) and

$$TA - FT = HC \tag{6.7}$$

$$TB - HD = G \tag{6.8}$$

Similarly for the error dynamics of (θ − f) to asymptotically approach zero, it is required that

$$MT + NC = K \tag{6.9}$$

$$ND + R = 0 \tag{6.10}$$

6.3 Unknown input observer with mixed outputs

In the case of the UIO, since the inputs are not available for state reconstruction, there is no freedom in selecting the observer matrices G and R and hence G = 0 and R = 0 in (6.8) and (6.10), respectively. This results in the constraints (6.8) and (6.10) taking the form

$$TB - HD = 0 \tag{6.11}$$

$$ND = 0 \tag{6.12}$$

For systems with mixed proper and strictly proper outputs, (6.12) takes the form

$$[N_1 \mid N_2]\begin{bmatrix} D_1 \\ \hdashline D_2 \end{bmatrix} = 0 \tag{6.13}$$

From (6.13) it is clear that $N_1 \equiv 0$. This implies there is no freedom in selecting the N matrix part corresponding to proper outputs. Since $D_2 \equiv 0$ (strictly proper outputs), N_2 can be freely chosen.

The key step in the development of the observer design algorithm is to formulate an eigenstructure assignment problem wherein the eigenvalues define the observer stability matrix F and the eigenvectors define the transformation matrix T. Towards this end, it is convenient to transform the system (6.1) into a canonical form wherein $C = [I_r \ \ 0]$. This structure can always be obtained through a state co-ordinate transformation of the form

$$T_0 = \begin{bmatrix} C_a^{-1} & C_a^{-1} & C_b \\ 0 & I_{n-r} & \end{bmatrix} \tag{6.14}$$

The conditioning of the matrix C_a for inversion is ensured by a permutation of state variables since C is full rank. Under this transformation, (6.1) can be written in partitioned form

$$\begin{aligned} \dot{\mathbf{x}} &= \begin{bmatrix} A_{11} & A_{12} \\ A_{21} & A_{22} \end{bmatrix} \mathbf{x} + \begin{bmatrix} B_1 \\ B_2 \end{bmatrix} \mathbf{u} \\ \mathbf{y} &= \begin{bmatrix} I_r & 0 \end{bmatrix} \mathbf{x} + D\mathbf{u}; \qquad f = \begin{bmatrix} K_1 & K_2 \end{bmatrix} \mathbf{x} \end{aligned} \tag{6.15}$$

where $A_{11} \in R^{r\times r}$, $K_1 \in R^{m\times r}$ and $B_1 \in R^{r\times s}$, and I_r is an rth-order identity matrix.

Multiplying the partitioned matrices in (6.7) yields

$$T_1A_{11} + T_2A_{21} - FT_1 = H \tag{6.16}$$

$$T_1A_{12} + T_2A_{22} - FT_2 = 0 \tag{6.17}$$

where $T_1 \in R^{p\times r}$ and $T_2 \in R^{p\times(n-r)}$.

Substituting for H from (6.16) into (6.11) in partitioned form yields

$$(T_1A_{11} + T_2A_{21} - FT_1)D - T_1B_1 - T_2B_2 = 0 \tag{6.18}$$

Define a *kernel* state space system with the quadruple (A_0, B_0, C_0, D_0), where

$$\begin{aligned} &A_{11}D = \tilde{A}_{11}; \quad A_{21}D = \tilde{A}_{21}; \quad F^T = \mathrm{Diag}(\lambda_i), \quad i = 1\text{–}p \\ &A_0 = A_{22}^T; \quad B_0 = A_{12}^T; \quad C_0 = \tilde{A}_{21}^T - B_2^T; \quad D_0 = \tilde{A}_{11}^T - B_1^T \end{aligned} \tag{6.19}$$

Using (6.16) and (6.17), the eigensystem assignment constraints of the *kernel* system can be written in vector form as

$$S_0(\lambda_i) = \left[\begin{array}{c|c|c} A_0 - \lambda_i I_{n-r} & B_{01} & B_{02} \\ \hline C_0 & D_{01} - \lambda_i D_1^T & D_{02} \end{array}\right]$$

and

$$S_0(\lambda_i)\left[\begin{array}{c} v_i \\ \hline w_i \end{array}\right] = 0, \quad i = 1 - p \tag{6.20}$$

with $D^T = [D_1^T \mid 0]$; D_{01} corresponds to the proper outputs and D_{02} corresponds to the strictly proper outputs. The vectors in (6.20) and their relation to the transformation matrix T are defined as

$$v_i \in R^{n-r}; \quad w_i \in R^r; \quad t_i \in R^n; \quad t_i = \left[\begin{array}{c} v_i \\ \hline w_i \end{array}\right]; \quad T^T = [t_1 \cdots t_p]$$

$$T_a^T = [w_1 \cdots w_p]; \quad T_b^T = [v_1 \cdots v_p]; \quad T = [\ T_a \mid T_b] \tag{6.21}$$

From (6.13), the N matrix takes the form $N = [N_1 \;\; N_2]$, $N_1 = 0$. The matrix N_1 corresponds to the proper outputs $\mathbf{y}_1$. Matrix N_1 has to be set to zero since input signal **u** through matrix D_1 is not available to the UIO. With this restriction on the N matrix, the constraint equation (6.9) in design co-ordinates reduces to

$$[M \mid [0 \quad N_2]]T_C = K, \quad \text{where } T_C = \left[\begin{array}{c|c|c} T_x & T_y & T_z \\ \hline 0 & I_{rs} & 0 \end{array}\right] \tag{6.22}$$

where $[0 \mid I_{rs} \mid 0] = C_2$ corresponds to strictly proper outputs $\mathbf{y}_2$ (in design co-ordinates).

Let

$$T_w = [T_x \mid T_z], \qquad T_w \in R^{p\times(n-rs)} \tag{6.23}$$

Equations (6.20) and (6.22) are the driving constraints for the design of the UIO. Equation (6.20) represents (n − r + s) linear equations in n-variables, and for a non-trivial solution to exist, r > s. In such a case, the null space of the matrix $S_0(\lambda_i)$ has the dimension (r − s). Thus, (r − s) elements in each normalised vector t_i can be freely selected. For (6.22) to have a solution for any K, the observer order has to be chosen as $p = (n - rs)$. In such a case, if $T_w \in R^{(n-rs)\times(n-rs)}$ in (6.23) is non-singular, the matrices M and N can be solved for any functional matrix K. Further if T_w is non-singular, the eigenvector matrix $T \in R^{p\times n}$ is full rank. Finally if the system has only proper outputs (rs = 0), a full-order observer (p = n) needs to be designed.

6.3.1 Necessary conditions for the existence of the observer

The eigenstructure constraint (6.20) can be written as

$$[S-\lambda_i E]t_i = 0, \quad i = 1\text{–}p \tag{6.24}$$

where

$$S = \left[\begin{array}{c|c|c} A_0 & B_{01} & B_{02} \\ \hline C_0 & D_{01} & D_{02} \end{array}\right], \qquad E = \left[\begin{array}{c|c|c} I_{n-r} & 0 & 0 \\ \hline 0 & D_1^T & 0 \end{array}\right] \tag{6.25}$$

$S, E \in R^{mp\times n}$; $mp = n - r + s$; $mp < n$.

The solution of (6.24) is closely related to the solution of the classical generalised eigenvalue problem $Sv = \lambda Ev$, where λ is a generalised eigenvalue and v is the corresponding generalised eigenvector. The Kronecker canonical form structure properties of the associated matrix pencil $[S - \lambda E]$ determine the existence of solution to (6.24) with arbitrary eigenvalues. These structural properties are briefly reviewed in Appendix D.

Essentially the Kronecker canonical form [15,16] of a rectangular (singular) matrix pencil has four structural blocks, namely (i) J – containing the finite eigenvalues, (ii) N – containing the infinite eigenvalues, (iii) L – singular block of right (or column) minimal indices and (iv) L^T – singular block of left (or row) minimal indices. These structural blocks influence the existence of the observer solution.

Theorem 6.1 (Necessary Condition): *The observer has a solution with arbitrary distinct eigenvalues, if the rectangular matrix pencil* $[S - \lambda E]$ *does not have*

Condition 1. L^T block ($L^T \in R^{mr\times nr}$, $mr > nr$) *such that* $nr \geq (n-r)$.
Condition 2. J block ($J \in R^{nf\times nf}$).

Proof: Assume the matrix pencil $[S - \lambda E]$ in (6.24) has L^T and J blocks. The matrix pencil $[S - \lambda E]$ can be transformed to a staircase canonical form (SCF) (Appendix D – (D.7)) using an equivalence transformation as

$$\lambda\bar{E} - \bar{S} = P(\lambda E - S)Q = \left[\begin{array}{c|c|c} \lambda B_r - A_r & 0 & 0 \\ \hline X_r & \lambda B_f - A_f & 0 \\ \hline X_1 & X_2 & \lambda\hat{B} - \hat{A} \end{array}\right] \tag{6.26}$$

where P and Q are unitary matrices. In (6.26), the $mr \times nr$ ($mr > nr$) matrix pencil partition $(\lambda B_r - A_r)$ contains the L^T structure of the pencil. The ($nf \times nf$) matrix pencil $(\lambda B_f - A_f)$ contains the finite generalised eigenvalues of the pencil and $(\lambda\hat{B} - \hat{A})$ contains the L and N blocks. The pencils X_r, X_1 and X_2 are non-zero.

Rewriting (6.24) in the canonical form of (6.26) yields

$$\left[\begin{array}{c|c|c} \lambda B_r - A_r & 0 & 0 \\ \hline X_r & \lambda B_f - A_f & 0 \\ \hline X_1 & X_2 & \lambda \hat{B} - \hat{A} \end{array}\right] \begin{bmatrix} \eta_r \\ \hline \eta_f \\ \hline \eta_c \end{bmatrix} = \begin{bmatrix} 0 \\ \hline 0 \\ \hline 0 \end{bmatrix} \tag{6.27}$$

The vector (η) is related to the eigenvector (t) through the equivalence transformation matrix Q as

$$t = Q\eta, \quad Q = \left[\begin{array}{c|c} Q_z & 0 \\ \hline 0 & I_{rs} \end{array}\right], \eta = \begin{bmatrix} \eta_r \\ \hline \eta_f \\ \hline \eta_c \end{bmatrix} \tag{6.28}$$

$$t, \eta \in R^n, \eta_r \in R^{nr}, \eta_f \in R^{nf}, \eta_c \in R^{nc}, Q_z \in R^{(n-rs)\times(n-rs)}$$

Condition 1. If the matrix pencil has an L^T block, from (6.27) it is seen that the matrix equation

$$[\lambda B_r - A_r][\eta_r] = 0 \tag{6.29}$$

Equation (6.29) constitutes an overdetermined set of linear equations (since mr > nr) and for all eigenvalue assignments has a solution $\eta_r \equiv 0$. This implies nr-components of vector v in (6.20) are zero. Further if $nr \geq (n - r)$, vector v in (6.20) is identically zero for any desired eigenvalue assignment. Hence, matrix T_w in (6.23) is singular, indicating that the observer does not exist.

Remarks:

1. For the case nr < (n − rs), the possibility of finding a solution exists. However, the presence of a L^T block drops the normal rank of the matrix $(S - \lambda E)$ for all eigenvalues. The normal rank of an $(m \times n)$ matrix is defined as min(m, n) and in the observer case it is m (m < n). Thus, the dimension of the null space of the matrix increases, allowing for more freedom in selection of the eigenvectors.
2. If the matrix pencil has a J block, choosing the observer eigenvalues corresponding to the finite eigenvalues of the pencil also results in the normal rank of the matrix $(S - \lambda E)$ dropping. This increases the dimension of the null space as in the case of the L^T block.

Condition 2. Let the matrix pencil has no L^T block and have only a J block; the SCF (6.27) reduces to

$$\left[\begin{array}{c|c} \lambda B_f - A_f & 0 \\ \hline X_2 & \lambda \hat{B} - \hat{A} \end{array}\right] \begin{bmatrix} \eta_f \\ \hline \eta_c \end{bmatrix} = \begin{bmatrix} 0 \\ \hline 0 \end{bmatrix} \tag{6.30}$$

From (6.30), it is seen that the matrix equation

$$[\lambda B_f - A_f][\eta_f] = 0 \tag{6.31}$$

has a non-trivial solution only for the nf-finite generalised eigenvalues corresponding to the matrix pencil in (6.31). Thus, for arbitrary eigenvalues, excluding those of the J block, vector v in (6.20) is identically zero. Hence, matrix Tw in (6.23) is singular, indicating that the observer does not exist.

Remark:

1. If $(n-r+s)=(n-1)$, there is no freedom in selection of eigenvector elements (t). In such a situation, Theorem 6.1 serves as *both necessary and sufficient* condition for the existence of the solution.

The observer design procedure for a system with mixed outputs can now be summarised as follows.

Algorithm 6.1:

Step 1. Construct the matrix pencil (6.24). Check if Theorem 6.1 is valid (*necessary condition*).

Step 2. If Theorem 6.1 is valid, for i = 1–p (p = n − rs), specify desired distinct eigenvalues λ_i and construct $N_p = N_s\{(S - \lambda_i E)\}$. $N_p \in R^{n\times(r-s)}$ is an $(r - s)$-dimensional subspace and $N_s\{\cdot\}$ indicates null space. Specify vector $X_{p_i} \in R^{(r-s)}$, which is a linear combination of the basis vectors of N_p (free-to-choose eigenvector elements), and compute the normalised eigenvector, $t_i = N_p * X_{p_i}$.

Step 3. Select vectors X_{p_i}, $i = 1$–p, in step 2 such that matrix T_w is non-singular (*sufficient condition*).

Step 4. Construct $F^T = \mathrm{Diag}(\lambda_i)$.
Construct the transformation matrix $T = [T_a \quad T_b]$ (6.21).
Compute M and N_2 from (6.22) and note that $N = [0 \mid N_2]$.

Step 5. Compute H from (6.16). G = 0 and R = 0 in (6.4).

Remark:

1. The choice of the eigenvalues (λ_i) and arbitrary eigenvector elements matrix X_p in step 2 determine the linear independence of the eigenvectors v_i and hence the numerical conditioning of transformation matrices T_w and T. Thus, the eigenstructure of (6.24) can be optimised to obtain well-conditioned transformation matrix T that results in low observer gains (H, M and N).

6.4 Unknown input observers with strictly proper outputs

If the system has only strictly proper outputs, rp = 0, D = 0 in (6.1) and the analysis of section 6.3 is still valid. However, the observer matrix N will now have full freedom in selection. It will be shown that this additional freedom can be utilised to construct a functional observer of order $m \le p \le (n - r)$. The analysis is primarily based on the approach used in Reference 4. Writing (6.9) in partitioned form (design co-ordinates), we have

$$M[T_1 \quad T_2] + N[I_r \quad 0] = [K_1 \quad K_2], \quad T_2, K_2 \in R^{m\times(n-r)} \tag{6.32}$$

Equation (6.32) leads to

$$MT_2 = K_2 \tag{6.33}$$

$$MT_1 + N = K_1 \tag{6.34}$$

Lemma 6.1 *Equation (6.33) has a solution if and only if*

$$\text{Rank}[T_2^T \quad K_2^T] = \text{Rank}[T_2^T] \tag{6.35}$$

This leads to a lower bound on the observer order as

$$p = \text{Rank}(K_2^T) = \sigma_k \tag{6.36}$$

The upper bound of $\sigma_k = m$. It is now possible to combine (6.20) and (6.32) to generate additional constraints on the construction of the transformation matrix T as

$$S_0(\lambda_i) = \begin{bmatrix} A_0 - \lambda_i I_{n-r} & B_0 \\ C_0 & D_0 \end{bmatrix}, \qquad S_0(\lambda_i)\begin{bmatrix} v_i \\ \hline w_i \end{bmatrix} = 0, \quad i = 1-p \tag{6.37}$$

The vectors in (6.37) are defined as follows:

$$\begin{gathered} v_i \in R^{n-r}; \quad w_i \in R^r; \quad t_i \in R^n; \quad t_i = \begin{bmatrix} v_i \\ \hline w_i \end{bmatrix}; \quad T^T = [t_1 \cdots t_p] \\ T_1^T = [w_1 \cdots w_p]; \ T_2^T = [v_1 \cdots v_p]; \quad T = [T_1 \mid T_2] \end{gathered} \tag{6.38}$$

The corresponding matrix pencil is

$$S = \begin{bmatrix} A_0 & B_0 \\ C_0 & D_0 \end{bmatrix}, \qquad E = \begin{bmatrix} I_{n-r} & 0 \\ 0 & 0 \end{bmatrix} \tag{6.39}$$

The matrices C_0 and D_0 take the form

$$C_0 = \begin{bmatrix} -B_2^T \\ \hline C_q \end{bmatrix}, \qquad D_0 = \begin{bmatrix} -B_1^T \\ \hline 0 \end{bmatrix} \tag{6.40}$$

where C_q is the additional *kernel* constraint matrix to satisfy the conditions of Lemma 6.1. The following relation dictates the number of kernel constraint equations q:

$$p + q = n - r \tag{6.41}$$

The constraint matrix $C_q \in R^{q \times (n-r)}$ is computed as

$$N_q = N_s\{(K_2^T)\}; \qquad C_q^T = N_q X_q, \quad X_q \in R^{(n-r-\sigma_k)\times q} \tag{6.42}$$

where $N_s\{\cdot\}$ indicates null space. $X_q \in R^{(n-r-\sigma_k)\times q}$ is a free-to-choose *kernel system* design parameter matrix, *provided* $q < (n - r - \sigma_k)$. X_q is chosen such that C_q^T is full rank. Note that the size of the X_q matrix is dependent on σ_k (6.36).

Equation (6.37) constitutes a set of $(n - r + s + q)$ linear equations in n-variables. For (6.37) to have a non-trivial solution, another lower bound, using (6.41), on the observer order can be derived as

$$p = n - 2r + s + 1 = \sigma_b \tag{6.43}$$

Combining the two lower bounds in (6.36) and (6.43), and noting that the upper bound on the functional observer is $p = (n - r)$ corresponding to the reduced-order observer, we have the lower bound for the order of the unknown input functional observer with strictly proper outputs as

$$p = \min\{(n-r), \max(\sigma_k, \sigma_b)\} \tag{6.44}$$

The design of the unknown input functional observer for strictly proper systems can now be summarised as follows.

Algorithm 6.2:

Step 1. Initialise the observer order from (6.44) or step 4 below.
Select X_q in (6.42) if *kernel* system constraint exists.

Step 2. Construct the matrix pencil (6.39). Check Theorem 6.1 is valid (*necessary condition*).

Step 3. If Theorem 6.1 is not valid, modify C_q (6.42) using X_q (if freedom exists) and go to step 2.

Step 4. If step 3 fails for all selections of X_q to meet the necessary condition of step 2, then if $p < (n - r)$, increment $p = p + 1$, and go to step 1.
If $p > n - r$; STOP (No solution)

Step 5. If Theorem 6.1 is valid, for $i = 1$–p, specify desired distinct eigenvalues λ_i and construct N_p.
$N_p \in R^{n\times(r-s-q)}$ is an $(r-s-q)$-dimensional subspace. Specify $X_{p_i} \in R^{(r-s-q)}$, which is a linear combination of the basis vectors of N_p (free-to-choose eigenvector elements), and compute the normalised eigenvector, $t_i = N_p * X_{p_i}$.

Step 6. Select vectors X_{p_i}, $i = 1-p$, in step 5 such that
(a) $T_2 \in R^{p\times(n-r)}$ is in the range space of K_2 (Lemma 6.1).
(b) $T \in R^{p\times n}$ is full rank.
The above conditions constitute *sufficient conditions* for solution to exist.

Step 7. Construct $F^T = \text{Diag}(\lambda_i)$. Construct the transformation matrix

$$T = [T_1 \quad T_2]$$

Step 8. Compute observer matrices H from (6.16), M from (6.33) and N from (6.34). Finally $G = 0$ and $R = 0$.

Remarks:

1. Each time p is incremented in step 4, the number of rows in C_q reduces, thus reducing the complexity of the assignment in step 5.
2. The procedure will terminate when $p = n - r$ ($q = 0$, $C_q = 0$) and the design corresponds to the reduced-order UIO.
3. The choice of the eigenvalues (λ_i) and arbitrary eigenvector elements matrix X_p in step 6 determine the linear independence of the eigenvectors v_i and hence the numerical conditioning of transformation matrices T_2 and T.

6.5 Known input observer

In the case of KIOs, since the inputs are available for constructing the observer, additional freedom in selection of observer matrices G and R exists ((6.8) and (6.10)). Algorithm 6.2 is directly applicable to the design of known input functional observers. The modifications to Algorithm 6.2 are

(i) $s = 0$;

(ii) $C_0 = C_q$ and $D_0 = 0$, in (6.37);

(iii) $G = TB - HD$ from (6.8) and $R = -ND$ from (6.10).

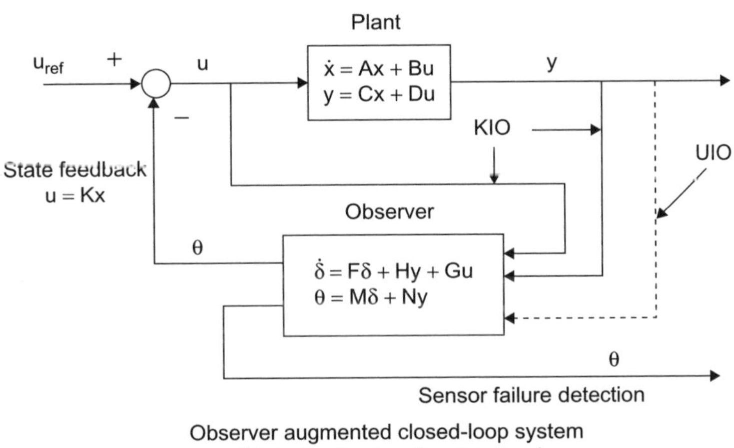

$$\begin{bmatrix} \dot{x} \\ \dot{\delta} \end{bmatrix} = \left[\begin{array}{c|c} A + BNC & BM \\ \hline HC + GNC & F + GM \end{array}\right] \begin{bmatrix} x \\ \delta \end{bmatrix} + \begin{bmatrix} B \\ G \end{bmatrix} u_{ref}$$

$$\theta = \begin{bmatrix} NC & M \end{bmatrix} \begin{bmatrix} x \\ \delta \end{bmatrix}$$

Figure 6.1 Schematic of functional observer controller

The constraints (6.11) and (6.12) required for the UIO design vanish due to the freedom in selection of G and R matrices (iii). If the system has only strictly proper outputs, $D = 0$ resulting in $G = TB$ and $R = 0$.

Figure 6.1 gives a schematic of the observer. The figure also includes the closed-loop state space system when a feedback control law system is implemented using the functional observer.

The feedback system given in Figure 6.1 is applicable to UIO. However, for KIO it is applicable only to strictly proper system ($D = 0$, $R = 0$). For KIO with mixed outputs ($D \neq 0$), the feedthrough matrix $R \neq 0$ (6.4), and the closed-loop system has to be constructed using this additional feedback loop.

6.6 Examples

Example 6.1: In this tutorial example the design of the UIO will be illustrated. Consider the dynamic system (6.1) with system matrices

$$A = \begin{bmatrix} -1 & 0 & 1 \\ 0 & -1 & -1 \\ -1 & 0 & 0 \end{bmatrix}, \quad B = \begin{bmatrix} 0 \\ 0 \\ 1 \end{bmatrix}, \quad C = \begin{bmatrix} 1 & 0 & 0 \\ 0 & 1 & 0 \end{bmatrix}, \quad K = \begin{bmatrix} 0 & 0 & 1 \end{bmatrix}$$

$$\text{Case 1:} \quad D = \begin{bmatrix} -1 & 1 \end{bmatrix}^T, \qquad \text{Case 2:} \quad D = \begin{bmatrix} 1 & 1 \end{bmatrix}^T$$
$$\text{Case 3:} \quad D = \begin{bmatrix} 1 & 0 \end{bmatrix}^T, \qquad \text{Case 4:} \quad D = \begin{bmatrix} 0 & 0 \end{bmatrix}^T \tag{6.45}$$

Case 1. The system has only proper outputs and hence $p = 3$ (full order). There is no freedom in selection of eigenvectors. The system matrix pencil (6.25) is

$$S = \begin{bmatrix} 0 & 1 & -1 \\ 0 & 1 & -1 \end{bmatrix}, \qquad E = \left[\begin{array}{cc|c} 1 & 0 & 0 \\ 0 & -1 & 1 \end{array}\right] \tag{6.46}$$

The Kronecker block structure is given by $J_1(0) \oplus J_1(-1) \oplus L_0$. The symbol $\oplus$ denotes summation of the canonical blocks (Appendix D). The matrix pencil has an SCF

$$\bar{S} = \begin{bmatrix} 2 & 0 & 0 \\ 0 & 0 & 0 \end{bmatrix}, \qquad \bar{E} = \left[\begin{array}{cc|c} -1 & -0.7071 & 0 \\ 1 & -0.7071 & 0 \end{array}\right] \tag{6.47}$$

The matrix pencil has a finite block J given by

$$A_f = \begin{bmatrix} 2 & 0 \\ 0 & 0 \end{bmatrix}, \qquad B_f = \begin{bmatrix} -1 & -0.7071 \\ 1 & -0.7071 \end{bmatrix} \tag{6.48}$$

The finite pencil has eigenvalues $\lambda_1 = 0$ and $\lambda_2 = -1$. Thus, from Theorem 6.1 an observer with arbitrary eigenvalues does not exist. The eigenvector has the form $t_i = [0 \;\; 0.7071 \;\; 0.7071]^T$, for all λ_i (except for λ_1 and λ_2), since the dimension of the null space of the matrix pencil is unity. This results in T_w being singular. However, for the eigenvalues of the finite pencil, the dimension of the null space of the pencil is two. Thus, an observer can be constructed by assigning the finite eigenvalues of the pencil with $\Lambda = [\lambda_1, \lambda_2, \lambda_{arbitrary}]$. Let the eigenvalues be $\Lambda = [0, -1, -2]$. Since the rank of the matrix pencil for finite eigenvalues of the pencil drops rank ($\lambda_1 = 0$ and $\lambda_2 = -1$), the free-to-choose eigenvector matrix is selected as $X_p = \begin{bmatrix} 1 & 1 \\ 0 & 0 \end{bmatrix}$ (selecting the first basis vector of the null space of the matrix $(S - \lambda_1 E)$ and $(S - \lambda_2 E)$). The resulting observer matrices are

$$F = \begin{bmatrix} 0 & 0 & 0 \\ 0 & -1 & 0 \\ 0 & 0 & -2 \end{bmatrix}, \quad H = \begin{bmatrix} 0 & 0 \\ 0\ 5774 & 0 \\ 0.7071 & 0.7071 \end{bmatrix},$$

$$M = \begin{bmatrix} -1 & 0 & 0 \end{bmatrix}, \quad N = \begin{bmatrix} 0 & 0 \end{bmatrix} \tag{6.49}$$

$$T = \begin{bmatrix} 0 & 0 & -1 \\ 0.7887 & 0.2113 & -0.5774 \\ 0.7071 & 0.7071 & 0 \end{bmatrix}$$

It should be noted that the N matrix is zero since system has all proper outputs.

Case 2. The system has only proper outputs and hence p = 3 (full order). The SCF satisfies condition of Theorem 6.1. Hence, a third-order observer with arbitrary eigenvalues exists. The system matrix pencil (6.25) is

$$S = \begin{bmatrix} 0 & 1 & -1 \\ -2 & -1 & -1 \end{bmatrix}, \quad E = \left[\begin{array}{cc|c} 1 & 0 & 0 \\ 0 & 1 & 1 \end{array}\right] \tag{6.50}$$

The matrix pencil is generic with an L_2 block. The corresponding SCF is

$$\bar{S} = \begin{bmatrix} -1.4142 & -2 & 0 \\ 0 & 0 & 1.4142 \end{bmatrix}, \quad \bar{E} = \begin{bmatrix} 1.4142 & 0 & 0 \\ 0 & -1 & 0 \end{bmatrix} \tag{6.51}$$

The observer has a solution since the matrix pencil does not have J or L^T block. For a choice of $\lambda = [-2, -1 \pm j1]$, the observer matrices are

$$F = \begin{bmatrix} -2 & 0 & 0 \\ 0 & -1 & -1 \\ 0 & 1 & -1 \end{bmatrix}, \quad H = \begin{bmatrix} -0.4472 & 0.8944 \\ 0.1923 & 0.4615 \\ 0.4615 & -0.1923 \end{bmatrix},$$

$$T = \begin{bmatrix} 0 & 0.8944 & 0.4472 \\ -0.7308 & 0.1923 & 0.6538 \\ 0.8462 & 0.4615 & 0.2692 \end{bmatrix} \tag{6.52}$$

$$M = \begin{bmatrix} -1.1180 & 1.6923 & 1.4615 \end{bmatrix}, \quad N = \begin{bmatrix} 0 & 0 \end{bmatrix}$$

It should be noted that the N matrix is zero since system has all proper outputs.

Case 3. This is a system with a mix of proper and strictly proper outputs. The order of the observer p = 2(p = n − rs). The system satisfies condition of Theorem 6.1. A second-order observer exists. The system matrix pencil (6.25) is

$$S=\begin{bmatrix}0 & 1 & -1\\ -2 & -1 & 0\end{bmatrix},\qquad E=\left[\begin{array}{cc|c}1 & 0 & 0\\ 0 & 1 & 0\end{array}\right] \tag{6.53}$$

The matrix pencil is generic with an L_2 block. The corresponding SCF is

$$\bar{S}=\begin{bmatrix}-1 & 2 & 0\\ 1 & 0 & -1\end{bmatrix},\qquad \bar{E}=\begin{bmatrix}1 & 0 & 0\\ 0 & -1 & 0\end{bmatrix} \tag{6.54}$$

The observer has a solution since the matrix pencil does not have J or L^T block. For a choice of $\lambda=[-1 \pm j1]$, the observer matrices are

$$F=\begin{bmatrix}-1 & -1\\ 1 & -1\end{bmatrix},\qquad H=\begin{bmatrix}0.4587 & 0.7331\\ 0.2744 & -0.1844\end{bmatrix},$$

$$T=\begin{bmatrix}-0.5487 & 0.1844 & 0.4587\\ 0.9175 & 0.7331 & 0.2744\end{bmatrix} \tag{6.55}$$

$$M=[1.6056 \quad 0.9603],\qquad N=[0 \quad -1]$$

It should be noted that the N matrix element corresponding to the proper output is zero.

Case 4. This is a system with only strictly proper outputs. From (6.44) observer order p = 1 and also corresponds to reduced-order observer. The system matrix pencil (6.25) is

$$S=\begin{bmatrix}0 & 1 & -1\\ -1 & 0 & 0\end{bmatrix},\qquad E=\left[\begin{array}{cc|c}1 & 0 & 0\\ 0 & 0 & 0\end{array}\right] \tag{6.56}$$

The Kronecker block structure is given by $J_1(\infty)\oplus N_1 \oplus L_0$. The corresponding SCF is

$$\bar{S}=\left[\begin{array}{c|cc}-1 & 0 & 0\\ \hline 0 & 1 & -1\end{array}\right],\qquad \bar{E}=\left[\begin{array}{c|cc}0 & 0 & 0\\ \hline 1 & 0 & 0\end{array}\right] \tag{6.57}$$

The matrix pencil has a J block where

$$A_f=-1,\qquad B_f=0 \tag{6.58}$$

The matrix pencil has an eigenvalue $\lambda_r=-\infty$. In general, the J block will not have an infinite eigenvalue. In this example since $B_f=0$, the matrix pencil has an infinite eigenvalue. The condition of Theorem 6.1 is violated. The solution to (6.37) has the form $t_i=[0 \mid 0.7071 \quad 0.7071]^T$ for all real eigenvalues. Thus, matrix T_w is singular and the observer does not exist.

Example 6.2: Observers play an important role in the fault diagnosis of aircraft sensors [14]. In particular, use of observers in estimating aircraft angle of attack and sideslip using accurate inertial rate and acceleration sensors provides analytical redundancy in the fault detection of flow angle sensors. The design of such an observer to estimate the sideslip angle of an aircraft using roll rate, yaw rate and lateral acceleration sensors will now be illustrated.

Consider the lateral dynamics of an aircraft in the form (6.1) with system matrices

$$A = \begin{bmatrix} -3.79 & 0.046 & -52.0 & 0 \\ -0.134 & -0.359 & 4.24 & 0 \\ 0.062 & -0.997 & -0.272 & 0.0462 \\ 1.0 & 0.0603 & 0 & 0 \end{bmatrix}, \quad B = \begin{bmatrix} 25 & 9.83 \\ 1.42 & -4.2 \\ 0.005 & 0.0503 \\ 0 & 0 \end{bmatrix} \tag{6.59}$$
$$C_1 = \begin{bmatrix} -0.125 & -0.0612 & -3.41 & -0.0015 \end{bmatrix}, \; D_1 = \begin{bmatrix} 1\ 0300 & -0.2660 \end{bmatrix}$$
$$K = \begin{bmatrix} 0 & 0 & 1 & 0 \end{bmatrix}$$

with $x = [p \quad r \quad \beta \quad \phi]^T$, $u = [\delta_a \quad \delta_r]$, $y_1 = n_y$ and $f = \beta$. The aircraft state variables are p – roll rate, r – yaw rate, β – sideslip angle, φ – bank angle, δ_a – aileron deflection, δ_r – rudder deflection and n_y – lateral acceleration (g-units). All angles and angular rates are in radian units. The observer design problem is to estimate the aircraft sideslip angle β using the inertial sensors (p, r, n_y, φ). It should be noted that the letter 'p' will be used to denote aircraft roll rate as well as order of the observer without causing any confusion.

Case A. UIO with strictly proper outputs: $[p \quad r \quad \varphi]$

From (6.44) the observer order is $p = 1$. The Kronecker block structure is given by $J_1(-0.1885) \oplus 2N_1 \oplus L_0$. The matrix pencil has a finite eigenvalue block with

$$A_f = 0.0804, \qquad B_f = -0.4267 \tag{6.60}$$

The matrix pencil has an eigenvalue $\lambda_0 = -0.1885$. Condition of Theorem 6.1 is violated. The solution to (6.37) has the form $t = [0 \quad 0 \quad 0 \quad 1]$ for all choices of λ, except for λ_0. Matrix T_w is singular, and the observer with arbitrary eigenvalues does not exist. However, for λ_0, (6.37) has a solution as $t = [0.99 \quad -0.0008 \quad 0.0102 \quad 0]$ with T_w non-singular. Hence, a first-order stable observer can be constructed with observer matrices

$$\begin{aligned} &F = -0.1885, \qquad H = [-0.0634 \quad 0.9987 \quad -0.0462] \\ &M = -1.0001, \qquad N = [0.0008 \quad -0.0102 \quad 0] \\ &T = [0.0008 \quad -0.0102 \quad -0.9999 \quad 0] \end{aligned} \tag{6.61}$$

Case B. UIO with mixed outputs: $y = [n_y \quad p \quad r]$

The observer order $p = 2$ $(p = n - rs)$. The matrix pencil has no J or L^T blocks. It is generic with the structure L_3. Theorem 6.1 is valid. For a choice of $\lambda = -1 \pm j2$, the observer matrices are

$$\mathrm{F} = \begin{bmatrix} -1 & -2 \\ 2 & -1 \end{bmatrix}, \qquad \mathrm{H} = \begin{bmatrix} 2.5065 & 0.2988 & -0.9125 \\ 3.1991 & -0.3615 & 3.4310 \end{bmatrix}$$

$$\mathrm{M} = \begin{bmatrix} 0.0220 & -0.2557 \end{bmatrix}, \qquad \mathrm{N} = \begin{bmatrix} 0 & 0.0261 & 0.0974 \end{bmatrix} \tag{6.62}$$

$$\mathrm{T} = \begin{bmatrix} 0.0815 & 0.3752 & 2.1378 & -0.0875 \\ 0.1091 & 0.4133 & -3.7273 & -0.0075 \end{bmatrix}$$

It should be noted that the element of N matrix corresponding to n_y output is zero.

Case C. Known input functional observer: $y = [p \quad r]$
From (6.43), with s = 0, the required observer order is p = 1. The matrix pencil of (6.25) has the form

$$\mathrm{S} = \left[\begin{array}{c|ccc} -0.2720 & 0 & -52.0 & 4.24 \\ \hline 0.0462 & 0 & 0 & 0 \\ 0 & 1 & 0 & 0 \end{array}\right], \qquad \mathrm{E} = \left[\begin{array}{c|ccc} 1 & 0 & 0 & 0 \\ \hline 0 & 1 & 0 & 0 \\ 0 & 0 & 0 & 0 \end{array}\right] \tag{6.63}$$

The Kronecker block structure is given by $J_1(\infty) \oplus N_2 \oplus L_0$.
The SCF is given by

$$\bar{\mathrm{S}} = \left[\begin{array}{c|ccc} 1 & 0 & 0 & 0 \\ \hline 0 & -0.0462 & 0 & 0 \\ 0 & 0.2720 & -52 & 4.24 \end{array}\right], \qquad \bar{\mathrm{E}} = \left[\begin{array}{c|ccc} 0 & 0 & 0 & 0 \\ \hline 1 & 0 & 0 & 0 \\ 0 & -1 & 0 & 0 \end{array}\right] \tag{6.64}$$

The matrix pencil has a J block where

$$A_f = 1, \qquad B_f = 0 \tag{6.65}$$

The matrix pencil has an eigenvalue $\lambda_0 = \infty$. The condition of Theorem 6.1 is violated. A first-order KIO does not exist.

In Algorithm 6.2, incrementing the order of the observer to p = 2 (step 4) results in a reduced-order observer (p = n − r). In this case, the *kernel system* constraints vanish (q, C_0 and $D_0 \equiv 0$). The resulting matrix pencil satisfies conditions of Theorem 6.1, and a reduced-order observer exists. The matrix pencil of (6.39) has the form

$$\mathrm{S} = \begin{bmatrix} -0.272 & 0 & -52.0 & 4.24 \\ 0.0462 & 0 & 0 & 0 \end{bmatrix}, \qquad \mathrm{E} = \begin{bmatrix} 1 & 0 & 0 & 0 \\ 0 & 1 & 0 & 0 \end{bmatrix} \tag{6.66}$$

The Kronecker block structure is given by $2L_1$, and the matrix pencil is generic.

The SCF is given by

$$\bar{\mathrm{S}} = \begin{bmatrix} 0 & -0.0462 & 0 & 0 \\ 0 & 0.272 & -52 & 4.24 \end{bmatrix}, \qquad \bar{\mathrm{E}} = \begin{bmatrix} 1 & 0 & 0 & 0 \\ 0 & -1 & 0 & 0 \end{bmatrix} \tag{6.67}$$

For a choice of $\lambda = -1 \pm j2$, and matrix $X_p = \begin{bmatrix} 1 & -7 \\ 0.5 & 1 \end{bmatrix}$, the observer matrices are

$$F = \begin{bmatrix} -1 & -2 \\ 2 & -1 \end{bmatrix}, \quad G = \begin{bmatrix} -4.0975 & -4.3835 \\ -2.3788 & -3.3601 \end{bmatrix}, \quad H = \begin{bmatrix} 0.1850 & -0.2356 \\ 0.2741 & 6.0206 \end{bmatrix}$$

$$M = [-0.0394 \quad -0.1521], \quad N = [-0.0259 \quad 0.0903], \quad R = [0 \quad 0]$$

$$T = \begin{bmatrix} -0.1981 & 0.5977 & 1.4833 & -0.1423 \\ -0.1187 & 0.4389 & -6.9588 & 0.0369 \end{bmatrix} \tag{6.68}$$

Case D. Known input functional observer: $y = [n_y \quad r]$
From (6.43), with s = 0, the observer order is p = 1. The matrix pencil (6.39) is given by

$$S = \left[\begin{array}{cc|cc} -1.8838 & 1 & 0.0063 & -0.2894 \\ \hline 0.0229 & 0 & -0.1607 & -0.0019 \\ -0.0120 & 1 & 0 & 0 \end{array}\right], \quad E = \left[\begin{array}{cc|cc} 1 & 0 & 0 & 0 \\ \hline 0 & 1 & 0 & 0 \\ 0 & 0 & 0 & 0 \end{array}\right] \tag{6.69}$$

The Kronecker block structure is given by $L_1 \oplus N_2$.

The SCF is given by

$$\bar{S} = \left[\begin{array}{cc|cc} -0.0120 & 1 & 0 & 0 \\ \hline -1.884 & 1 & 0.0076 & -0.2894 \\ -0.0077 & -0.0081 & 0.1606 & 0.0042 \end{array}\right], \quad \bar{E} = \left[\begin{array}{cc|cc} 0 & 0 & 0 & 0 \\ \hline 1 & -0.0081 & 0 & 0 \\ -0.0081 & -1 & 0 & 0 \end{array}\right] \tag{6.70}$$

A KIO with p = 1 meets the condition of Theorem 6.1. For a choice of $\lambda = -6$, $X_p = 1$ (no eigenvector selection freedom) and $X_q = 1$ (no *kernel system* design freedom), the observer matrices are

$$\begin{aligned} &F = -6, \quad H = [-0.0588 \quad 5.7149], \quad M = -0.5247, \\ &N = [-0.2777 \quad 0.5052] \quad G = [3.1272 \quad -3.5507], \\ &R = [0.2860 \quad -0.0739] \quad T = [0.0662 \quad 0.9953 \quad -0.1012 \quad 0.0008] \end{aligned} \tag{6.71}$$

In summary, the sideslip angle (β) of an aircraft can be estimated with a first-order known input functional observer using yaw rate and lateral acceleration sensors. A second-order UIO exists using roll rate, yaw rate and lateral acceleration sensors. UIO, with arbitrary eigenvalues, using only strictly proper outputs does not exist. The example illustrates the utility of the properties of matrix pencils in determining the existence of appropriate-order observers. The example also highlights that the lateral acceleration sensor, which is called a surrogate signal for sideslip, is essential in the estimation of sideslip angle with the lowest order observers.

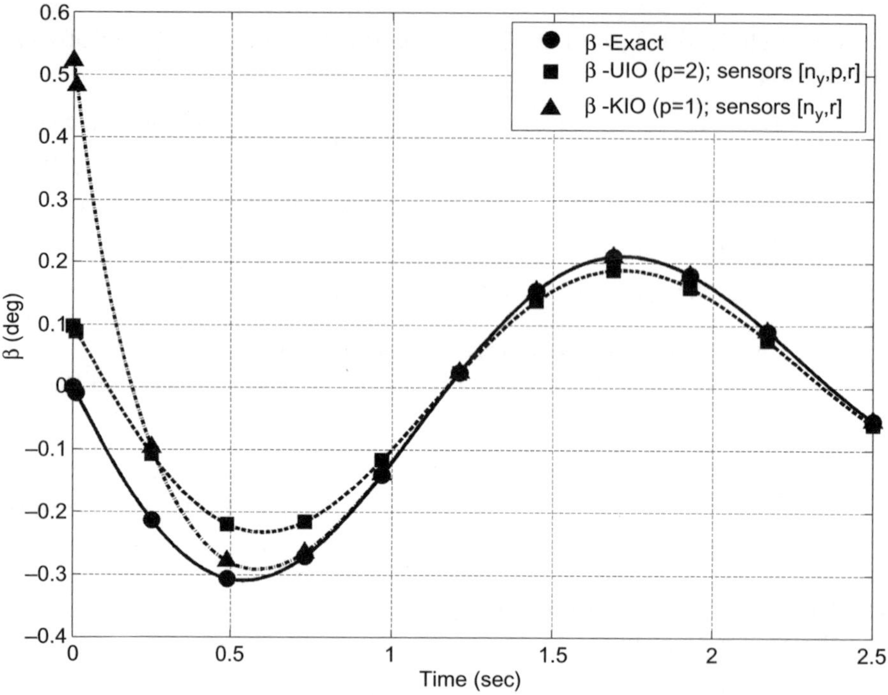

Figure 6.2 Sideslip response to yaw rate disturbance (r(0) = 1 deg/sec)

Figure 6.2 demonstrates the asymptotic convergence characteristics of the observers in estimating sideslip angle (β), for an initial yaw rate disturbance (r(0) = 1 deg/sec).

6.7 Summary

In this chapter a unified approach to the design of observers has been detailed. The algorithm developed is applicable to all categories of observers studied in the literature. The key concept of eigenstructure assignment developed in Chapter 2 is used in the construction of the observer. The observer eigenvalues define the observer stability matrix F, and the eigenvectors define the transformation matrix T. Defining the stability matrix in the modal canonical form facilitates formulation of a computationally simple observer design algorithm. Further this modal canonical observer structure also enjoys advantages in real-time implementation since (i) the stability matrix F has a minimum parameter representation and (ii) only a bank of first-order (real eigenvalues) or second-order (complex conjugate eigenvalues) filters needs to be implemented. Existence conditions of the observer with arbitrary eigenvalues are derived using the Kronecker canonical structure properties of the resulting singular rectangular matrix pencils. The computationally stable SCF is used to extract these structural properties. However, construction of the SCF is only required to check the existence of the solution for a specified observer problem and is not needed for design optimisation of the observer.

For a system with n-states, s-inputs, r-outputs and m-functionals, the minimum observer order, for different categories of observers, is

1. UIO with mixed (proper and strictly proper) outputs: $p = (n - rs)$, where rs is the number of strictly proper outputs. As a corollary, if the observer has only proper outputs, $p = n$.
2. UIO with strictly proper outputs: $p = \min\{(n-r), \max(\sigma_k, \sigma_b)\}$, where $\sigma_b = n - 2r + s + 1$, and σ_k is the rank of the functional's submatrix in the design canonical form.
3. KIO: $p = \min\{(n-r), \max(\sigma_k, \sigma_b)\}$, where $\sigma_b = n - 2r + 1$, and σ_k is the rank of the functional's submatrix in the design canonical form.

The two primary applications of observers are (i) implementing a state variable feedback control law using only the available sensor set with the observer acting as a dynamic compensator and (ii) analytically estimating a sensor response in the sensor FDI schemes. The functional observer design methodology discussed in this chapter aids in arriving at the lowest observer order to meet the given estimation problem.

Finally an unknown input functional observer that estimates a state variable feedback control law could be looked upon as the design of a dynamic output feedback compensator with guaranteed stability of the closed-loop system as well as that of the compensator. If the order of this observer is comparable to that of the dynamic compensator design based on eigenstructure assignment discussed in Chapter 4, the observer-based design is to be preferred since the stability of the compensator is also guaranteed.

References

1. KAILATH, T.: *Linear Systems* (Englewood Cliffs, NJ, Prentice-Hall, 1980)
2. LUENBERGER, D.G.: 'Observers for multi-variable systems', *IEEE Trans. Automat. Control*, 1966, **AC-11**, pp. 190–197
3. FORTMAN, T.E., and WILLAMSON, D.: 'Design of low-order observers for linear feedback control laws', *IEEE Trans. Automat. Control*, 1972, **AC-17**, pp. 301–308
4. SRINATHKUMAR, S.: 'Minimal order observers for linear functions of the state vector', *Electron. Lett.*, 1978, **14**, (9), pp. 275–277
5. TSUI, C.C.: 'A new algorithm for the design of multifunctional observers', *IEEE Trans. Automat. Control*, 1985, **30**, (1), pp. 89–93
6. ALDEEN, M., and TRINH, H.: 'Reduced-order linear functional observer for linear systems', *IEE Proc. Control Theory Appl.*, 1999, **146**, (5), pp. 399–405

7. DAROUACH, M.: 'Existence and design of functional observers for linear systems', *IEEE Trans. Automat. Control*, 2000, **45**, (5), pp. 940–943
8. HOU, M., and MULLER, P.C.: 'Design of observers for linear systems with unknown inputs', *IEEE Trans. Automat. Control*, 1992, **37**, (6), pp. 871–875
9. SYRMOS, V.L.: 'Computational observer design techniques for linear systems with unknown inputs using the concepts of transmission zeros', *IEEE Trans. Automat. Control*, 1993, **38**, (5), pp. 790–794
10. LIN, S.F., and WANG, A.P.: 'Unknown input observers designed by eigenstructure assignment', *SAMS*, 2002, **42**, (3), pp. 415–428
11. LIN, S.F., and WANG, A.P.: 'Reduced order unknown input observers designed by eigenstructure assignment', *SAMS*, 2001, **39**, (4), pp. 497–506
12. TRINH, H., and HA, Q.: 'Design of linear functional observers for linear systems with unknown inputs', *Int. J. Syst. Sci.*, 2000, **31**, (6), pp. 741–749
13. FRANK, P.M.: 'Fault diagnosis in dynamic systems using analytical and knowledge-based redundancy – a survey and some new results', *Automatica*, 1990, **26**, (3), pp. 459–474
14. HAJIYEV, C., and CALISKAN, F.: *Fault Diagnosis and Reconfiguration in Flight Control Systems* (Kluwer Academic Publishers, Boston, 2003)
15. KAGSTROM, B.: 'Singular Matrix Pencils (Section 8.7)', In BAI, Z., DEMMEL, J., DONGARRA, J., RUHE, A., and VAN DER VORST, H. (eds.). *Templates for the Solution of Algebraic Eigenvalue Problems: A Practical Guide* (SIAM, Philadelphia, 2000)
16. VAN DOOREN, P.: 'The computation of Kronecker's canonical form of a singular pencil', *Lin. Algebra Appl.*, 1979, **27**, pp. 103–140

Chapter 7
Model following control systems

7.1 Introduction

Model following control is based on the principle of matching the output responses of a system with that of a defined 'reference' model. Two forms of model following control have been studied in the literature. The 'implicit' model following (IMF) uses only a feedback control structure. The 'explicit' model following (EMF) has a combination of feedforward and feedback structures. The EMF control requires comparison of the model and plant outputs, implying real-time simulation of the model. The IMF control does not have this additional computational requirement. The EMF is to be preferred if the model following has to occur in the presence of unknown disturbances. Erzberger [1] postulated the concept of perfect model following (PMF) wherein the plant outputs exactly satisfy the reference model's differential equation. However, the conditions of PMF severely restrict the selection of the reference models and are hard to satisfy in practical control system design. This has led to development of model following schemes where the error between the plant and model outputs (due to model and plant initial conditions mismatch) is allowed to approach zero as time increases.

Early studies on the synthesis of such model following control systems using quadratic optimal control theory can be found in References 2–4. The major limitation of this approach is the trial and error method of selecting the weighting matrices in the performance index to achieve good model matching. Alternate formulation of the feedforward control problem has also been studied in Reference 5. A command generator tracker (CGT) for a randomly forced model has been studied in Reference 6, and the computation of the feedforward matrices is accomplished by solving linear algebraic matrix equations. The study also identifies the deficiency of PMF control laws. O'Brien and Broussard [7] provide a comprehensive solution scheme for the general CGT problem with model inputs. Broussard and Berry [8] have established the relation between IMF and eigenstructure assignment. Finally more recently the concept of non-linear dynamic inversion, which is also a form of EMF control, is being extensively studied for flight control applications [9]. However, the problem formulation and mathematical analysis are completely different in the context of the discussions in the present chapter. The relative merits of these approaches will be further discussed in Chapter 11.

In this chapter an algorithm for the CGT problem is developed, using the relation established in Reference 8, as an eigenstructure assignment procedure. For completeness' sake, model following control problem as originally posed in Reference 2, which corresponds to the general command tracker problem without inputs, is also discussed, and the equivalence between IMF and EMF structures is established. Finally the constraints of perfect IMF problem [1] and its relation to eigenstructure assignment constraints of Chapter 2 are highlighted.

7.2 Command generator tracker

The design of a CGT with inputs is considered in this section. Consider a state variable plant to be controlled as

$$\dot{\mathbf{x}} = \mathbf{A}\mathbf{x} + \mathbf{B}\mathbf{u} \tag{7.1}$$

$$\mathbf{y} = \mathbf{C}\mathbf{x} + \mathbf{D}\mathbf{u} \tag{7.2}$$

and a reference model that the plant has to track as

$$\dot{x}_m = A_m x_m + B_m u_m \tag{7.3}$$

$$y_m = C_m x_m + D_m u_m \tag{7.4}$$

where $\mathbf{x} \in R^n, x_m \in R^p, u_m \in R^q, \mathbf{u}, \mathbf{y}, y_m \in R^m$ (the sizes of the plant inputs, plant outputs and model outputs are equal). The matrices A, B, C, D, A_m, B_m, C_m and D_m are of compatible dimensions. It is assumed that both the plant and the model are controllable and observable. Further the plant/model control and output matrices are full rank. Define the plant state and control vectors as a linear combination of the model states and controls as follows:

$$\mathbf{x} = Tx_m + Mu_m + \delta x \tag{7.5}$$

$$\mathbf{u} = Gx_m + Nu_m + \delta u \tag{7.6}$$

$$\delta y = y - y_m \tag{7.7}$$

where δx, δu and δy are the respective errors. The problem is to determine T, G, M and N such that the error in (7.7) decays to zero as time increases.

Substituting for **x** and **u** from (7.5) and (7.6) into (7.1), eliminating $\dot{x}_m$ term from (7.3) and considering only step inputs for u_m ($\dot{u}_m = 0$) results in

$$\delta\dot{x} = (AT + BG - TA_m)x_m + (AM + BN - TB_m)u_m + A\,\delta x + B\,\delta u \tag{7.8}$$

Using similar substitutions results in

$$\delta y = (CT + DG - C_m)x_m + (CM + DN - D_m)u_m + C\,\delta x + D\,\delta u \tag{7.9}$$

If the controller parameters T, G, M and N are selected such that

$$AT + BG = TA_m \tag{7.10}$$

$$AM + BN = TB_m \tag{7.11}$$

$$CT + DG = C_m \tag{7.12}$$

$$CM + DN = D_m \tag{7.13}$$

then (7.8) and (7.9) take the form

$$\begin{aligned} \delta\dot{x} &= A\,\delta x + B\,\delta u \\ \delta y &= C\,\delta x + D\,\delta u \\ \delta x(0) &= \mathbf{x}(0) - Tx_m(0) \end{aligned} \tag{7.14}$$

For $\delta u = 0$, δx and δy tend to zero, as time increases, if the plant system matrix A is stable.

Equations (7.10)–(7.13) can be written in compact form as

$$\begin{bmatrix} A & B \\ C & D \end{bmatrix} \begin{bmatrix} T & M \\ G & N \end{bmatrix} = \begin{bmatrix} TA_m & TB_m \\ C_m & D_m \end{bmatrix} \tag{7.15}$$

Equation (7.15) is the driving constraint for the solution to exist. It can now be solved by formulating an equivalent eigenstructure assignment problem. This approach is also used in Reference 8 to establish the equivalence between IMF and eigenstructure assignment. However, in Reference 8, an additional alternate computation scheme to solve (7.15) is also provided.

Assuming the eigenvalues of the model are distinct (not a restriction if the model is a designer's choice), apply a coordinate transformation to the model dynamics (7.3) and (7.4) such that

$$\begin{aligned} T_m^{-1} A_m T_m &= \hat{A}_m = \mathrm{Diag}(\lambda_i), \quad i = 1\text{–}p \\ T_m^{-1} B_m &= \hat{B}_m, \qquad C_m T_m = \hat{C}_m \end{aligned} \tag{7.16}$$

The matrices T and G in (7.15) are also transformed to

$$\hat{T} = TT_m, \qquad \hat{G} = GT_m \tag{7.17}$$

The matrices D_m, M and N remain invariant. Let

$$\begin{aligned} &B_m = \hat{T}\hat{B}_m, \quad \hat{T} = [t_1 \cdots t_p], \quad \hat{G} = [g_1 \cdots g_p], \quad M = [m_1 \cdots m_q], \quad N = [n_1 \cdots n_q] \\ &\hat{C}_m = [\hat{c}_{m_1} \cdots \hat{c}_{m_p}], \quad B_m = [b_{m_1} \cdots b_{m_q}], \quad D_m = [d_{m_1} \cdots d_{m_q}] \end{aligned} \tag{7.18}$$

Then (7.15) can be written in vector form as

$$\begin{bmatrix} A - \lambda_i I & B \\ C & D \end{bmatrix} \begin{bmatrix} t_i \\ g_i \end{bmatrix} = \begin{bmatrix} 0 \\ \hat{c}_{m_i} \end{bmatrix}, \quad i = 1\text{–}p \tag{7.19}$$

$$\begin{bmatrix} A & B \\ C & D \end{bmatrix}\begin{bmatrix} m_i \\ n_i \end{bmatrix}=\begin{bmatrix} \tilde{b}_{m_i} \\ d_{m_i} \end{bmatrix}, \quad i=1\text{–}q \tag{7.20}$$

Lemma 7.1

Condition 1. Equation (7.19) has a solution if and only if the model's eigenvalues (λ_i) are not a zero of the plant (for LHS to have full rank).

Condition 2. Equation (7.20) has a solution if and only if the plant has no zeros at the origin (for LHS to have full rank).

Remark: Condition 1 can be satisfied if the model's eigenvalues are designer's choice. Condition 2, however, is purely a plant characteristic. These existence conditions have also been established in Reference 7, and in addition a general solution using model input derivatives and compensator states is provided if the conditions of Lemma 7.1 are violated. These extensions are not pursued here.

The computation of the EMF control law can now be summarised as follows:

Algorithm 7.1:

1. If condition 2 of Lemma 7.1 is satisfied, solve for m_i and n_i from (7.20)
   Else Stop (No solution)
2. Select the model eigenvalues (λ_i) to satisfy condition 1 of Lemma 7.1.
3. Solve for $\hat{T}$ and $\hat{G}$ from (7.19).
4. Compute T and G from (7.17).
5. Using $\hat{T}$ compute $\tilde{B}_m$ in (7.18) and solve for M and N from (7.20).

If the plant is stable, the model following control law takes the form

$$u = Gx_m + Nu_m \tag{7.21a}$$

If a state feedback law is employed to augment the stability of the plant, then

$$\delta u = K_s\, \delta x \tag{7.21b}$$

From (7.5), (7.6) and (7.21b), the model following control law becomes

$$\mathbf{u} = (G - K_sT)x_m + (N - K_sM)u_m + K_s\mathbf{x} \tag{7.22}$$

If an output feedback law with proper outputs is employed to augment the stability of the plant, then

$$\delta u = K_o\, \delta y \tag{7.23a}$$

Following the analysis in section 4.3, the equivalent output control law of (7.23a) without proper outputs (D = 0) is given by

$$K_e = PK_o \tag{7.23b}$$

where $P = [I_m - K_oD]^{-1}$. Accordingly from (7.11) and (7.23b), the model following control law becomes

$$u = (G - K_e CT)x_m + (N - K_e CM)u_m + K_e Cx \tag{7.24}$$

It should be noted that the feedforward matrices are modified when stability augmentation is employed.

7.3 Tunable command generator tracker

The CGT problem formulation discussed in section 7.2 tacitly assures *only the steady-state tracking of the output variables.* However, the transient response mismatch could be significant. Further since the number of output variables that can be tracked is equal to the number of inputs (usually less than the number of state variables), there can be both transient and steady-state errors in state variables not included in the output. The solution scheme as described in Algorithm 7.1 indicates that for a specified plant, model and output variables, the feedforward matrices (G and N) are uniquely determined. Thus, there is no apparent design freedom that can be used to improve the transient response. This appears, at the outset, to be a major limitation of the CGT scheme. However, a closer examination of the feedforward control structure reveals that if the (p*m) elements of the model output matrix C_m are used as free design parameters, fine-tuning of the transient response of selected *response variables* $\mathbf{y_r} = C_r x_m$ can be achieved by optimising a time response objective function of the form

$$J = \min_{C_m}[\text{rms}\{\mathbf{y_p}(t) - \mathbf{y_r}(t)\}] \tag{7.25}$$

where $\mathbf{y_p}$ is selected plant outputs. The idea here is to optimally blend the model state variables (linear combination of model states) to minimise tracking errors of selected model and plant response variables. This concept leads to the design of a *tunable* CGT *controller*. This modified design scheme actually maps the inherent design freedom available in selecting the feedforward gain matrix G to a meaningful selection of the model output matrix C_m to improve the model tracking performance.

7.4 Explicit and implicit model following control

A special case of the CGT problem is when the order of the model and the plant is the same. This has been studied as IMF and EMF problems in the literature. For the present analysis, initially assume the matrices D in (7.2), B_m in (7.3) and D_m (7.4) are identically zero ($u_m = 0$). Then (7.15) reduces to

$$\begin{bmatrix} A & B \\ C & 0 \end{bmatrix} \begin{bmatrix} T \\ G \end{bmatrix} = \begin{bmatrix} TA_m \\ C_m \end{bmatrix} \tag{7.26}$$

From (7.26) it is evident that $T \in R^{n \times n}$. Equation (7.26) can be solved for T and G using the method outlined in section 7.2. If a stabilising state variable feedback K (yet to be determined) is also used, the composite feedforward and feedback control law (EMF) takes the form

$$u = (G - KT)x_m + Kx \tag{7.27}$$

If the model states are initialised using $\mathbf{x}(0) = Tx_m(0)$ (7.14), then $y \equiv y_m$ for all $t > 0$. If the control law is to be implemented using only a feedback loop (IMF), (7.27) can be solved for the equivalent implicit feedback gain matrix K as

$$K = GT^{-1} \tag{7.28}$$

Use of this feedback gain matrix in (7.27) implies the feedforward from the model states is eliminated and only a state feedback around the plant remains. This control law is called the IMF control law. This establishes the equivalence between the IMF and EMF control structures. It is to be noted that the feedback gain matrix K of (7.28) assigns the poles of the model stability matrix (A_m) to the plant as is evident from the solution of T and G satisfying (7.10).

The combined plant and model system using the EMF control law of (7.22) takes the form

$$\begin{bmatrix} \dot{x} \\ \dot{x}_m \end{bmatrix} = \begin{bmatrix} A & BG \\ 0 & A_m \end{bmatrix} \begin{bmatrix} x \\ x_m \end{bmatrix} + \begin{bmatrix} BH \\ B_m \end{bmatrix} u_m \tag{7.29}$$

The interconnect matrix H in (7.29) is computed as

$$H = N - KM \tag{7.30}$$

The closed-loop system of IMF control law that is equivalent to the EMF controller in (7.22) and (7.28) is given by

$$\dot{x} = (A + BK)x + BHu \tag{7.31}$$

7.5 Perfect implicit model following control

The model following control problem considered in Reference 1 is as follows. Given a state variable plant as in (7.1) and (7.2) with $D = 0$ and $\mathbf{x} \in R^n$, $\mathbf{u} \in R^m$, $\mathbf{y} \in R^r$, it is required to find a state feedback control law $\mathbf{u} = K\mathbf{x}$ such that the output $\mathbf{y}$ *exactly* satisfies the differential equation

$$\dot{\mathbf{y}} = A_m \mathbf{y} \tag{7.32}$$

where A_m is the desired model dynamics. It is shown in Reference 1 that if the matrix CB is full rank and if

$$[(CB)(CB)^+ - I_r](A_m C - CA) = 0 \tag{7.33}$$

where superscript '+' indicates pseudo-inverse, then the state feedback control law

$$\mathbf{u} = (CB)^{+}(A_mC - CA)\mathbf{x} \tag{7.34}$$

will exactly satisfy (7.32). As observed in Reference 6, control law in (7.34) often results in (i) large control gains and (ii) destabilising the plant (those plant modes not involved in the model following). Thus, design methods discussed in sections 7.2 and 7.3 are to be preferred. This would imply accepting error decaying to zero as time increases.

Consider the special case when the order of the model is same as that of the plant and $C_m = C$. From the analysis of section 7.2, for PMF of all states ($\mathbf{y} = \mathbf{x}$ in (7.31)), we have $T = I_n$ and $K = G$ with the PMF constraint

$$A + BK = A_m \tag{7.35}$$

where K is the state feedback gain matrix. The restriction on the selection of the reference model A_m to satisfy (7.35) can be derived using the following analysis.

Partition the plant matrices in (7.35) as

$$A = \left[\begin{array}{c|c} A_{11} & A_{12} \\ \hline A_{21} & A_{22} \end{array}\right], \qquad B = \left[\begin{array}{c} B_1 \\ \hline B_2 \end{array}\right], \quad A_{11}, B_1 \in R^{m\times m} \tag{7.36}$$

with B_1 non-singular (always possible with permutation of state variables since B is full rank). Apply a coordinate transformation T_0 to the plant, model and gain matrix K such that

$$\{\hat{A}, \hat{A}_m, \hat{B}, \hat{K}\} \Rightarrow \{T_0^{-1}AT_0, T_0^{-1}A_mT_0, T_0^{-1}B, T_0K\} \tag{7.37}$$

where

$$T_0 = \left[\begin{array}{c|c} I_m & 0 \\ \hline B_2B_1^{-1} & I_{n-m} \end{array}\right] \quad \text{and} \quad B = \left[\begin{array}{c} B_1 \\ \hline 0 \end{array}\right] \tag{7.38}$$

In the new coordinates (7.36) in partitioned form can be written as

$$\left[\begin{array}{c|c} \hat{A}_{11} & \hat{A}_{12} \\ \hline \hat{A}_{21} & \hat{A}_{22} \end{array}\right] + \left[\begin{array}{c} B_1 \\ \hline 0 \end{array}\right]\hat{K} = \left[\begin{array}{c|c} \hat{A}_{m_{11}} & \hat{A}_{m_{12}} \\ \hline \hat{A}_{m_{21}} & \hat{A}_{m_{22}} \end{array}\right] \tag{7.39}$$

It is clearly seen that PMF can be achieved if the model sub-matrices in (7.39) are chosen such that

$$[\hat{A}_{21} \quad \hat{A}_{22}] = [\hat{A}_{m_{21}} \quad \hat{A}_{m_{22}}] \tag{7.40}$$

and the feedback gain matrix is given by

$$\hat{K} = B_1^{-1}[(\hat{A}_{m_{11}} - \hat{A}_{11}) \ (\hat{A}_{m_{12}} - \hat{A}_{12})] \tag{7.41}$$

Following the analysis of Chapter 3 (3.1b), an alternate way of expressing the model selection constraint (7.40) is that the distinct eigenvalues (λ_i) and eigenvectors (v_i) of the model stability matrix $\hat{A}_m$ and the plant sub-matrices must satisfy the constraint

$$[\hat{A}_{21} \mid \hat{A}_{22} - \lambda_i I_{n-m}] v_i = 0, \quad i = 1 - n \tag{7.42}$$

If (7.42) is satisfied, from the analysis of Chapter 3, a feedback gain matrix K always exists satisfying (7.40) that will ensure PMF. Another way of interpreting (7.42) is that the eigenvalues (λ_i) and eigenvectors (v_i) of the model stability matrix $\hat{A}_m$ must lie in the corresponding subspaces defined by the plant stability matrix $\hat{A}$.

Since T is the identity matrix for PMF, if the plant states are initialised as $\mathbf{x}(0) = x_m(0)$, the plant state trajectories will track the model states perfectly for any initial condition of the model (7.14). PMF would also occur for model inputs, if B_m is in the range space of B.

7.6 Examples

Example 7.1: A tutorial example is considered to illustrate the features of the model following algorithms discussed in this chapter. The example is specially constructed to meet the special case where the model order is same as the plant order and the PMF conditions are satisfied. Consider the state space description of a plant, which has to follow a reference model given by

Plant: (A, B), model: (A_m, B_m)

$$A = \begin{bmatrix} -0.5 & 3 & 0 \\ -3 & -0.5 & 0 \\ 0 & -4 & -0.5 \end{bmatrix}, B = \begin{bmatrix} 1 & -1 \\ 1 & 1 \\ -1 & 1 \end{bmatrix}, A_m = \begin{bmatrix} -1 & 0 & 0 \\ 1 & -0.5 & 1 \\ 0.5 & -1 & -0.5 \end{bmatrix},$$

$$B_m = \begin{bmatrix} -1 & 1 \\ 1 & 1 \\ 0.9 & -1.1 \end{bmatrix}$$

Plant and model outputs

$$c = \begin{bmatrix} 0 & 1 & 0 \\ 0 & 0 & 1 \end{bmatrix}, \quad c_m = c, \quad d = 0, \quad d_m = 0 \tag{7.43}$$

The plant eigenvalues are located at [−0.5, −0.5 ± j3] and the model eigenvalues are at [−1.0, −0.5 ± j1]. Since the model order is same as that of the plant, applying the coordinate transformation (7.38) to the plant and model stability matrices reveals that the plant and model satisfy the PMF condition (7.40). The model control distribution matrix B_m is however not in the range space of B.

Case 1. Explicit model following design
Applying Algorithm 7.1 results in the following EMF controller matrices:

$$T = \begin{bmatrix} 1 & 0 & 0 \\ 0 & 1 & 0 \\ 0 & 0 & 1 \end{bmatrix}, G = \begin{bmatrix} 1.75 & -1.5 & 0.5 \\ 2.25 & 1.5 & 0.5 \end{bmatrix}, M = \begin{bmatrix} 0.2 & 0.2 \\ 0 & 0 \\ 0 & 0 \end{bmatrix}, N = \begin{bmatrix} 0.35 & 1.35 \\ 1.25 & 0.25 \end{bmatrix} \tag{7.44}$$

The plant steering control inputs are given by $u = Gx_m + Nu_m$ (7.21a).

Case 2. Implicit model following controller
Using (7.28) and (7.30) the equivalent IMF state feedback and control interconnect matrices are

$$K = \begin{bmatrix} 1.75 & -1.5 & 0.5 \\ 2.25 & 1.5 & 0.5 \end{bmatrix}, \quad H = \begin{bmatrix} 0 & 1 \\ 0.8 & -0.2 \end{bmatrix} \tag{7.45}$$

The feedback matrix K will assign the model eigenvalues and eigenvectors to the closed-loop plant stability matrix $A_c = A + BK$ implying $A_c \equiv A_m$. As noted earlier, if B_m is in the range space of B, the IMF control law will perfectly track the model. In this example B_m does not satisfy this condition. Figure 7.1 shows

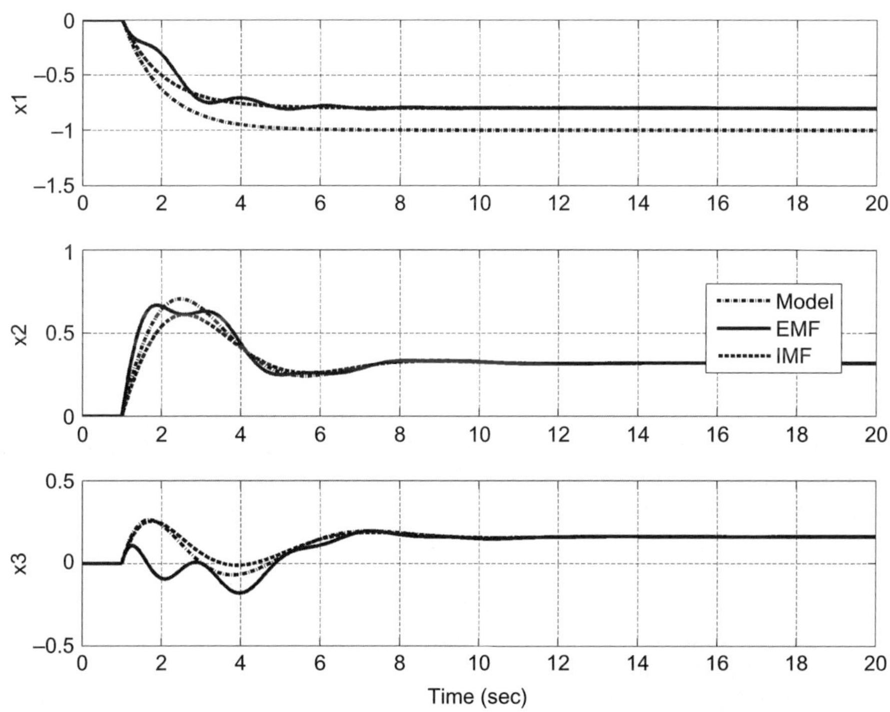

Figure 7.1 Step response characteristics of EMF and IMF controllers

the step response for the first model input. It is seen that the plant and model outputs track, in steady state, using both EMF and IMF controllers. There is a mismatch in the steady-state response state of x1 as it is not a tracked output. It is also seen that there are substantial mismatch errors in the transient response characteristics. This is the limitation of the basic model following control problem formulation as observed earlier. The tunable model following controller formulation (7.25) is a method to fine-tune this transient response error. This concept is further illustrated in example 7.3.

Example 7.2: The aircraft landing flare problem considered in Reference 6 is used to illustrate the use of the EMF controller design. The state variable model matrices for the longitudinal dynamics of the aircraft (C-8 Augmentor Wing STOL) are given by

$$A = \begin{bmatrix} -0.0397 & -0.28 & -0.282 & 0 & 0 \\ 0.135 & -0.538 & 0.538 & 0.0434 & 0 \\ 0 & 0 & 0 & 1 & 0 \\ 0.0207 & 0.441 & -0.441 & -1.41 & 0 \\ -0.017 & 1.92 & 0 & 0 & 0 \end{bmatrix}, \quad B = \begin{bmatrix} -0.0052 & -0.102 \\ 0.031 & 0.037 \\ 0 & 0 \\ -1.46 & -0.066 \\ 0 & 0 \end{bmatrix} \tag{7.46}$$

The state variables are forward velocity (u (ft/sec)), flight path angle (γ (deg)), pitch attitude (θ (deg)), pitch rate (q (deg/sec)) and altitude (h (ft)), respectively. The control variables are elevator (δ_e (deg)) and engine nozzle angle (δ_T (deg)), respectively. The output variables are $y = [u \quad h]^T$.

The command model is given by

$$\begin{aligned} \dot{u}_m &= 0; \qquad u_m(0) = 0 \\ \dot{h}_m &= -\left[\frac{1}{\tau}\right] h_m; \qquad \tau = 0.2778; \quad h_m(0) \text{ specified} \\ y_m &= \begin{bmatrix} u_m \\ h_m \end{bmatrix} \end{aligned} \tag{7.47}$$

The aircraft trim states at the reference condition are $V_0 = 110$ ft/sec, $\gamma_0 = -1.0$ deg, $\theta_0 = 1.3$ deg, $\delta_e = -9.6$ deg, $\delta_T = -55.4$ deg [6]. The command model generates a smooth flare trajectory, which the aircraft has to follow. For simulation purposes the aircraft and command model states are initialised with approach trim conditions as $u(0) = 0$, $\gamma(0) = -6.5$ deg, $\theta(0) = -5.8$ deg, $q(0) = 0$, $h(0) = 45$ ft, $u_m(0) = 0$, $h_m(0) = 45$ ft [6].

Figure 7.2 gives a block schematic of EMF control for the landing flare problem. Note that the model has no inputs and only the initial condition on the aircraft and model initiates the simulation.

The control law is given by

$$u = Gx_m + Kx \tag{7.48}$$

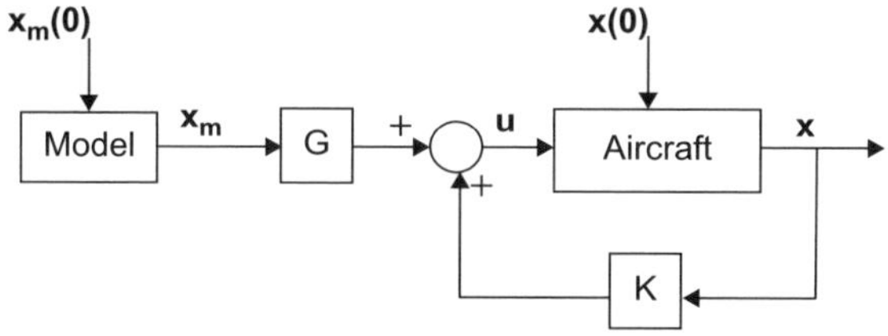

Figure 7.2 Schematic of explicit model following controller

The model following control law is computed for two cases: (i) basic aircraft (K = 0) and (ii) state feedback augmented aircraft (K ≠ 0). From Theorem 2.1, all closed-loop eigenvalues can be assigned and in addition five eigenvector elements can be arbitrarily chosen. In this case, the eigenvector freedom is utilised to minimise the gain matrix (K) norm. This results in minimum control effort feedback solution. The resulting feedback and command path gain matrices are

Basic aircraft:

$$G = \begin{bmatrix} 0.0834 & -0.0664 \\ 0.3269 & 0.7295 \end{bmatrix}; \qquad K = 0$$

Augmented aircraft:

$$G = \begin{bmatrix} -0.1382 & -0.0961 \\ -0.5937 & 1.3538 \end{bmatrix};$$

$$K = \begin{bmatrix} 0.3977 & 0.6375 & 0.6748 & 0.4873 & 0.1862 \\ 0.8790 & -0.1995 & -0.1609 & 0.0236 & -0.6731 \end{bmatrix} \tag{7.49}$$

Figure 7.3 shows the aircraft eigenvalues for the two cases. The state feedback is designed to increase the phugoid mode damping and speed up the heave mode while retaining the short period mode close to that of basic aircraft. The advantages of this pole assignment will be evident when the simulation results shown in Figures 7.4 and 7.5 are studied.

Figure 7.4 shows the tracking of the flare trajectory for the two control laws of (7.49). It is seen that with K = 0 (basic aircraft), the low phugoid damping of the aircraft introduces substantial tracking errors. The augmented aircraft shows good tracking characteristics. The heave mode eigenvalue ($\lambda = -0.5$) also results in better tracking performance. Lower values of this mode, even though resulting in lower feedback gains, increase tracking error close to ground (h = 0).

Figure 7.5 shows the tracking error in altitude and forward speed. The model demands no change in forward velocity ($u_m = 0$). Figure 7.6 shows the state variable responses, and Figure 7.7 indicates the elevator and engine nozzle control activity. The control activity is not too different between the control laws. It should be noted that the control activity shown is an increment over

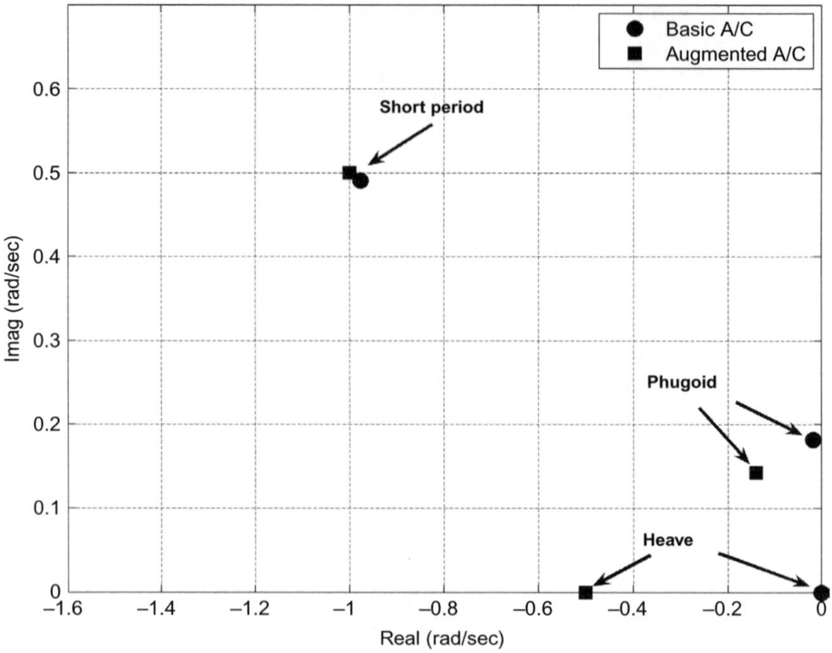

Figure 7.3 Eigenvalues of basic and augmented aircraft

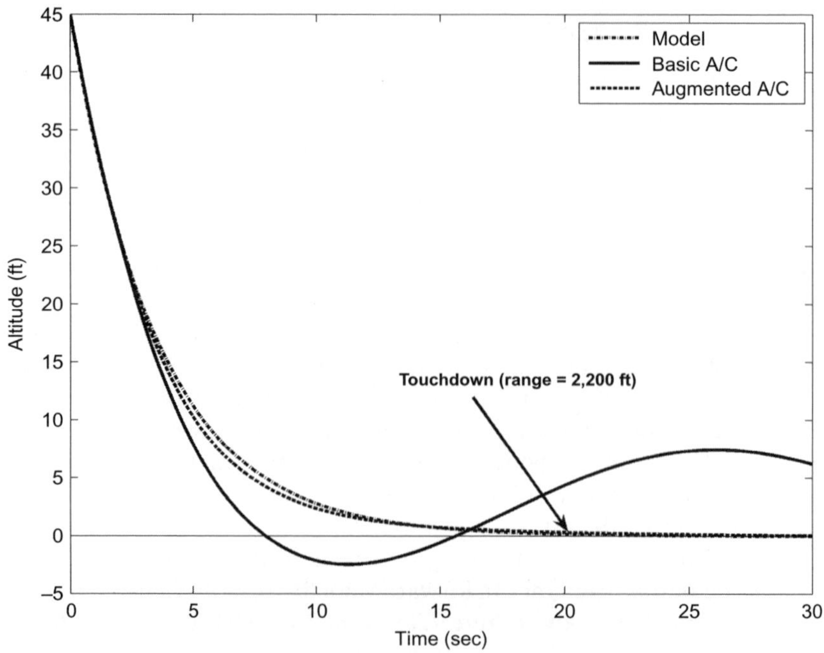

Figure 7.4 Aircraft flare trajectory

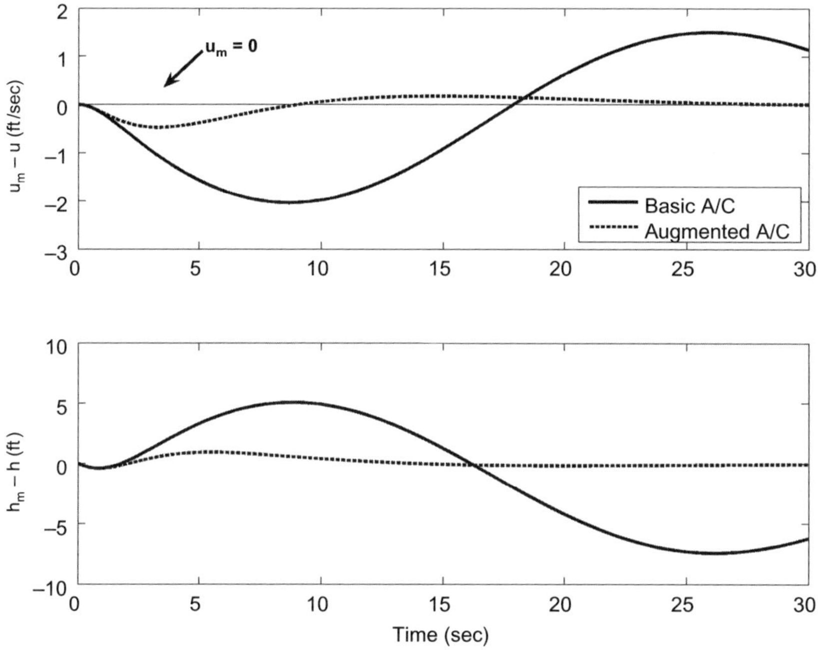

Figure 7.5 Model tracking errors

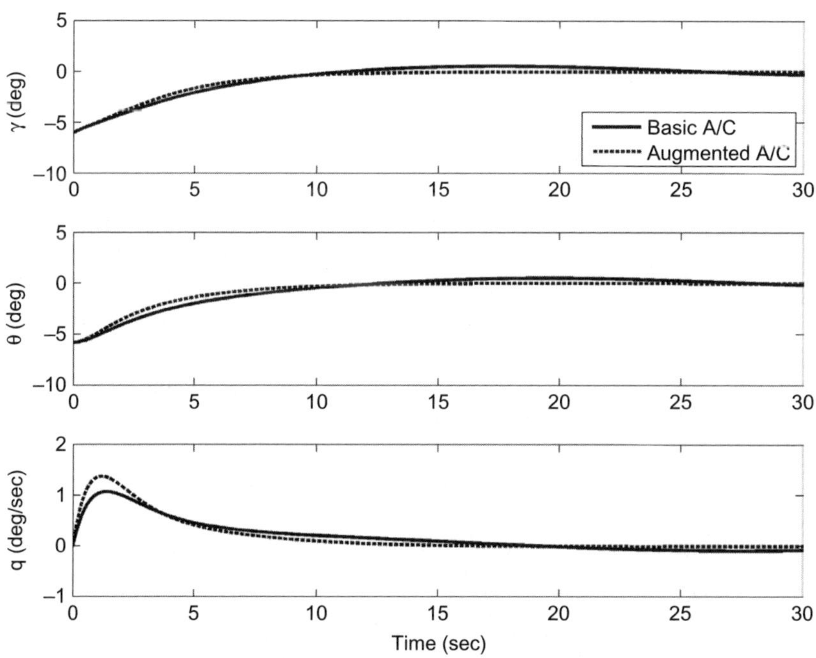

Figure 7.6 State variable response

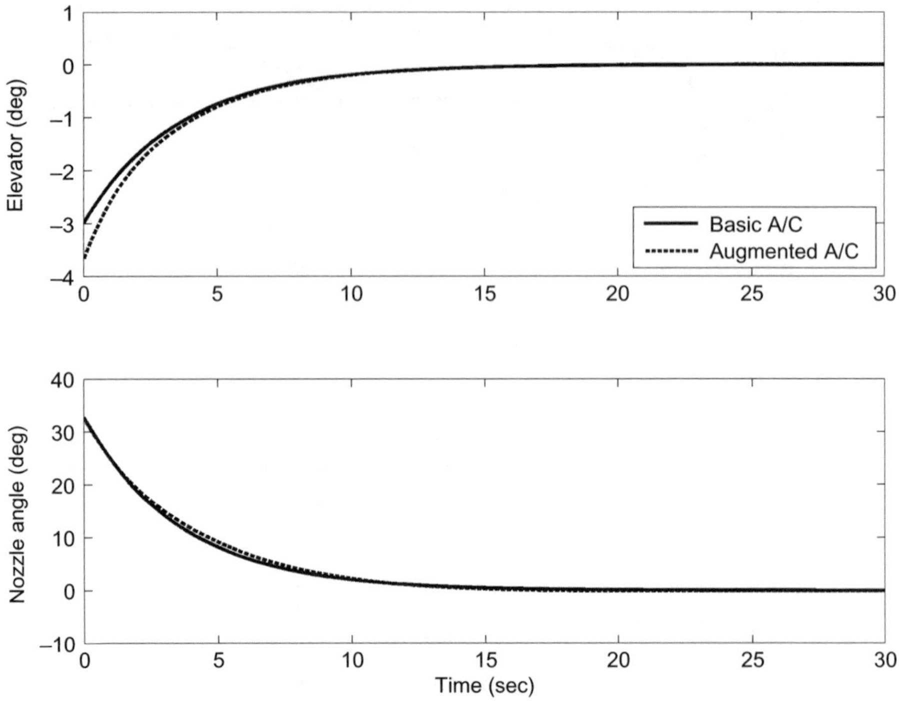

Figure 7.7 Control actuator response

the trim settings at the reference condition. An alternative solution, based on quadratic synthesis [6], has similar results shown here and leads to an interesting comparison of the two different approaches.

Example 7.3: In this example, the merits of the tunable CGT design of section 7.3 are illustrated. The longitudinal state variable model of the F-8C aircraft at a reference condition of M = 0.6, h = 20,000 ft (Appendix B – Table B.2a) is considered as the plant. The problem is to design an EMF control law such that the plant follows the model specified as F-8C aircraft at a reference condition M = 0.219, h = 0 (Appendix B – Table B.2c). The aircraft has two control inputs and thus two outputs can be tracked. The plant and the model order are the same in this case. The plant outputs selected for tracking are angle of attack and pitch attitude. Table 7.1 lists the plant and model eigenvalues.

Table 7.1 F-8C aircraft longitudinal eigenvalues

Mode	Plant (FC1)	Model (FC3)
Short period	$-0.8326 \pm j2.5141$	$-0.4297 \pm j0.7933$
	$\omega_n = 2.65, \zeta = 0.31$	$\omega_n = 0.90, \zeta = 0.48$
Phugoid	$-0.0056 \pm j0.0698$	$-0.0181 \pm j0.1689$
	$\omega_n = 0.07, \zeta = 0.08$	$\omega_n = 0.17, \zeta = 0.11$

The following designs are derived for comparing the resulting solutions:

1. Nominal design (NOM): Choose the model outputs (C_m) as angle of attack and pitch attitude (same as plant outputs) and apply Algorithm 7.1. The resulting controller matrices are

$$C_m = \begin{bmatrix} 1 & 0 & 0 & 0 \\ 0 & 0 & 1 & 0 \end{bmatrix}$$

$$G = \begin{bmatrix} 0.5411 & -0.0132 & 0.0164 & -0.0024 \\ -5.822 & -0.0339 & -0.098 & 0.0146 \end{bmatrix}, \quad N = \begin{bmatrix} 1.6649 & 0.1571 \\ -9.1284 & -0.7534 \end{bmatrix}$$

$$T = \begin{bmatrix} 1 & 0 & 0 & 0 \\ 0 & 1 & 0 & 0 \\ 0 & 0 & 1 & 0 \\ -40.3347 & 36.4634 & 52.956 & 1.0176 \end{bmatrix}, M = \begin{bmatrix} 0 & 0 \\ 0 & 0 \\ 0 & 0 \\ 5.3385e+3 & 0.818e+3 \end{bmatrix} \tag{7.50}$$

2. Optimal design (OPT): Using the tunable CGT algorithm of section 7.3, compute the model output matrix C_m to minimise the error between the plant and model state variable responses for an elevator doublet input. The following objective function is used for the optimisation:

$$J = \min_{C_m} \left[\text{rms}\left(\sum_{i=1}^{n} w_i e_i \right) \right], \qquad e_i = x_i(t) - x_{mi}(t) \tag{7.51}$$

where w_i are error weighting constants and n is the number of state variables. The resulting controller matrices are

$$C_m = \begin{bmatrix} 1.3923 & -0.1389 & -0.1773 & -0.0009 \\ 0.0336 & -0.0322 & 0.9353 & 0.0007 \end{bmatrix}$$

$$G = \begin{bmatrix} 0.842 & 0.328 & -0.1552 & -0.0043 \\ -8.7567 & -1.7161 & 1.4456 & 0.0291 \end{bmatrix}, N = \begin{bmatrix} 0.3418 & -0.1208 \\ -1.2886 & 0.8979 \end{bmatrix}$$

$$T = \begin{bmatrix} 1.3923 & -0.1389 & -0.1773 & -0.0009 \\ 0.0177 & 0.9803 & -0.0211 & -0.0001 \\ 0.0336 & -0.0322 & 0.9353 & 0.0007 \\ -6.1053 & 1.5260 & 5.7036 & 1.0377 \end{bmatrix}, M = \begin{bmatrix} 0 & 0 \\ 0.058 & 0.0117 \\ 0 & 0 \\ 29.2161 & -171.1521 \end{bmatrix} \tag{7.52}$$

Figure 7.8 shows the state variable time response for both the NOM and the OPT. It is clear from the responses that the OPT substantially reduces the transient response errors. Table 7.2 shows the steady-state tracking error of the OPT. The NOM perfectly tracks the model in steady state. A trade-off between transient and steady-state response errors can be accomplished by the addition of steady-state tracking requirement in the objective function (7.51). This example

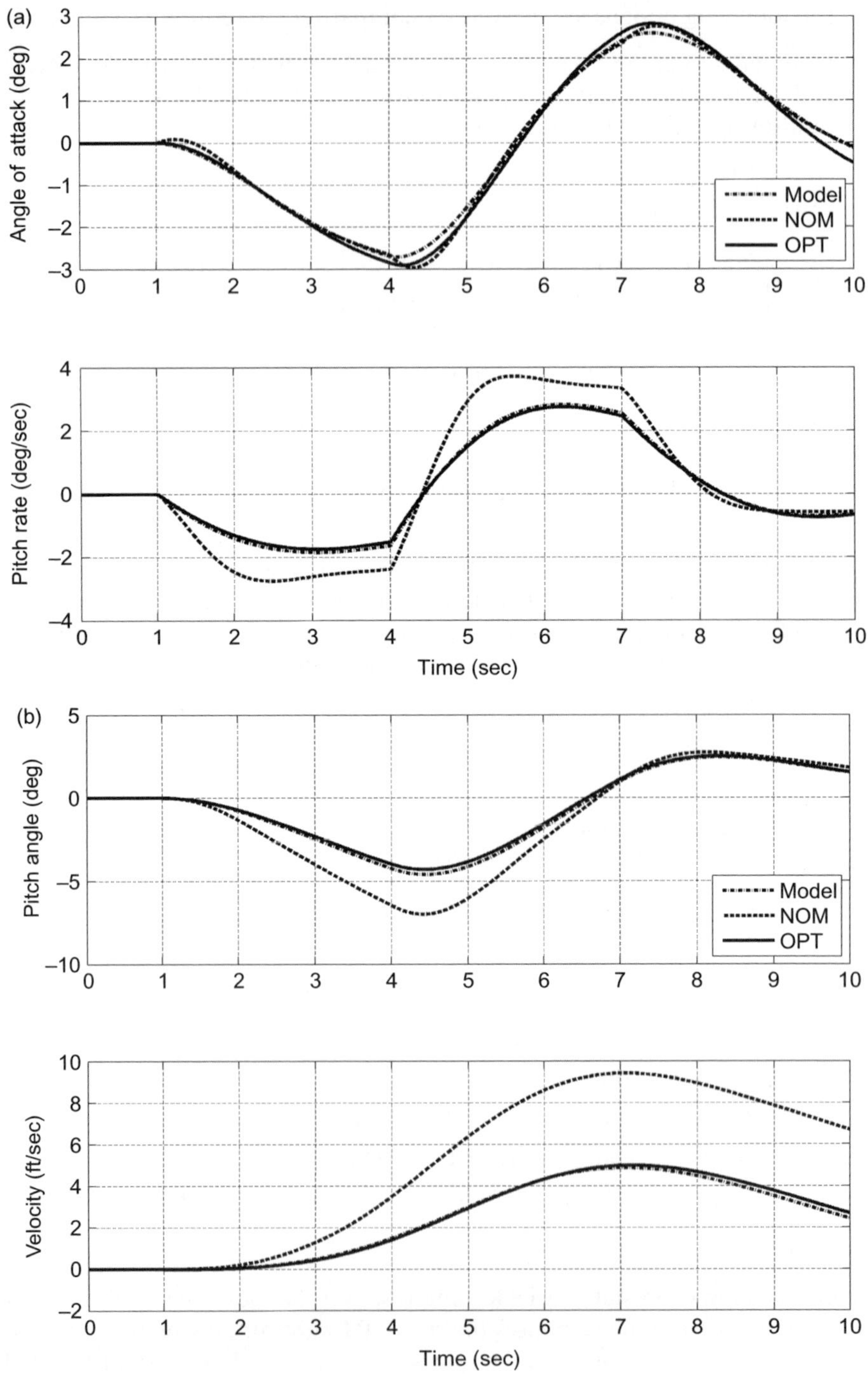

Figure 7.8 State variable responses (elevator doublet input) (a) Angle of attack and pitch rate (b) Pitch attitude and aircraft velocity

illustrates that tunable CGT algorithm brings in flexibility in the design of the EMF controller.

Table 7.2 Steady-state responses for elevator step input

Variable	Steady-state response	
	Nominal design	**Optimal design**
α (deg)	−3.5386	−5.1591
q (deg/sec)	0	0
θ (deg)	−4.4185	−3.4930

7.7 Summary

In this chapter the model following control problem has been formulated as an eigenstructure assignment problem. For the special case where the plant and model have the same number of states, both EMF and IMF control schemes can be adopted. The equivalence between the two schemes has been established. For this special case, the constraints on the plant and model stability matrices for PMF are also defined. In cases where model selection flexibility exists, these constraints can be used to advantage to improve tracking performance. The command tracker problem formulation at the outset tacitly assures only steady-state tracking of the output variables. This leads to the possibility of unacceptable transient response errors. This is especially true if the model and plant orders are the same. This appears to be a major limitation of the scheme. However, the key concept of considering the elements of the *model* output matrix as free design parameters, as proposed in this chapter, alleviates this problem. This allows fine-tuning of transient/steady-state mismatch errors in *selected* response variables by formulating a suitable optimisation problem. This *tunable CGT design* scheme substantially enhances the capability of the controller to optimise model-matching errors.

References

1. ERZBERGER, H.: 'On the Use of Algebraic Methods in the Analysis and Design of Model-Following Control Systems', July 1968, NASA TN D-4663
2. TYLER, J.S.: 'The characteristics of model following systems as synthesised by optimal control', *IEEE Trans. Automat. Control*, 1964, **AC-9**, (4), pp. 485–498
3. ASSEO, S.J.: 'Application of optimal control to perfect model-following', *J. Aircraft*, 1970, **7**, (4), pp. 308–313
4. KREINDLER, E., and ROTHSCHILD, D.: 'Model-following in linear-quadratic optimization', *AIAA J.*, 1976, **14**, (7), pp. 835–842

5. DAVISON, E.J.: 'The feedforward control of linear multivariable time-invariant systems', *Automatica*, 1973, **9**, pp. 561–573
6. TRANKLE, T.L., and BRYSON, Jr, A.E.: 'Control logic to track the outputs of a command generator or randomly forced target', *Proceedings of the AIAA Guidance, Navigation and Control Conference*, 1977, Hollywood, Florida, pp. 45–52
7. O'BRIEN, M.J., and BROUSSARD, J.R.: 'Feed forward control to track the output of a forced model', *Proceedings of the 1978 IEEE Conference on Decision and Control*, 1979, San Diego, California
8. BROUSSARD, J.R., and BERRY, P.W.: 'The relationship between implicit model following and eigenvalue eigenvector placement', *IEEE Trans. Automat. Control*, 1980, **AC-25**, (3), pp. 591–594
9. ENNS, D., BUGAJSKI, D., HENDRICK, R., and STEIN, G.: 'Dynamic inversion: an evolving methodology for flight control design', *Int. J. Control*, 1994, **59**, (1), pp. 71–91

Chapter 8

Flight control system design guidelines

8.1 Introduction

In Chapters 1–7, control/estimation algorithms based on eigenstructure control theory have been detailed to cover the normally used feedback control structures such as state/output feedback, dynamic feedback, observer-based dynamic feedback and explicit/implicit model following controllers. While these algorithms can be used in any control design situation, in this book their utility for the design of flight control systems is highlighted in Chapters 9–12.

With the development of high reliability sensors, computers and actuators, it is now possible to build flight vehicle control systems with extraordinary performance. Indeed the full authority feedback systems can substantially mask the basic airframe dynamic characteristics and consequently their performance limitations. These technology developments in turn have brought into focus the role of multivariable control system design methods to evolve complex multi-loop systems using multiple sensors and control effectors. Multivariable control techniques are now being demonstrated in experimental flight research programs. It is thus reasonable to expect that these methods will be used in flight vehicle control design of operational aircraft as increased sophistication in airframe design and performance is sought.

It should be emphasised that the design of the dynamic performance of an aircraft over its full flight and manoeuvre envelope is a non-trivial task. The control of the dynamics of the aircraft can be broadly classified as (i) small control input response, addressing tracking/precision flying, and (ii) large control input manoeuvres such as rapid gross target acquisition. The controller structure includes feedback and command path elements, which are tuned to shape both the small and large input responses of the aircraft. For the small input case, the non-linear aircraft dynamics is considered as a continuum of equilibrium states in the operational envelope of the aircraft and linear perturbation models about these equilibrium points are used for design. Scheduling these fixed-point linear designs as a function of appropriate air data parameters, which define the aircraft state, results in the full flight envelope control logic. The large-amplitude control design on the other hand addresses the optimum use of the control power available while ensuring that the structural, aerodynamic and actuator limits of the aircraft are not exceeded. These control elements are primarily in the form of non-linear limiters that are fine-tuned using piloted aircraft real-time simulation.

The linear control design and analysis methods, as discussed in Chapters 9–11, are applicable *only* to the small control input response requirements. Nevertheless, it should be emphasised that this is a *non-trivial* part of the total control law design. It plays an important role in ensuring the safe operation of the flight vehicle under precision flying tasks that are associated with demanding pilot workload. There are instances of inadequate small perturbation linear designs triggering the so-called Pilot Involved Oscillations (PIO).

To evolve a flight control design, a number of performance requirements have to be addressed. These are broadly categorised as (i) providing a satisfactory pilot–vehicle interface referred to as *handling qualities* (HQ) [1] and (ii) the conventional feedback control requirements of providing adequate stability margins, good transient response and disturbance rejection characteristics. In this chapter, these performance requirements, as applicable to the design examples in Chapters 9–11, are briefly reviewed. Comprehensive control system performance requirements for typical fly-by-wire aircraft flight control law design, including the effects of sensor and actuator dynamic characteristics, are detailed in Reference 9.

8.2 Flight vehicle handling qualities requirements

The requirements of satisfactory pilot–vehicle interaction in high workload precision flying tasks, such as close formation flying, air to air refuelling, etc. play a very important and dominant role in the design of flight control systems. Based on extensive flight test experience, on a range of aircrafts, over the past few decades, *quantitative* HQ specifications to aid control law design have been derived. For aircraft systems, these have been documented in MIL-HDBK-1797 [1], MIL-F-8785C [2] and MIL-F-9490D [3]. The corresponding rotorcraft design standard is documented in ADS-33E-PRF [4]. A number of HQ metrics have been postulated to aid the flight control engineer to evolve good PIO-resistant designs. For aircraft systems, the definition and the rationale behind these metrics are well documented in Reference 1. In this section the relevant HQ specifications applicable to the case studies in Chapters 9–11 will be briefly reviewed.

The aircraft HQ specifications defined in Reference 1 (Appendix C) imply:

> a general framework that permits tailoring each requirement according to: (i) the kind of aircraft (class), (ii) the job to be done (flight phase category) and (iii) how well the job must be done (level).

The level definition is derived from the Cooper–Harper pilot-rating scale [5].

The aircraft control law designs studied in Chapters 9 and 10 are required to meet the HQ specifications corresponding to ***class IV*** (high manoeuvrability aircraft) operating in flight phase ***category A*** (non-terminal flight phase requiring rapid manoeuvring and precision tracking) with ***level 1*** (satisfactory) performance. The performance specifications for class IV, category A and level 1 aircraft are summarised below. The numbers in brackets refer to MIL-HDBK-1797 paragraphs.

8.3 Lateral–directional aircraft handling qualities requirements

1. Roll mode time constant shall be less than 1.0 sec (4.5.1.1).
2. If the spiral mode is unstable, the minimum time to double shall be greater than 12 sec. If the spiral mode is stable, the time to half shall be greater than 10 sec (4.5.1.2).
3. The aircraft shall not exhibit a coupled roll–spiral (lateral phugoid) mode in category A flight phases (4.5.1.3).
4. Following a step roll control command, held until the bank angle change is at least 90 deg, the roll rate at the first minimum following the first peak shall be of the same sign and not less than 60% of the roll rate at the first peak (4.5.1.4).

   Remarks: The p_{osc}/p_{av} requirement associated with this requirement, primarily directed towards configurations with high magnitudes of roll-to-sideslip ratio for the Dutch roll mode, $\left|\frac{\varphi}{\beta}\right|_d$ (typically 3.5–5), combined with low Dutch roll damping ζ_d has been generally ignored for current airplanes for category A operations because the suggested levels of ζ_d in (4.6.1.1) defined in requirement 5 below effectively eliminate this requirement. The parameter $\left|\frac{\varphi}{\beta}\right|_d$ is computed by evaluating the transfer function ratio $R(s) = \frac{\varphi(s)/\delta_a(s)}{\beta(s)/\delta_a(s)}$ at the Dutch roll frequency (ω_d). An approximation to the roll-to-sideslip ratio for the Dutch roll mode is given by [6]

$$\left|\frac{\varphi}{\beta}\right|_d \cong \left|\frac{L'_\beta}{N'_\beta}\right| \left\{\frac{1+[(N'_\beta(L'_r)^2)/(L'_\beta)^2]}{1+[(L'_p)^2/N'_\beta]}\right\}^{1/2} \tag{8.1}$$

5. The Dutch roll natural frequency (ω_d) and damping (ζ_d) shall satisfy (4.6.1.1): $\zeta_d > 0.4$; $\zeta_d\omega_d > 0.4$ rad/sec; $\omega_d > 1$ rad/sec. When $\omega_d\left|\frac{\varphi}{\beta}\right|_d$ is greater than 20 rad/sec, the minimum $\zeta_d\omega_d$ shall be increased to $\Delta\varsigma_d\omega_d = \omega_d^2\left|\frac{\varphi}{\beta}\right|_d$.
6. Turn co-ordination – large roll inputs (4.6.2):
   (i) For category A, following larger step roll control commands, the parameter β_k, which is the ratio of the sideslip increment $\Delta\beta$ to the parameter k ($\beta_k = \Delta\beta/k$), should be less than the values specified below.

The roll command shall be held fixed until the bank angle has changed at least to 90 deg. The sideslip excursion $\Delta\beta$ is the maximum change in sideslip occurring within 2 sec or the damped period of the Dutch roll T_d

$$\left(T_d = \frac{2\pi}{\omega_d \sqrt{1-\zeta_d^2}} \text{ sec} \right) \text{ whichever is greater:}$$

$$\beta_k < 6\,\text{deg (adverse sideslip or positive } \beta \text{ for a right roll)}$$
$$\beta_k < 2\,\text{deg (proverse sideslip or negative } \beta \text{ for a right roll)}$$

where $k = (\varphi_t/30)$ and φ_t is the bank angle at time $t = 1$ sec.

(ii) For category A turn co-ordination, the lateral acceleration at the pilot station be less than ± 0.15 g for a 60 deg/sec roll to 60 deg bank [7]. The bounds shall be reduced depending on the maximum roll rate attained using the relation

$$n_{y_{max}} = 0.15 \times \left(\frac{p_{max}}{60} \right)$$

7. Departure resistance metrics [8]: At high angle of attack conditions the aircraft is susceptible to depart from controlled flight. Typical lateral divergence characteristics are nose slice, wing drop or a wing rock motion. The following metrics are generally used to assess the degree of departure resistance the aircraft possesses:

$$\text{(a) } c_{n_\beta}\text{dyn} = c_{n_\beta} \cos\alpha_0 - \left(\frac{I_z}{I_x} \right) c_{l_\beta} \sin\alpha_0$$

$$\text{(b) Lateral control divergence parameter (LCDP)} = c_{n_\beta} - c_{l_\beta} \left(\frac{c_{n_{\delta a}}}{c_{l_{\delta a}}} \right) \quad (8.2)$$

where α_0 is the trim angle of attack, I_x is the roll inertia, I_z is the yaw inertia and the other variables are the non-dimensional lateral stability and control derivatives (Appendix A). The departure resistance metrics in (8.2) have the units of rad^{-1}.

For departure resistance both $c_{n_\beta}\text{dyn}$ and LCDP should be positive. Further the industry practice is $c_{n_\beta}\text{dyn} > 0.1\,\text{rad}^{-1}$.

8. Gibson's PIO resistance criterion [9]: With the aircraft trimmed at straight and level flight, compute the transfer function of the bank angle to lateral stick force. From the Nichol's plot of the frequency response, compute the magnitude of the average phase rate, φ_{avg} (deg/Hz), defined as

$$|\varphi_{avg}| = \frac{\varphi_{2f_c} - \varphi_{f_c}}{f_c} \quad (8.3)$$

where f_c is the frequency in hertz at which the phase angle is −180 deg. To show that the aircraft has no PIO tendencies, it is required that φ_{avg} meets the bounds given in Table 8.1. See also Figure 8.6 for the phase rate criterion template.

Table 8.1 Average phase rate criterion

HQ level	**f_c (Hz)**	**φ_{avg} (deg/Hz)**
Level 1*	0.65–1.8	<55
Level 1	0.48–1.8	<85
Level 2	0.38–1.8	<145

9. Control margins: Feedback designs attempt to achieve good dynamic response and also exhibit good disturbance rejection properties while respecting the stability margins discussed earlier. These design objectives usually result in high-gain systems requiring high control power. The total control actuator displacement and rate excursions have to be budgeted to cater to (i) turbulence disturbance, (ii) sensor noise excitation, (iii) stabilisation, (iv) recovery, (v) inter-axis coupling compensation and (vi) manoeuvring. Margin levels on control deflection and rate recommended in Reference 1 (5.1.11.5) for turbulence (Appendix A) and sensor noise are shown in Table 8.2.

Table 8.2 Control margins for actuator displacement and rate

Disturbance	**Control margin**	
	Displacement	**Rate**
σ_g	$3\sigma_\delta$	$3\sigma_{\dot{\delta}}$
σ_s	$3\sigma_\delta$	$2\sigma_{\dot{\delta}}$

(i) σ_g: RMS severe turbulence intensity, (ii) σ_s: RMS sensor noise disturbance, (iii) σ_δ: RMS actuator displacement response and (iv) $\sigma_{\dot{\delta}}$: RMS actuator rate response.

8.4 Longitudinal aircraft handling qualities requirements

The safe control of the pitch axis of an aircraft is extremely important. A number of HQ criteria have been developed towards this end. In this section the definitions of the criteria used in assessing the designs in Chapter 10 will be highlighted.

'The short-term pitch response characteristics are universally regarded to be extremely important – so important that controversy over both the form and the substance of requirements has continued for some years. Many issues still remain to be resolved' [1]. Many HQ criteria are based on the dominant short-period mode characteristics. While these criteria can be applied to conventional aircraft without feedback augmentation, they cannot be directly applied to highly augmented aircraft with additional higher order dynamics. For this purpose, the concept of lower order equivalent systems (LOES) has been developed.

8.4.1 Lower order equivalent system

The rationale behind this concept is to derive an equivalent short-period model for the high-order system (HOS) in a specified frequency range. Typically a low-order equivalent pitch rate and normal load factor frequency response to pilot control force input are simultaneously matched with the higher order system response. In the present book only a low-order model for the pitch rate (q)/pilot control force (F) will be used. The transfer function takes the form

$$\frac{q(s)}{F(s)} = \frac{K_\theta(s + 1/T_{\theta 2})e^{-\tau_\theta s}}{s^2 + 2\zeta_{sp} + \omega_{sp}^2} \qquad (8.4)$$

where K_θ is the gain, $1/T_{\theta 2}$ is fixed as the basic aircraft pitch rate zero, τ_θ is a time delay parameter and ω_{sp} and ζ_{sp} are the short-period natural frequency and damping of the LOES model. The low-order model parameters in (8.4) are optimised to get a model match, both in magnitude and phase, of the high-order model of the augmented aircraft typically in the frequency range of 0.1–10 rad/sec. The error bounds for both magnitude and phase are also specified [1]. The equivalent time delay parameter τ_θ is a measure of delay between pilot control input and pitch attitude response as perceived by the pilot. Thus, larger time delays result in poor pilot ratings. The allowable delays are given in Table 8.3 [1].

Table 8.3 Allowable time delay parameter

Level	Allowable τ_θ (sec)
1	0.10
2	0.20
3	0.25

8.4.2 Control anticipation parameter

The control anticipation parameter (CAP) is defined as the ratio of initial pitch acceleration to steady-state change in load factor (computed using a short-period approximation of the aircraft) for a step control input:

$$\text{CAP} = \frac{\ddot{\theta}_0}{\Delta n_{z_{ss}}} (\text{rad/sec}^2)/\text{g} \qquad (8.5)$$

An approximation for CAP in terms of LOES model parameters is given by

$$\mathrm{CAP} = \frac{\omega_{sp}^2}{n/\alpha}, \quad \text{where } \frac{n}{\alpha} \approx \frac{V}{g}\frac{1}{T_{\theta 2}} \tag{8.6}$$

An alternative definition of CAP based on direct computation from step input response of the HOS is given by

$$\mathrm{CAP}' = \frac{\dot{q}_{\mathrm{maxHOS}}}{n_{z_{ss}}} \tag{8.7}$$

Figure 8.1 gives the CAP HQ level boundaries for category A flight phases. It is emphasised that CAP parameters are derived using the LOES model.

8.4.3 Bandwidth criterion

The bandwidth of the open-loop pitch attitude (θ) response to pilot control input (F) is defined as

$$\omega_{BW} = \min(\omega_{BWg}, \omega_{BWp}) \tag{8.8}$$

where ω_{BWg} is the frequency at which the gain margin is 6 dB and ω_{BWp} is the frequency at which the phase margin is 45 deg. The phase delay τ_p is given by

$$\tau_p = \frac{-(\varphi_{2\omega_{180}} + 180)}{57.3(2\omega_{180})} \text{ sec} \tag{8.9}$$

where ω_{180} is the frequency corresponding to −180 deg phase and $\varphi_{2\omega_{180}}$ is the phase angle at $2\omega_{180}$. The ω_{BW} and τ_p cross plot, with boundaries defined for different HQ levels for category A flight phase, is given in Figure 8.2.

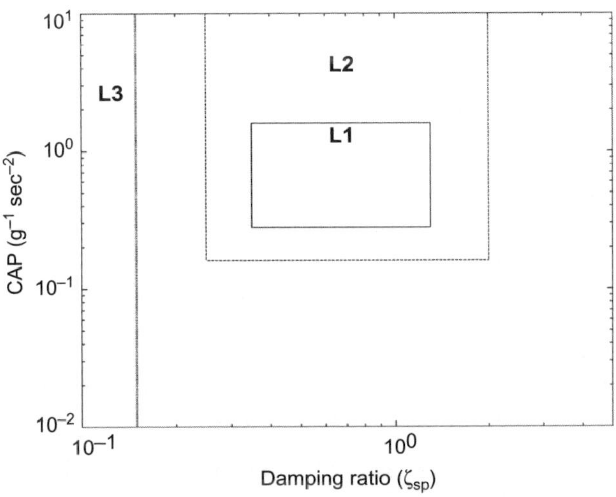

Figure 8.1 CAP chart for category A flight phase

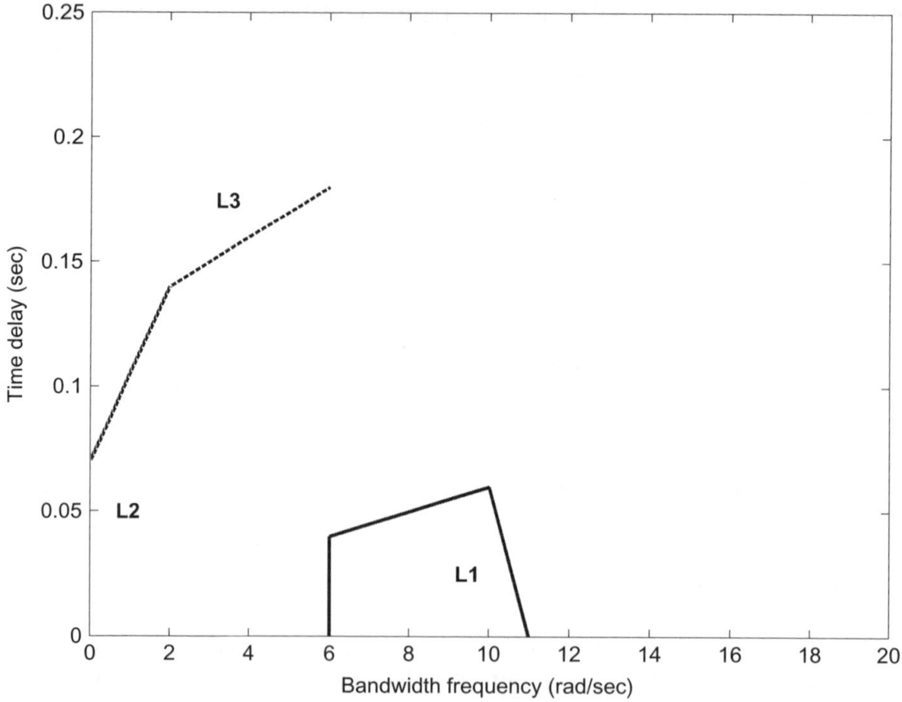

Figure 8.2 Bandwidth criterion boundaries for category A flight phase

8.4.4 Gibson's longitudinal handling qualities criteria

Gibson [10,11] has developed a set of criteria based on pitch attitude time and frequency response characteristics. The time domain criterion is based on the relationship between pitch attitude dropback (db), steady-state pitch rate (q_{ss}) and flight path angle time delay (t_γ). The analytical relations for these parameters are given by

$$t_\gamma = \left(\frac{2\zeta_{sp}}{\omega_{sp}}\right), \qquad dbq = \left(\frac{db}{q_{ss}}\right), \qquad dbq = (T_{\theta_2} - t_\gamma) \tag{8.10}$$

In (8.10), dbq is the ratio of dropback to steady-state pitch rate (q_{ss}). For good precision control, the value of dbq should lie in the range of 0–0.25. The relationship between dbq and ratio of peak/steady-state pitch rate (q_m/q_{ss}) is shown in Figure 8.3 where the satisfactory regions are also indicated.

The Gibson's frequency domain criterion is based on the frequency response $\theta(j\omega)/F(j\omega)$, where θ is the pitch attitude and F is the pilot stick force. The gain and phase are plotted as in Figure 8.4 to ensure PIO resistance; the frequency response curve should meet the following criteria:

1. In Figure 8.4, the attitude gain at 180 deg phase should be <-25 dB.
2. The frequency response curve should lie within the level 1 boundary of Figure 8.5. While plotting the frequency response curve, the response

must be displaced vertically until the curve passes through the 0 dB and −110 deg point.

3. Meet the average phase rate criterion of Table 8.1 as shown in Figure 8.6.

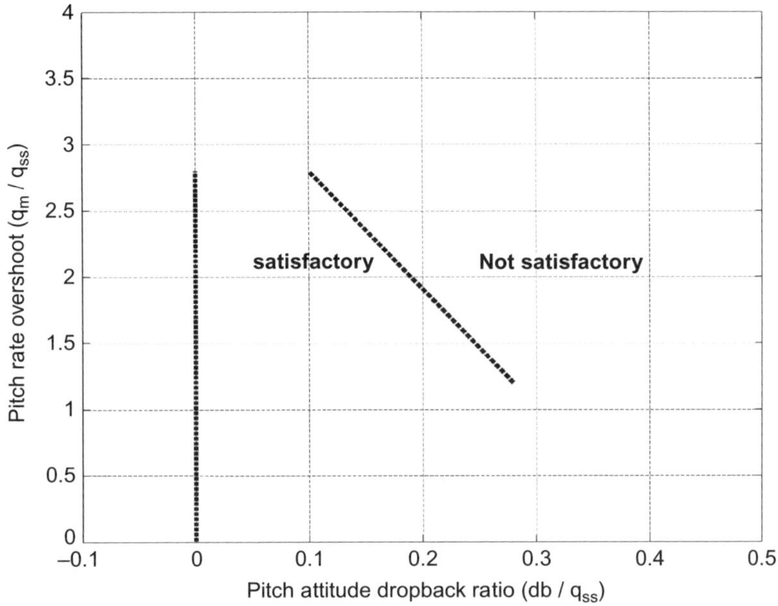

Figure 8.3 Boundaries of dbq *and* q_m/q_{ss} *for precision tracking*

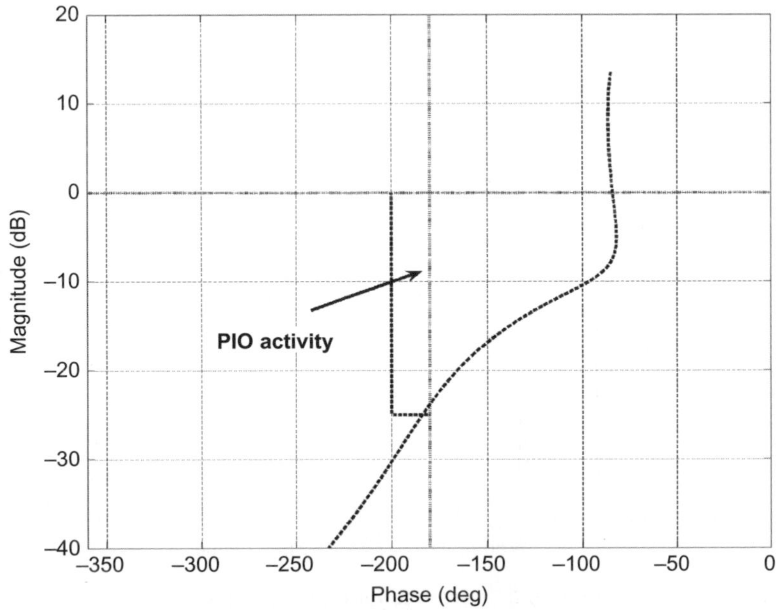

Figure 8.4 Pitch axis frequency response criterion 1

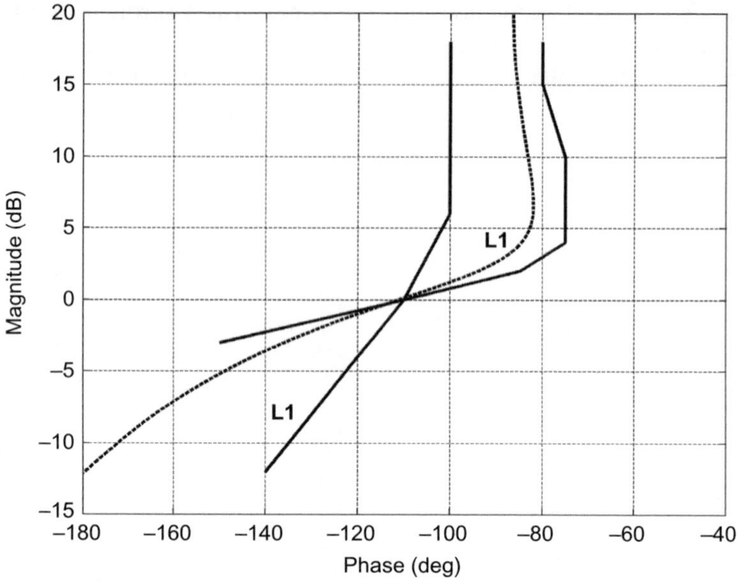

Figure 8.5 Pitch axis frequency response criterion 2

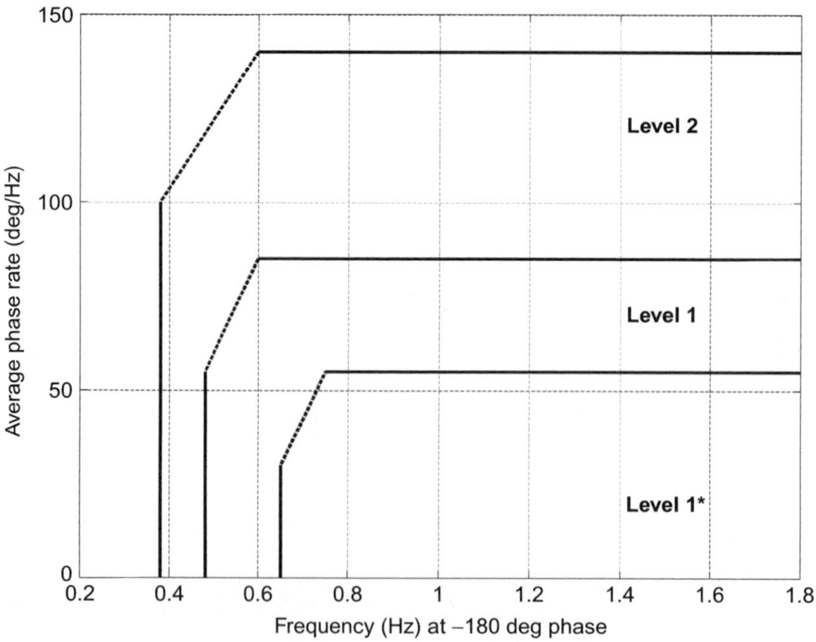

Figure 8.6 Average phase rate criterion

A brief synopsis of these criteria is given in Reference 1 – Appendix A, pp. 244–253. A summary of pertinent comments given in Reference 1 is reproduced as follows for information:

1. Negative attitude dropback (i.e. overshoot) was usually associated with sluggish, unpredictable response both in flight path control and in tracking, leading sometimes to overdriving PIO.
2. Attitude dropback from 0 to about 0.25 sec was excellent for fine tracking and was associated with comments typified by the nose following the stick.
3. Increasing attitude dropback with large pitch rate overshoot led to abrupt response and bobbling, from slight tendency to continuous oscillations, in tracking tasks. Sometimes this was called PIO, but it did not cause concern for safety.
4. Attitude dropback had little effect within the range tested upon gross manoeuvring without target, landing approach or flight refuelling, provided it was not negative.
5. CAP up to 3.6 rad/sec^2/g was satisfactory for gross manoeuvring without a target, but was unsatisfactory above 2 rad/sec^2/g for the landing approach, above 1 rad/sec^2/g for fine tracking and below 0.28 rad/sec^2/g for any task.

8.5 Rotorcraft handling qualities requirements

The rotorcraft HQ specifications are defined in the document ADS-33E-PRF [4]. These HQ requirements can be broadly classified into (i) small/medium-amplitude short-term response and (ii) response shaping to pilot control input. The rationale for the development of these specifications is well documented in References 12 and 13.

A relatively high-gain inner-loop feedback controller augments the short-term response and improves disturbance rejection properties. The principal short-term response requirements are (i) dynamic stability, (ii) bandwidth and (iii) inter-axis cross coupling. The dynamic stability specifications translate into the Phugoid and Dutch roll modes having desirable natural frequency and damping. The bandwidth requirement defines the short-term attitude response of the vehicle to pilot control especially in terms of the phase delay. These specifications also vary with mission task element (MTE) with target acquisition and tracking tasks having stringent requirement. The inter-axis coupling specification is the most important design requirement. Most helicopters, without some form of control augmentation, have significant inter-axis coupling resulting in high pilot workload. The ADS definition for inter-axis coupling is as follows (section 3.4.5.4):

> The pitch due to roll (q/p) and roll due to pitch (p/q) coupling for Target Acquisition and Tracking shall not exceed the limits specified in Figure 8.7. The average q/p and average p/q are derived from ratios of pitch and roll frequency responses. Specifically, average q/p is defined as the magnitude of pitch-due-to-roll control input (q/δ_{lat}) divided by roll-due-to-roll control input (p/δ_{lat}) averaged between the bandwidth (θ/δ_{lon}) and neutral-stability (phase $= -180$ deg) frequencies of the pitch-due-to-pitch control inputs (q/δ_{lon}). Similarly average

p/q is defined as the magnitude (p/δ_{lon}) divided by (q/δ_{lon}) averaged between the roll axis bandwidth (ϕ/δ_{lat}) and neutral-stability frequencies.

Figure 8.7 gives the ADS template categorising the pitch/roll inter-axis coupling. The response shaping is essentially an autopilot function, which is achieved by having a low-bandwidth outer-loop forward path/feedback controller. The response shaping requirements are defined based on usable cue environment (UCE) and MTE. These are categorised as '*response types*'. Typical response types are (i) rate, (ii) attitude command (AC), (iii) attitude command/attitude hold (ACAH) and (iv) translational rate command with position hold (TRCPH). The response types are briefly defined in Table 8.4 [12].

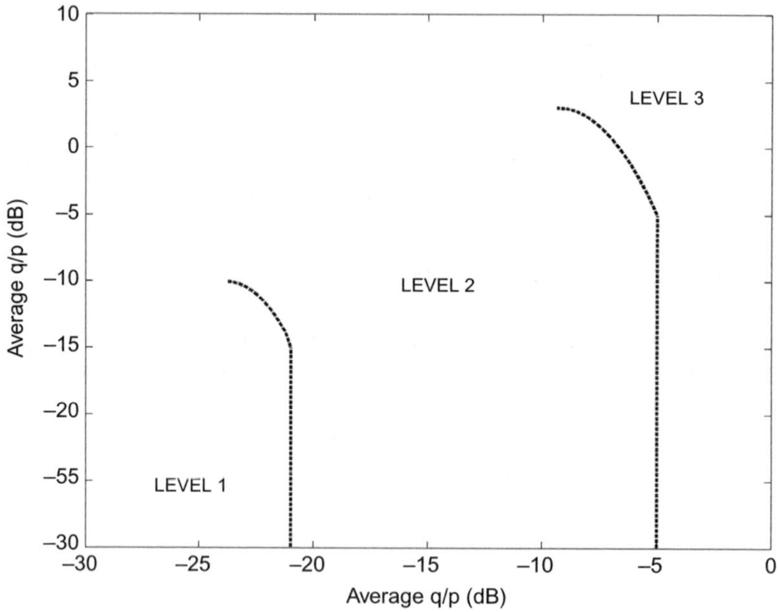

Figure 8.7 Requirements for pitch due to roll and roll due to pitch coupling for aggressive agility (ADS-33E-PRF)

Table 8.4 Typical response types

Response type	Brief definition
Rate	Attitude diverges away from trim for at least 4 sec following a step cockpit control input
Attitude command (AC)	Constant cockpit control force input must produce proportional angular displacement
Attitude hold (AH)	Attitude must be held following a pulse command input
Translational rate command with position hold (TRCPH)	Constant controller force input must result in constant translational rate. The rotorcraft must hold position if the force on the cockpit controller is zero

8.6 Control system performance specifications

8.6.1 Single-loop stability margins

For single input/output systems, the robustness of the feedback system to model uncertainties is defined using the classical gain and phase margin criteria. However, for multi-input/output systems many alternative metrics can be postulated to measure the robustness of the system. For military aircraft applications [3 (3.1.3.6.1)], robust stability has to be guaranteed in terms of gain and phase margins. In multiple-loop systems, variations shall be made with all gain and phase values in the feedback paths held at nominal values except for the path under investigation.

The above conventional stability margin definition requires satisfying independently the gain and phase margins. However, aircraft industry practice dictates a more stringent requirement in that the gain and phase margins must be simultaneously satisfied by ensuring that the loop transfer function does not intersect an exclusion boundary of ±6 dB gain and ±35 deg phase. An approximate analytical function (ellipse) of the exclusion boundary is also defined as follows:

$$\left(\frac{g}{6.0}\right)^2+\left(\frac{\varphi+180}{35}\right)^2=1 \qquad (8.11)$$

where g is the gain in decibels and φ is the phase in degrees. Figure 8.8 shows the exclusion boundary limits.

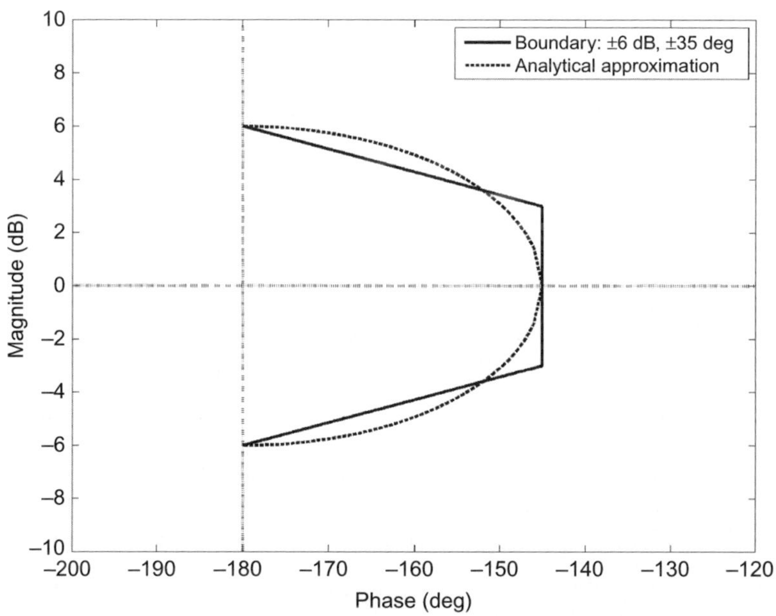

Figure 8.8 Stability margin boundaries

8.6.2 Multivariable stability margins

An alternative multivariable stability criterion is defined [14], wherein the closed-loop system should be able to withstand the application of simultaneous *independent* gain (κ_i) and phase offset (φ_i) at the input of each control channel without becoming unstable. The perturbation matrix κP has the form $P = \mathrm{diag}(\kappa_1 e^{-j\varphi_1}, \ldots, \kappa_m e^{-j\varphi_m})$, where m is the number of control channels. The stability must be evaluated at the combined gain and phase offsets (at least at the boundary values) of the region shown in Table 8.5.

Table 8.5 Gain and phase perturbations for stability analysis

κ – Gain offset (dB)	**φ – Phase offset (deg)**
−4.5	0
−2.5	30
+2.5	30
+4.5	0

For modelling the phase offset effect φ, an equivalent time delay τ (sec) has to be computed. The closed-loop system matrix with the perturbations added takes the form

$$\hat{A} = A + B\,\mathrm{diag}(\kappa_1 e^{-\tau_1 s}, \ldots, \kappa_m e^{-\tau_m s})K \tag{8.12}$$

where A and B are the plant matrices and K is the equivalent state feedback matrix. It is required that the closed-loop system matrix $\hat{A}$ be stable.

A scalar equivalent criterion for (8.12) is to add a scalar gain and phase delay perturbation of the form $\rho e^{-\tau_d s}$ in all channels with the resulting closed-loop system matrix

$$\hat{A} = A + \rho e^{-\tau_d s} BK \tag{8.13}$$

required to be stable.

In order to test the system stability under worst case gain and offset conditions given in Table 8.5, the following procedure is suggested [14]:

1. Apply a gain perturbation of $\rho = 4.5$ dB ($\varphi = 0$). Determine the eigenvalue ω_c, which is closest to the imaginary axis (lowest phase margin).
2. Compute the time delay $\tau_d = (0.6/\omega_c)$ sec, which adds an additional phase offset of $\varphi = 33$ deg at ω_c.
3. Check the closed-loop stability with $\rho = 2.5$ dB and τ_d using (8.13).

Remark: A multivariable root locus, with gain and phase offset perturbations in (8.13), helps in identifying all root locus branches, which are migrating towards the imaginary axis. However, it should be remembered that the rules governing the mapping of the multivariable root locus trajectories towards multivariable zeros are not the same as those of the single input/output system.

8.6.3 *Modal robustness metrics*

Eigensystem robustness metrics, as discussed in Chapter 5, provide direct insight into the sensitivity of the evolving designs since the synthesis is performed in a modal control framework. A good scalar robustness metric of the eigensystem is $\kappa_2(V)$, which is defined as the *condition number* of the eigenvector matrix V, with bounds $1 \le \kappa_2(V) \le \infty$. The lower bound of unity indicates the ideal conditioning of the modal matrix with eigenvectors being mutually orthogonal. A vector of robustness metric called mode *condition numbers* is also defined in Chapter 5. These metrics help in determining the sensitivity of the individual modes to parameter perturbation.

An alternative measure of determining the linear dependence between eigenvectors is to define the angles between eigen-subspaces [15] as follows.

Let X and Y be orthogonal bases of the given subspaces with $X \in R^{n\times r}$ and $Y \in R^{n\times m}$, and without loss of generality, let $(r + m) \le n$ and $r \le m$. Let θ_i $(i = 1 - r)$ be the ith angle of inclination. Then

$$\cos q_i = \sigma_i, \quad \text{where } \sigma_i \text{ is the ith singular value of } M = X^T Y \tag{8.14}$$

8.6.4 *Multivariable zeros*

A system with n-states, m-inputs and r-outputs (m, r < n) is called square if m = r [16]. For a square controllable, observable, proper system, the sets of system zeros, invariant zeros and transmission zeros coincide, and they can be defined as follows.

For a square, proper state space system defined by the triple (A, B, C), let

$$z(s) = \rho_0(s)|G(s)| = \begin{vmatrix} sI_n - A & B \\ C & 0 \end{vmatrix} \tag{8.15}$$

where z(s) is the zero polynomial, ρ_0 is the characteristic polynomial and $|\cdot|$ is the determinant operator. Then the complex number λ is a finite zero of the system if and only if it is a root of the zero polynomial z(s).

8.6.5 *Allowable feedback gain magnitudes*

Aircraft industry experience indicates that feedback gains from inertial rate and attitude sensors of the order of 0.1 deg/deg/sec, 0.1 deg/deg, respectively, are considered as low gains with gains approaching 1 deg/deg/sec, 1 deg/deg, respectively, for full authority high-gain systems. The guideline for maximum feedback gains from acceleration sensors is dictated by the accuracy of the sensors, and for aircraft applications typical values are 4 deg/m/sec^2 (0.68 rad/g) for lateral acceleration sensors.

8.7 Summary

In this chapter, the basic specifications for design of flight control systems have been outlined. It should be pointed out that historically, flight vehicle HQ guidelines evolved as a basis for the design of a new aircraft with desirable dynamic characteristics using only low authority stability augmentation systems. With the modern full authority high-gain fly-by-wire flight control systems, most of these specifications can easily be achieved. However, other design problems associated with such high-gain/high-bandwidth feedback systems, namely (i) stability margins, (ii) interaction of structural modes with rigid body dynamics and (iii) actuator rate saturation during large-amplitude manoeuvres in the presence of severe atmospheric turbulence, etc. take centre stage in the control law design process. Brief specifications/guidelines to meet these practical control system design requirements have also been outlined. The guidelines detailed in this chapter will be used for the design studies in Chapters 9–11. Additional useful flight control design practices have also been well documented in Reference 17.

References

1. ANONYMOUS: 'Flying Qualities of Piloted Aircraft', 1997, MIL-HDBK-1797
2. ANONYMOUS: 'Military Specifications – Flying Qualities of Piloted Aircraft', 1980, MIL-8785C
3. ANONYMOUS: 'Flight Control Systems – Design, Installation and Test of Piloted Aircraft, General Specifications', 1975, MIL-F-9490D
4. ANONYMOUS: 'Aeronautical Design Standard – Performance Specification: Handling Qualities Requirements for Military Rotorcraft', March 2000, ADS-33E-PRF
5. COOPER, G.E., and HARPER, Jr, R.P.: 'The Use of Pilot Rating in the Evaluation of Aircraft Handling Qualities', April 1969, NASA TN D-5153
6. CHALK, C.R., NEAL, T.P., HARRIS, T.M., PRITCHARD, F.E., and WOODCOCK, R.J.: 'Background Information and User Guide for MIL-F-8785B (ASG), Military Specification – Flying Qualities of Piloted Airplanes', August 1969, AFFDL-TR-69-72
7. HARTMANN, G.L., HAUGE, J.A., and HENDRICK, R.C.: 'F-8C Digital CCV Flight Control Laws', February 1976, NASA CR-2629
8. WEISSMAN, R.: 'Status of design criteria for predicting departure characteristics and spin susceptibility', *J. Aircraft*, 1975, **12**, (12), pp. 989–993
9. MUIR, E.: 'HIRM Design Challenge Problem Description' In MAGNI, J.-F., BENNANI, S., and TERLOUW, J. (eds.): *Robust Flight Control – A Design Challenge*. Lecture Notes in Control and Information Sciences, vol. 224 (Springer-Verlag, Berlin, 1997)
10. GIBSON, J.C.: 'Handling qualities for unstable aircraft', *Proceedings of the ICAS Congress*, 1986, London, pp. 433–445, paper ICAS-86-5.3.4

11. GIBSON, J.C.: *Development of a Methodology for Excellence in Handling Qualities Design for Fly-by-Wire Aircraft* (Delft University Press, Delft, The Netherlands, 1999)
12. KEY, D.L.: 'A new handling qualities specification for U.S. military rotorcraft', *Proceedings of the Helicopter Handling Qualities and Control Conference*, Royal Aeronautical Society, London, 1988
13. HOH, R.H.: 'Dynamic requirements in the new handling qualities specification for U.S. military rotorcraft', *Proceedings of the Helicopter Handling Qualities and Control Conference*, Royal Aeronautical Society, London, 1988
14. MUIR, E.: 'The Robust Inverse Dynamics Estimation Approach' In MAGNI, J.-F., BENNANI, S., and TERLOUW, J. (eds.): *Robust Flight Control – A Design Challenge*. Lecture Notes in Control and Information Sciences, vol. 224 (Springer-Verlag, Berlin, 1997)
15. KLEMA, V.C., and LAUB, A.J.: 'The singular value decomposition: its computation and some applications', *IEEE Trans. Automat. Control*, 1980, **AC-25**, pp. 164–176
16. OWEN, D.H.: *Feedback and Multivariable Systems* (Peter Peregrinus Ltd., London, 1978)
17. ANONYMOUS: 'Flight Control Design – Best Practices', December 2000, NATO RTO Technical Report 29

Chapter 9

Aircraft lateral–directional handling qualities design

9.1 Introduction

Analytical studies on flight control system design using eigenstructure assignment technique have been reported widely. Formulating aircraft handling qualities design, as an eigenstructure assignment problem, was first proposed in Reference 1. Eigenstructure assignment wherein selected feedback gains can be forced to zero is used in Reference 2 to design an aircraft lateral control system. Many other studies have been undertaken, highlighting specific features of the method such as (i) gain suppression, (ii) modal robustness, (iii) command following such as pitch/yaw pointing and (iv) pseudo-control, in various aircraft control design situations [3, and references therein]. In an effort to bridge the gap between analytical studies, based on modern control theory, and their infusion into the aircraft industry design practices, a major study of developing practical flight control laws has been conducted [4]. The design study has been directed both on a civil and a military aircraft configuration. Eigenstructure assignment technique has been explored in the civilian aircraft design study. Assignment of right and left eigenvectors to achieve both output and input decoupling has been explored in Reference 5. Application of eigenstructure techniques to the design of a high-performance aircraft to improve system robustness is reported in Reference 6. These extensive analytical studies have established the maturity of the eigenstructure theory to be used as a practical design tool for flight control law development.

The handling qualities specifications defined in terms of modes with desired time constants, natural frequency and damping easily translate into eigenvalue specifications. However, translation of specifications such as turn co-ordination to equivalent eigenvector shapes is not straightforward. The approach has been to seek a mode decoupling structure between the roll and yaw axes. Thus, the problem of defining an ideal eigenvector structure for design is still elusive. In Reference 7 using literal transfer function expressions, relationship between handling qualities metrics and eigenvector element ratios is derived and guidelines are developed to select eigenvector shapes to meet handling qualities criteria. Attempts have also been made to define ideal models for the lateral–directional dynamics of aircraft [8] and also helicopters [9,10].

A successful flight test demonstration of the lateral–directional controller design for the A320 aircraft, using eigenstructure assignment, has been conducted [11]. A major flight test program, on the F18-HARV aircraft, using eigenstructure assignment based lateral–directional controller has been successfully completed [12,13]. The design has been accomplished to operate the aircraft safely at angle of attack as high as 60 deg. Both the above flight test programs however required the aircraft attitude sensor information for feedback control. These flight demonstration programs have given a strong impetus to the aircraft industry to adapt these design procedures for production aircraft programs. However, for a production aircraft, there are still some concerns regarding the availability of attitude sensors with sufficient redundancy, for use as primary feedback sensors.

In this chapter, the lateral–directional handling qualities of an aircraft will be formulated as an eigenvalue/eigenvector assignment problem. The direct relationship between eigenvector elements and their influence on response shaping as postulated in Chapter 2 will be highlighted in this design study. The design objectives specified in Chapter 8, which cover most of the aircraft industry design requirements, will be used to generate design solutions. These requirements, invariably being conflicting in nature, result in an iterative design process. Thus, the major emphasis in this chapter will be to establish a *design process*, which provides enough physical insight to the iterative performance optimisation process.

9.2 Control problem formulation

A state variable model can be derived to represent the lateral–directional dynamics of an aircraft (Appendix A). Appendix B defines F-8C aircraft model constants for three flight conditions, representing low angle of attack (FC1, $\alpha = 3.45$ deg), high angle of attack (FC2, $\alpha = 15.45$ deg) and power approach (FC3, $\alpha = 7.48$ deg) flight. The design for high angle of attack flight condition will be studied in this chapter.

The desired handling qualities specifications in section 8.3 can be broadly summarised as requiring the aircraft to behave like two decoupled sub-systems with (i) roll rate (p) and bank angle (φ) dominantly exhibiting the roll subsidence and spiral modes, respectively, and (ii) yaw rate (r) and sideslip (β) motions containing the Dutch roll mode. It is apparent that these specifications tend to suggest a mode decoupling design using eigenstructure modification. However, the main aim of the design is to achieve good turn co-ordination. This implies minimum yaw coupling in roll entries and exits using aileron control. The turn co-ordination specification can be quantified in many equivalent ways, e.g.: (i) zero sideslip angle, (ii) zero lateral acceleration (n_y) at the c.g., (iii) turn rate (r) consistent with bank angle and speed ($r = g\varphi/V_0$), (iv) zero lateral acceleration at the cockpit (ball in the middle), (v) velocity vector roll and (vi) zero stability axis yaw rate ($r - p\alpha = 0$). The last two criteria become more relevant when the aircraft is rolling at high angle of attack conditions (coning motion).

As discussed in Chapter 8, in addition to handling qualities issues, use of high-gain feedback systems to improve performance brings with it attendant problems of stability, control actuator rate saturation, etc., which have to be adequately addressed.

9.3 Feedback sensor considerations

For production class military aircraft programs with full authority flight control systems, it is required that the feedback sensors be quadruple or at least triplex redundant. This requirement originates from flight control system reliability considerations.

Ideally, to assign all the rigid body modes, a state feedback (SF) control law utilising the maximum freedom in gain selection is desirable. This implies availability of p, r, β and φ sensors. Among these, externally mounted sideslip sensors are not preferred due to damage vulnerability. Flush-mounted flow direction (α, β) sensors are being developed to overcome this limitation [14]. From cost considerations multiple redundant attitude sensors are also not preferred. Synthesis of sideslip rate ($\dot{\beta}$) feedback signal used in the F18-HARV program again requires attitude (θ, φ) information [15].

In summary, the industry practice is to avoid use of flow sensors (α, β) and attitude sensors (θ, φ) as *primary* feedback variables. Towards this end, a set of feedback signals for control purposes, using p, r and n_y sensors, is derived based on the following flight mechanics analysis.

Consider the side force equation in polar co-ordinates

$$\dot{\beta} = \frac{\bar{q}\,SC_Y}{mV_0} + p\sin\alpha - r\cos\alpha + \frac{g}{V_0}\sin\varphi\cos\theta \tag{9.1}$$

where $\bar{q}$ is the dynamic pressure, α is the angle of attack, S is the aircraft wing area, C_Y is the aerodynamic side force coefficient, m is the mass of the aircraft and V_0 is the trim aircraft velocity. The lateral acceleration is given by $n_y = \frac{\bar{q}\,SC_Y}{m}$. The side force coefficient C_Y can be expanded as

$$C_Y = C_{Y_\beta}\beta + C_{Y_\delta}\delta \tag{9.2}$$

Using small angle approximations and neglecting force term due to control (δ) in (9.2), (9.1) reduces to

$$\dot{\beta} = Y_\beta\beta + p\alpha - r + \frac{g}{V_0}\varphi \tag{9.3}$$

where $Y_\beta = \frac{\bar{q}\,SC_{Y_\beta}}{mV_0}$. From (9.3) an approximation for $\dot{\beta}$, neglecting the gravity term, can be derived as

$$\dot{\beta}_e = \left(\frac{-s}{s+p_W}\right) r_s, \quad r_s = r - p\alpha \quad \text{and} \quad p_W = -Y_\beta \tag{9.4}$$

where r_s is the stability axis yaw rate and s is the Laplace operator. The lateral acceleration is a good surrogate signal to derive sideslip information. Passing the acceleration signal through a low-pass filter generates an approximation for sideslip as

$$\beta_e = \frac{p_L}{s+p_L} \beta_{n_y} \tag{9.5}$$

where

$$\beta_{n_y} = \frac{n_y}{Y_\beta} \tag{9.6}$$

Thus, the primary feedback variables for the lateral/directional controller design are (i) body axis roll rate (p), (ii) body axis yaw rate (r), (iii) sideslip estimate (β_e) and (iv) sideslip rate estimate ($\dot{\beta}_e$). The angle of attack signal (α) required to construct the stability axis yaw rate (r_s) is derived from the flow sensor or its estimate ($\hat{\alpha}$) derived from normal acceleration.

9.4 Aircraft eigenstructure assignment

For eigenstructure assignment, as indicated earlier, at least four independent feedback variables must be identified to assign all the rigid body modes and also have maximum flexibility for eigenstructure modification. Towards this end, the following control law structures will be investigated:

(1) State feedback (SF), with variables p, r, β and φ.
(2) Dynamic output feedback (DF), with variables p, r, β_e and $\dot{\beta}_e$.

The SF structure will be used to (i) generate good benchmark solutions and (ii) characterise the relationship between eigenstructure modification and consequent response variation. The DF structure meets the industry's preferred sensor set requirement and forms the basis for generating practical design solutions, which meet all the performance requirements.

Remark: A reduced feedback variable set p, β_e and $\dot{\beta}_e$ can also be investigated to eliminate yaw rate feedback, which is likely to affect steady-state turn performance. The consequent reduction in design freedom will entail assigning only roll and Dutch roll modes. This controller structure will be further discussed in section 9.7. However, design studies reveal that as long as the feedback gains from yaw rate are kept reasonable in the DF control law structure, good steady-state turn rate performance can be achieved.

9.4.1 Synthesis of mode decoupled eigenvectors

For the eigenstructure synthesis formulation, the mode decoupling requirement alluded to in section 9.1 leads to a desirable closed-loop eigenstructure as shown in Table 9.1. However, the structure of the Dutch roll mode eigenvector shape (real eigenvector pair) shown in Table 9.1, requiring the decoupling of the bank angle variable, is not necessarily the best for achieving good turn co-ordination as later analysis will reveal.

Table 9.1 Aircraft mode decoupling eigenstructure

State variable component	Roll mode	Dutch roll		Spiral
		Real	Complex	
p	1	zd	0	1
r	zr	zx	zy	zs
β	0	x	x	0
φ	x	0	0	x

From Theorem 2.1, two eigenvector entries, in each mode, can be arbitrarily selected with one entry used for scaling. Thus, in Table 9.1, the ratio of eigenvector components corresponding to the roll rate and yaw rate variables can be arbitrarily chosen, while the sideslip and bank angle components get determined from (2.17) and (2.19). Further the selection of p and r eigenvector components is possible since the submatrix of the control distribution matrix (B) corresponding to these variables is non-singular as required by Theorem 2.1.

For the roll mode, the parameter zr is chosen to make β component zero. The x-mark for eigenvector components in Table 9.1 indicates that there is no freedom in their selection. Similarly for the spiral mode, zs is chosen to make β-component zero. These selections decouple the roll and spiral modes from the sideslip response. The complex pair eigenvalues of the Dutch roll mode have four eigenvector elements to be selected including scaling. The special structure chosen for these elements in Table 9.1 has some advantages, which will be discussed in later sections. Thus, zd, zx and zy are the free parameters to be selected to decouple the Dutch roll mode from the roll rate and bank angle response. The eigenstructure postulated in Table 9.1, in addition to meeting some handling qualities requirements, tends to make the eigenvectors linearly independent and consequently improves the modal robustness of the augmented system as discussed in Chapter 5. Further, reduction of Dutch roll mode contamination in the roll rate response and sideslip excursions can be accomplished by using the aileron to rudder interconnect (ARI) (Appendix A). Use of ARI, which modifies the control distribution matrix (B matrix), does not alter the range space of B. *Hence, the eigenstructure assignment is invariant with respect to ARI gain.*

In the following sections, we will explore how best these ideal eigenstructure modifications can be achieved, within the constraints of feedback gain magnitude limits and other specifications of Chapter 8.

An example will be used to illustrate the synthesis of the mode decoupling eigenstructure specified in Table 9.1. The F-8C aircraft *design model* (Appendix B) will be used for the study. The high angle of attack flight condition, exhibiting significant mode coupling, will be used for illustration. The aircraft modes and eigenvectors are listed in Table 9.2.

Table 9.2 F-8C eigenvector characteristics

State variable component	Eigenvalues			
	λ_R (Roll) (−0.7665)	**λ_S (Spiral) (−0.0968)**	**λ_D (Dutch roll) (−1.0329 ± j2.8998, ω_d = 3.08 rad/sec, ζ_d = 0.336)**	
p	−0.5852	−0.1023	−0.9475	0
r	−0.1100	0.0219	−0.0060	−0.0209
β	0.0214	0.0065	0.0314	0.0752
φ	0.8031	0.9945	0.1017	0.2911

Flight condition-FC2 (α = 15.45 deg).

Examination of the eigenvector matrix elements reveals: (i) bank angle is the dominant variable in both the roll and spiral modes and (ii) the Dutch roll mode dominantly appears in the roll rate and bank angle variables. It should be noted that the desired eigenstructure modification of Table 9.1 is dependent on selection of both the eigenvalue and the associated eigenvector freedom. Thus, it is imperative that the mode coupling characteristics are assessed in the range of assignable eigenvalues and an optimum selection made.

9.4.1.1 Roll mode modification

The aircraft has a slow roll mode ($\lambda_R = -0.7665$), and the associated eigenvector has strong coupling between roll rate and bank angle as seen in Table 9.2. The modification required is to speed up the mode and modify the eigenvector shape as defined in Table 9.1. This requires that roll mode be dominated by the roll rate variable and also be decoupled from sideslip response. The eigenvector is synthesised using the fundamental eigenvector constraint relationship defined in (2.21) as

$$w_i = C_i z_i, \quad \text{with } C_i = [\lambda_i I_{n-m} - F]^{-1}[G + \lambda_i H], \quad i = 1\text{–}n \tag{9.7}$$

The respective matrices in (9.7) corresponding to flight condition FC2 are

$$F = \begin{bmatrix} -0.2510 & 0.0498 \\ 0 & 0 \end{bmatrix}, G = \begin{bmatrix} 0.2677 & -0.9663 \\ 1.0 & 0.276 \end{bmatrix}, H = \begin{bmatrix} 0.0009 & -0.0103 \\ 0 & 0 \end{bmatrix}$$

For the roll mode, the constraint (9.7) for the assignment in Table 9.1 takes the form

$$\begin{bmatrix} \beta_r \\ \varphi_r \end{bmatrix} = \begin{bmatrix} c_{11} & c_{12} \\ c_{21} & c_{22} \end{bmatrix} \begin{bmatrix} 1 \\ zr \end{bmatrix} \tag{9.8}$$

Figure 9.1 gives the variation in the yaw rate (zr) and bank angle (φ_r) eigenvector components as a function of roll mode eigenvalue (λ_R) for making the sideslip component (β_r) zero. As the roll mode becomes faster, the dominance of φ_r decreases as desired. For a specified roll mode eigenvalue of $\lambda_R = -4.0$, the mode coupling relationship of (9.7) for the eigenvector assignment of Table 9.1 is given by

$$\begin{bmatrix} \beta_r \\ \varphi_r \end{bmatrix} = \begin{bmatrix} -0.0671 & 0.2448 \\ -0.2500 & -0.0690 \end{bmatrix} \begin{bmatrix} 1 \\ zr \end{bmatrix} \tag{9.9}$$

For β_r to be zero, zr = 0.2742, resulting in $\varphi_r = -0.2689$. This results in the eigenvector $x_R = [0.9335 \quad 0.2560 \quad 0 \quad -0.2510]^T$, with the eigenvector normalised to unit length. The bank angle coupling has been reduced, and sideslip response decoupling has been achieved.

9.4.1.2 Spiral mode modification

From Figure 9.1 it is seen that as the spiral mode is decreased, φ_s component becomes dominant, which is the desired result. For a spiral mode of $\lambda_S = -0.01$, the mode coupling relationship in (9.7) for the desired assignment of Table 9.1 is

$$\begin{bmatrix} \beta_s \\ \varphi_s \end{bmatrix} = \begin{bmatrix} -19.5486 & -9.7115 \\ -100.0 & -27.6 \end{bmatrix} \begin{bmatrix} 1 \\ zs \end{bmatrix} \tag{9.10}$$

For β_s to be zero, zs = −2.0129, resulting in $\varphi_s = -44.4433$ (Figure 9.1). This results in the eigenvector $x_S = [0.0225 \quad -0.0452 \quad 0 \quad -0.9987]^T$, with the eigenvector normalised to unit length. The roll rate component in the eigenvector has been reduced, and sideslip response has been decoupled from the slow spiral mode.

9.4.1.3 Dutch roll mode modification

In general, the eigenstructure synthesis of the complex Dutch roll mode is not as straightforward as those of roll and spiral modes. This is due to the difficulty in visualising the structure of a pair of real eigenvectors that will achieve the desired response shaping. However, the structure defined in Table 9.1 is easily realised by constructing the coupling matrix using the constraint relationship (2.19). The aircraft Dutch roll mode damping is improved by shifting the pole to $\lambda_D = -2.1 \pm j2.1424$ ($\omega_d = 3.0$ rad/sec, $\zeta_d = 0.7$). For this complex mode assignment, the corresponding mode coupling constraint takes the form

$$\begin{bmatrix} \beta_x \\ \varphi_x \\ \beta_y \\ \varphi_y \end{bmatrix} = \begin{bmatrix} -0.0613 & 0.2092 & 0.0657 & -0.2592 \\ -0.2333 & -0.0644 & 0.2380 & 0.0657 \\ -0.0657 & 0.2592 & -0.0613 & 0.2092 \\ -0.2380 & -0.0657 & -0.2333 & -0.0644 \end{bmatrix} \begin{bmatrix} zd \\ zx \\ 0 \\ zy \end{bmatrix} \tag{9.11}$$

Setting zd = 1 and solving for zx and zy from (9.11) to make $\varphi_x = 0$ and $\varphi_y = 0$ results in zx = −3.6232, zy = 0, $\beta_x = -0.8192$ and $\beta_y = -1.005$. The resulting eigenvector pair normalised to unit length (2.4a) corresponding to the Dutch roll mode eigenvector is synthesised as

$$x_D = \begin{bmatrix} 0.2515 & 0 \\ -0.9113 & 0 \\ -0.2060 & -0.2528 \\ 0 & 0 \end{bmatrix} \begin{matrix} p \\ r \\ \beta \\ \varphi \end{matrix} \tag{9.12}$$

It is seen that the roll rate component has been reduced compared with the basic aircraft. The complex mode is dominant in the yaw rate and sideslip variable as desired. The achieved eigenstructure is summarised in Table 9.3.

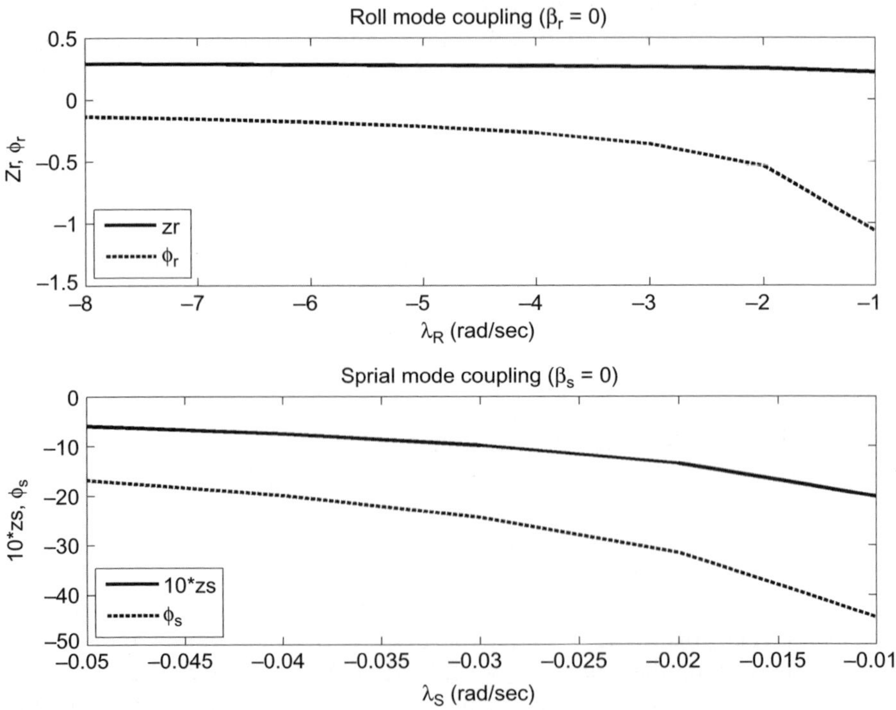

Figure 9.1 Real mode eigenvector coupling characteristics

Table 9.3 Eigenstructure assignment for mode decoupling

State variable component	Eigenvalues			
	Roll (–4.0)	Spiral (–0.01)	Dutch roll (–2.1 ± j2.1424, ω_d = 3.0 rad/sec, ζ_d = 0.7)	
p	0.9335	−0.0225	−0.2517	0
r	0.2560	0.0452	0.9112	0
β	0.0	0.0	0.2060	0.2528
φ	−0.2510	0.9987	0.0001	0.0001

Flight condition-FC2 (α = 15.45 deg).

The significant improvement in the modal robustness characteristics (Chapter 5) due to this eigenstructure assignment of Table 9.3 is listed in Table 9.4.

Table 9.4 Modal robustness characteristics

F-8C aircraft		Augmented aircraft	
Eigenvalue	Mode condition number (σ)	Eigenvalue	Mode condition number (σ)
−0.7665	8.6481	−4.00	1.0357
−0.0968	6.4071	−0.01	1.0370
−1.0329 ± j2.8998	14.3213	−2.1 ± j2.1424	4.1858

Flight condition-FC2 (α = 15.45 deg).

An SF control law assigning the eigenstructure of Table 9.3 to the system

$$A_c = A + B * K \tag{9.13}$$

is given by

$$\begin{bmatrix} \delta_a \\ \delta_r \end{bmatrix} = K \begin{bmatrix} p \\ r \\ \beta \\ \varphi \end{bmatrix}, \quad K = \begin{bmatrix} -0.1196 & -0.0002 & 2.1632 & -0.0086 \\ -0.0535 & 0.9270 & -1.4911 & -0.0492 \end{bmatrix} \tag{9.14}$$

All angular rotation and rate gains in (9.14) are in radian units. From (9.14) it is clear that the solution requires unacceptably high sideslip feedback gains (gain magnitudes greater than unity). The cause can be traced to the requirement to *fully* decouple the Dutch roll mode from bank angle response and also reduce the coupling to roll rate variable. Iterative designs relaxing this requirement can be attempted to reduce the feedback gain magnitudes while accepting lower robustness. This approach however will not be pursued here. An alternative and more effective design methodology of using optimisation techniques to realise good turn co-ordination control laws will be outlined in the subsequent sections.

9.5 Aircraft eigenstructure optimisation

As stated earlier, an important design goal of aircraft lateral–directional feedback augmentation is to achieve good turn co-ordination. However, there are additional performance objectives that need to be satisfied as detailed in Chapter 8. There are many ways these requirements can be postulated as a performance index for optimisation. Table 9.5 lists some of the candidate objective functions for system optimisation including turn co-ordination.

In case of SF, the design variables available for optimising the performance are (i) rigid body modes (λ_R, λ_S and λ_{DR}), (ii) free eigenvector elements zr, zs, zd, zx and zy (Table 9.1) and (iii) aileron–rudder interconnect gain g_x. Among the above design parameters, the Dutch roll mode (λ_{DR}) with the associated free eigenvector elements (zd, zx and zy) combined with ARI gain g_x strongly influences the minimisation process. Thus, the roll and spiral modes and the associated eigenvector parameters zr and zs are fixed to the values computed from (9.9) and (9.10), respectively, to ensure full decoupling of the sideslip response from roll and spiral modes. In case of DF, the washout filter root $\lambda_W = -p_W$ and the n_y lag filter root $\lambda_L = -p_L$ are additional variables available for optimisation. Table 9.6 summarises the design variables used along with their limits.

Table 9.5 Objective functions for performance optimisation

Performance index	Objective
$J_1 = \min\limits_{\omega_{min} \le \omega \le \omega_{max}} \left\{ \text{avg} \left\| \frac{\beta(s)}{p(s)} \right\|_{s=j\omega} \right\}$	Turn co-ordination (frequency domain) $\frac{\beta(s)}{p(s)}$: Ratio of aileron transfer functions
$J_2 = \min\left(\frac{\beta_{rms}}{p_{rms}}\right)_{\text{aileron doublet input}}$	Turn co-ordination (time domain)
$J_3 = \sum\limits_{i=1}^{m}\sum\limits_{j=1}^{n} w_{ij}k_{ij}$	Weighted feedback gain constraint
$J_4 = \min(\beta_{ss})_{\text{aileron step input}}$	Steady-state sideslip
$J_5 = [\text{Avg}(\beta)]_{\text{aileron pulse (right roll)}} > 0$	Ensure adverse sideslip (to eliminate PIO tendencies)
$J_6 = \text{Real}(\lambda_i) < 0$, for all i	Ensure closed-loop stability

Table 9.6 Design variable limits

Design variable (λ_{dr})	Lower limit (ω_d = 1 rad/sec, $\zeta_d = 0.5$)	Upper limit (ω_d = 5 rad/sec, $\zeta_d = 0.9$)
zd	$-\infty$	∞
zx	$-\infty$	∞
zy	$-\infty$	∞
g_x	−1.5	0.20
λ_W	−1.0	−0.30
λ_L	−4.0	−3.33

The ω_d lower limit is set by the handling qualities requirements of Chapter 8, and the upper limit is based on engineering judgment. It is advisable not to shift the natural frequency (ω_d) too far away from the basic aircraft value. The ζ_d lower limit is set higher than the handling qualities requirement. The ARI gain upper limit is set such that under worst combination of roll and yaw control derivatives estimate (L_δ, N_δ), (i) augmented L_{δ_a} does not reverse sign (roll reversal) and (ii) aileron proverse yaw characteristic is retained (N_{δ_a} does not reverse sign).

9.5.1 State feedback design

SF design offers maximum flexibility for eigenstructure modification. Thus, the design provides a *benchmark* on the best achievable performance. In addition, it offers direct insight into the level of augmentation required in the stability derivatives to achieve a desired performance. These insights can be appropriately used in the iterative re-optimisation of the dynamics of the augmented aircraft to meet all the, often conflicting, performance requirements. From a turn co-ordination perspective, examination of (9.3) reveals that SF to the rudder channel has the property of minimising an ideal turn co-ordination parameter $\left(\dot{\beta} = Y_\beta \beta + p\alpha - r + \frac{g}{V_0}\varphi\right)$.

An example will now be presented for optimising turn co-ordination using state variable feedback. The *design model* (Appendix B) will be used for this purpose.

The *objective function* for optimisation is chosen as (Table 9.5)

$$J = J_1 + J_3 \tag{9.15}$$

The frequency range for J_1 is chosen as ω_{min} = 0.1 rad/sec and ω_{max} = 10 rad/sec. One acceptable design resulted in the SF gain matrix and control interconnect gain

$$K = \begin{bmatrix} -0.2587 & 0.4805 & 0.0 & -0.0351 \\ -0.0938 & 0.6798 & -0.0101 & -0.0464 \end{bmatrix}, \quad g_x = -0.9171 \tag{9.16}$$

Even though the design model has n_y-lag filter and r-washout filter states, the SF influences only the rigid body dynamics. Thus, only the rigid body modes are modified. Figure 9.2 shows the ratio of sideslip/roll rate frequency response magnitude ratio. The figure indicates that the controller has achieved excellent reduction in the response ratio over the bandwidth frequency range.

The aileron doublet response in Figure 9.3 corroborates this observation. Further notice that the roll rate response of the closed-loop system is excited only by the roll mode, indicating that the Dutch roll mode contamination has been eliminated. The sideslip response has been significantly reduced.

At high angle of attack conditions, the ARI gain has a very strong influence on achieving good turn co-ordination performance. The variation of ARI gain (g_x) on the turn co-ordination properties is shown in Figure 9.4. The variation of the numerator zero (ω_φ, ζ_φ) of the roll rate to aileron transfer function and the average closed-loop (β/p) ratio (Figure 9.1) as a function of ARI gain reveals that for the optimum gain $g_x = -0.9171$ (9.16), the average (β/p) ratio is a minimum and the numerator zero (ω_φ, ζ_φ) practically cancels the Dutch roll mode (ω_d, z_d). Thus, both good turn co-ordination and decoupling of the Dutch roll mode from the roll rate response have been achieved.

In a turn coordinated feedback-augmented aircraft, the rudder control is generally used to intentionally generate sideslip motion such as during crosswind landing. Figure 9.5 shows the sideslip and bank angle excursions for a rudder doublet input. It is seen that for the basic aircraft, a sideslip motion of approximately 2 deg (peak to peak) results in a large bank angle excursion (−30 deg).

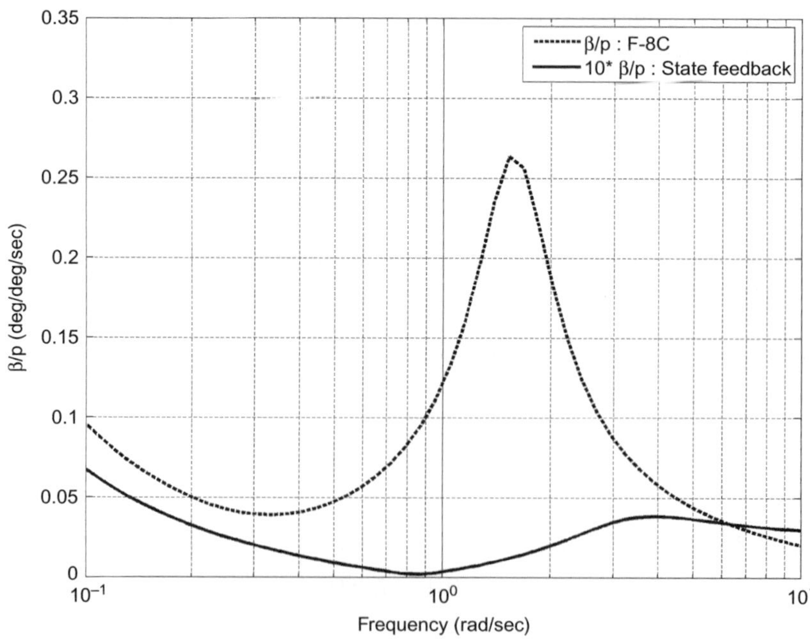

Figure 9.2 Magnitude ratio of sideslip to roll rate transfer functions (aileron input)

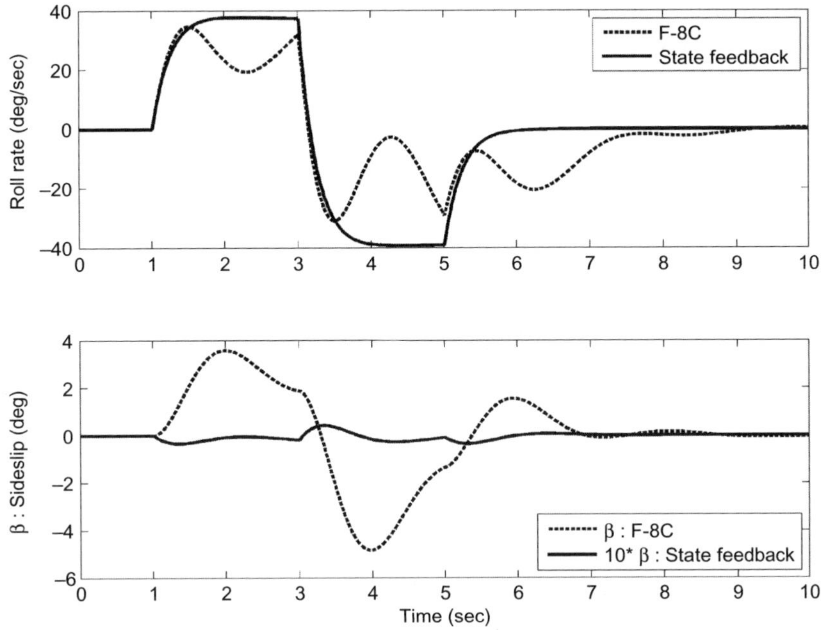

Figure 9.3 Aileron doublet command response

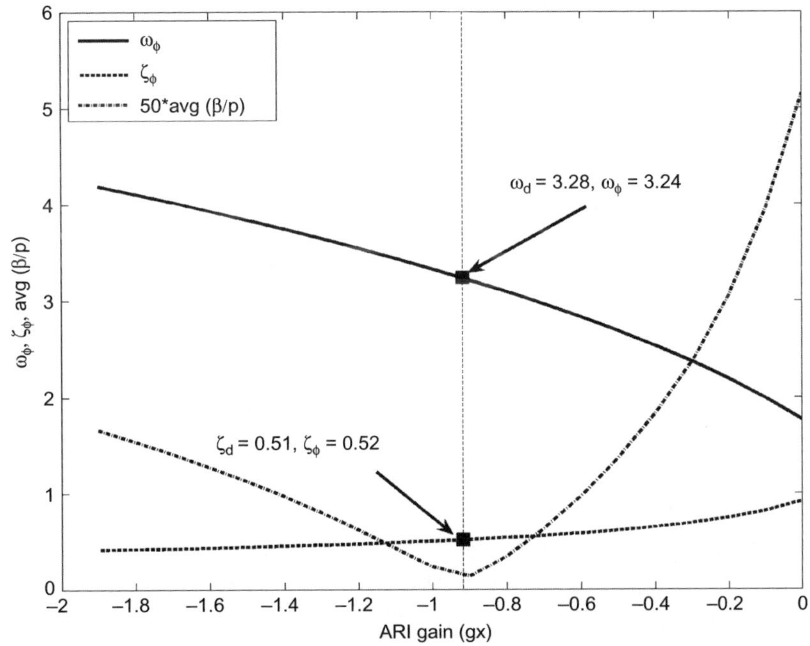

Figure 9.4 Aileron–rudder interconnect characteristics

Figure 9.5 also shows that by scaling the design variable zd (Table 9.1), from its optimum value ($\rho = 1$ in Figure 9.4) to half its value ($\rho = 0.5$), bank angle deviations of the closed-loop system can be reduced while the sideslip response remains invariant. *Another remarkable property is that the response characteristics of the feedback system to aileron input remain invariant with respect to scaling of zd.* This invariance property allows fine-tuning of the bank angle response to rudder input without affecting turn co-ordination. Thus, using the special eigenvector structure of Table 9.1, independent control of two important dynamic characteristics of the aircraft is possible.

The feedback gain variation, as a function of scaling parameter (ρ), is shown in Table 9.7. The only gain that undergoes significant change is the gain from sideslip to aileron (k_β). This results in the additional augmentation of only the effective dihedral (c_{l_β}) and static directional stability (c_{n_β}) of the form

$$\hat{c}_{l_\beta} = c_{l_\beta} + k_\beta c_{l_{\delta a}}$$
$$\hat{c}_{n_\beta} = c_{n_\beta} + k_\beta c_{n_{\delta a}} \tag{9.17}$$

Table 9.8 lists the key parameters that undergo change due to the scaling (ρ) of the Dutch roll eigenvector parameter zd. The F-8C aircraft has close to neutral static directional stability at this high angle of attack ($\alpha = 15.45$ deg). As ρ decreases, static stability increases while the effective dihedral decreases. However, there is no loss in departure resistance as indicated by the criteria c_{n_β} dyn and LCDP (8.2). Thus, by modifying the scale factor of the Dutch roll mode eigenvector design parameter zd, the bank angle excursions due to rudder inputs can be controlled without affecting either the turn co-ordination or the closed-loop stability characteristics as indicated by the departure resistance parameters. As a corollary, changes in sideslip to aileron feedback gain do not alter turn coordination characteristics (see Table 9.7a).

Table 9.7 Feedback gain variations with scaling parameter (ρ)

ρ	p	R	β	φ
(a) Aileron channel gains*				
1.0	−0.2587	0.4805	0.00	−0.0351
0.7	−0.2512	0.4554	0.64	−0.0334
0.5	−0.2282	0.3764	1.18	−0.0290
(b) Rudder channel gains*				
1.0	−0.0938	0.6798	−0.0100	−0.0464
0.7	−0.0921	0.6741	−0.0200	−0.0461
0.5	−0.0915	0.6717	−0.0347	−0.0460

*All gains are in radian units.

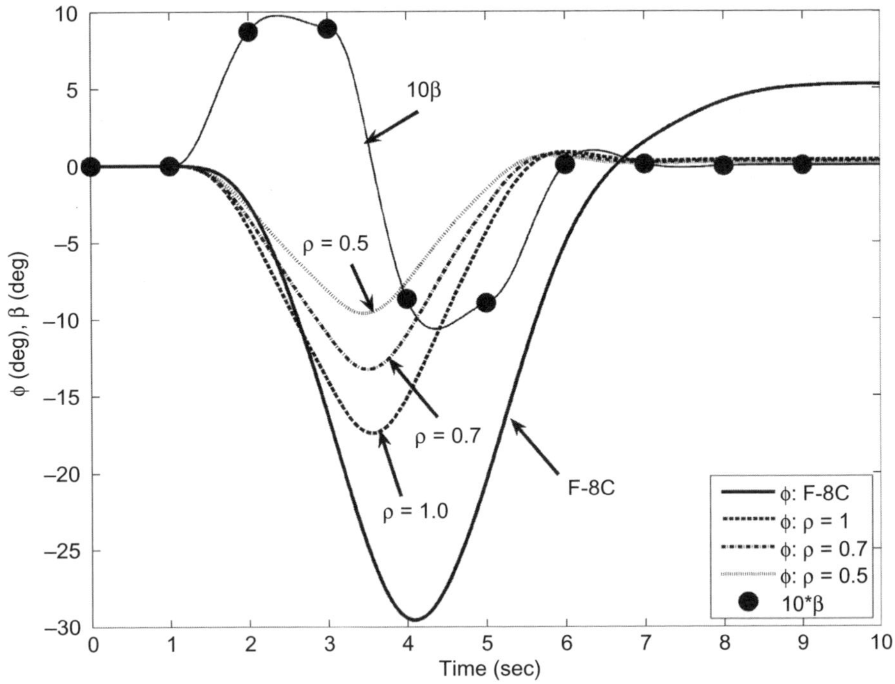

Figure 9.5 Effect of Dutch roll eigenvector scaling (ρ) on rudder input response

Table 9.8 Key aircraft parameter variations with scaling parameter (ρ)

ρ	c_{n_β}*	c_{l_β}*	c_{n_β} dyn*	LCDP*	k_β	φ_{peak}
F-8C	−0.002	−0.102	0.267	0.0674	–	−30.00
1.0	−0.001	−0.102	0.268	0.2800	0.00	−17.40
0.7	0.080	−0.073	0.271	0.2820	0.64	−13.30
0.5	0.148	−0.049	0.270	0.2810	1.18	−9.60

*All derivatives in radian^{-1} units.
k_β – Sideslip to aileron gain (rad/rad).
φ_{peak} – Peak bank angle (deg) (Figure 9.5).

9.5.2 Dynamic output feedback design

The DF controller has two additional state variables corresponding to λ_W and λ_L. The resulting sixth-order *design model* is given in Appendix B. Since all states are not used for feedback, output feedback algorithms of Chapter 4 will be used. Using Theorem 4.1, all rigid body modes (four eigenvalues) along with one entry in each eigenvector can be assigned as in the SF case. However, the migrations of the unassigned lag roots (λ_W and λ_L) are not under control. Alternately using Theorem 4.4, five eigenvalues can be assigned with a loss of freedom in eigenvector selection. In the present study algorithm of Theorem 4.1 will be used for assigning all the rigid body modes. A constrained optimisation procedure will ensure that unassigned roots (λ_W and λ_L) will be restricted to lie in the stable region. The use of independent feedback gains from both the rate (k_g) and washed out rate (k_w) of roll and yaw rate signals results in the composite dynamic feedback as shown below:

$$\hat{k}_g = k_g + \frac{k_w s}{s + p_W} \tag{9.18}$$

leading to the relation

$$\hat{k}_g = \frac{k_t(s + z_W)}{s + p_W}, \quad \text{where } k_t = (k_g + k_w), \quad z_W = p_W\left(\frac{k_g}{k_t}\right) \tag{9.19}$$

This results in either a lead or a lag filter depending on the zero (z_w) location. The filter can be either minimum phase or non-minimum phase. There are two zeros $\lambda_a = -z_a$ and $\lambda_b = -z_b$, corresponding to the aileron and rudder channels, respectively, *for each of roll rate and yaw rate feedback paths*. Again constraints on (i) minimum phase filters (λ_a, $\lambda_b < 0$) and (ii) the ratio of (z_W/p_W) can be imposed in the optimisation procedure to restrict the feasible solution space. The ratio constraint emerges from the requirement to limit signal quantisation noise amplification effects in digital controllers using fixed-point arithmetic.

9.6 Aircraft performance assessment

The SF design example in section 9.5.1 used only the design model for optimisation and performance assessment. However, the control system has additional dynamics such as (i) actuator lags, (ii) computation delays associated with digital controller, (iii) lags associated with structural filters in sensor channels and (iv) a lead filter in the aileron channel to inhibit the hardware modes from coalescing with the rigid body modes. The hardware-related dynamics would influence the final performance of the feedback system. A *truth model* (Appendix B) is constructed including all these additional dynamic elements. The truth model is used for computing the performance index of the optimising function. Additional dynamics in the form of (i) a second-order secondary servo actuator in each channel and (ii) a second-order Dryden turbulence model (Appendix A) is added to the truth model to assess the actuator rate requirements for (i) full stick roll rate commands and (ii) turbulence disturbance inputs.

Figure 9.12 gives a schematic of the DF control law with feedback loops to the aileron channel and a command path (inset). The feedback loops shown are identical to the rudder channel also. However, the command path design for the rudder channel is not relevant to the present design study. The aileron command path details shown in Figure 9.12 (inset) are relevant to the assessment of Gibson's pilot involved oscillation (PIO) criterion to be discussed subsequently.

The following steps are involved in the design optimisation:

Step 1. Assign the roll mode and associated decoupled eigenvector using (9.9).
Step 2. Assign the spiral mode and associated decoupled eigenvector using (9.10).
Step 3. Initiate a design with an initial guess of the design parameters from Table 9.6, using the *design model.*
Step 4. Compute the specified *objective function* selected from Table 9.5, using the *truth model.*
Step 5. Refine the design parameters using an optimisation procedure, to minimise the *objective function.*
Step 6. Assess all the key performance attributes. Re-optimise if some attributes are not satisfactory.

Remark: The use of two models in steps 3 and 4 is intentionally done: (i) to retain the direct relationship of the design parameters (Table 9.6) to the eigenstructure modification process as revealed by the *design model* and (ii) to compute an accurate *objective function* capturing the effects of unmodelled dynamics using the *truth model.*

Based on this optimisation procedure, both SF and DF controllers are designed to meet all the handling qualities and control system design requirements outlined in Chapter 8. The performance results are summarised in the following sections.

9.6.1 Feedback design characteristics

Table 9.9 summarises the feedback gains and optimum design parameters for both SF and DF designs. Table 9.10 summarises the modal robustness characteristics for *design* and *truth* models. Table 9.11 lists the steady-state response for unit aileron and rudder inputs. The key observations about the solutions are:

(1) The Dutch roll mode damping is adequate.
(2) SF and DF designs have (ω_d, z_d) and (ω_φ, z_φ) pairs close to each other, indicating good Dutch roll mode decoupling from roll rate response for aileron input.
(3) All the feedback gains are well within the limits specified in Chapter 8.
(4) The SF design modal condition numbers based on *truth model* are better than the DF design.
(5) The ratio of z_W/p_W of all the dynamic filters is less than unity. The roll rate to rudder feedback channel zero is non-minimum phase. Since the z_W/p_W ratio is less than unity, no attempt is made to force the non-minimum phase constraint.
(6) The steady-state sideslip response to aileron input is very low for both control laws, indicating that the spiral mode is decoupled.

Table 9.9 Feedback gains and design parameters

	p	r	β	φ	β_e	$\dot{\beta}_e$
(a) Aileron channel gains*						
SF	−0.2703	0.5225	0.0008	−0.0359	–	–
DF	−0.1855	−0.3173	–	–	−0.1299	−0.4144
(b) Rudder channel gains*						
SF	−0.1545	0.9581	−0.0088	−0.0577	–	–
DF	0.1042	−0.2193	–	–	−0.0788	−0.5131

	λ_L	λ_W	g_x	ω_d	ζ_d	ω_φ	ζ_ι
(c) Design parameters (based on *design model*)							
SF	−3.30	−1.00	−0.76	3.31	0.70	3.09	0.70
DF	−3.37	−0.629	−0.66	2.36	0.53	2.43	0.48

*All gains are in radian units except β_e gains that are in rad/g.

Table 9.10 Modal robustness of feedback system

Mode	**Model**		σ_d^*	σ_t^*
	Design	**Truth**		
(a) State feedback (SF)				
λ_R	−4.00	−4.1750	9.90	7.04
λ_S	−0.01	−0.0095	1.99	2.46
λ_W	−1.00	−1.0000	1.45	1.44
λ_L	−3.33	−3.3300	2.88	2.11
$[\omega_d, \zeta_d]$	[3.31, 0.70]	[4.10, 0.75]	14.68	18.46
$[\omega_\varphi, \zeta_\varphi]$	[3.09, 0.70]	[3.91, 0.82]	–	–
(b) Dynamic output feedback (DF)				
λ_R	−4.000	−3.5900	17.13	10.07
λ_S	−0.010	−0.0097	1.22	1.62
λ_W	−0.738	−0.7620	2.11	2.43
λ_L	−5.105	−5.2200	21.75	13.71
$[\omega_d, \zeta_d]$	[2.36, 0.53]	[2.55, 0.59]	9.06	11.58
$[\omega_\varphi, \zeta_\varphi]$	[2.43, 0.48]	[2.63, 0.54]	–	–

Table 9.10 Concluded

Feedback channels	**Control channels**	
	Aileron	**Rudder**
(c) Dynamic filter zeros (DF)		
Roll rate	−0.194	0.160
Yaw rate	−0.273	−0.188

*Mode condition numbers (Chapter 5): σ_d − design model and σ_t − truth model (with no secondary servo dynamics – Table B.4).

Table 9.11 Steady-state step response

Configuration	**Aileron (+1 deg)***				**Rudder (+1 deg)***			
	p	**r**	**β**	**φ**	**p**	**r**	**β**	**φ**
F-8C	−0.72	2.6	0.52	57.18	2.67	−9.66	−0.39	−203.74
F-8Cx†	−2.73	9.88	0.812	210.74	2.67	−9.66	−0.39	−203.74
SF	−5.12	18.55	0.00	387.70	4.79	−17.37	0.36	−361.07
DF	−3.71	13.44	0.05	281.00	3.57	−12.93	0.50	−267.64

*All responses are in degrees.
†F-8C with ARI (g_x = −0.7663).

Figure 9.6a–e summarises the response of the system to an aileron pulse command. The pulse input is applied through a command path filter with a time constant of 0.25 sec. The roll rate response of both SF and DF design is not corrupted by the Dutch roll mode as required. The SF design sideslip response is not coupled with the Dutch roll mode either and also shows the desirable feature of a slow build-up of the sideslip. The sideslip response of the DF design has Dutch roll mode component. The response of the basic F-8C is included to show the excellent improvement in both roll rate and sideslip response achieved using feedback. The sideslip excursion has been reduced substantially. The adverse sideslip response ensures no PIO tendencies. The sideslip response is also not contaminated by the spiral mode so that the, near zero, steady-state sideslip is quickly attained. From Figure 9.6c it is seen that the aircraft turn rate capability, given by $r_t = \frac{g\varphi}{V}$ (deg/sec), is not affected by feedback. For the DF design, the aileron actuator response shows an overshoot and consequently peak aileron actuator rate is slightly higher.

Figure 9.7 shows the feedback system response to a rudder pedal input. The SF design shows a larger bank angle excursion compared with the DF design. This is to be expected since the sideslip sensor feedback gains are very low (Table 9.9). Further reduction in bank angle response can be achieved, *without affecting*

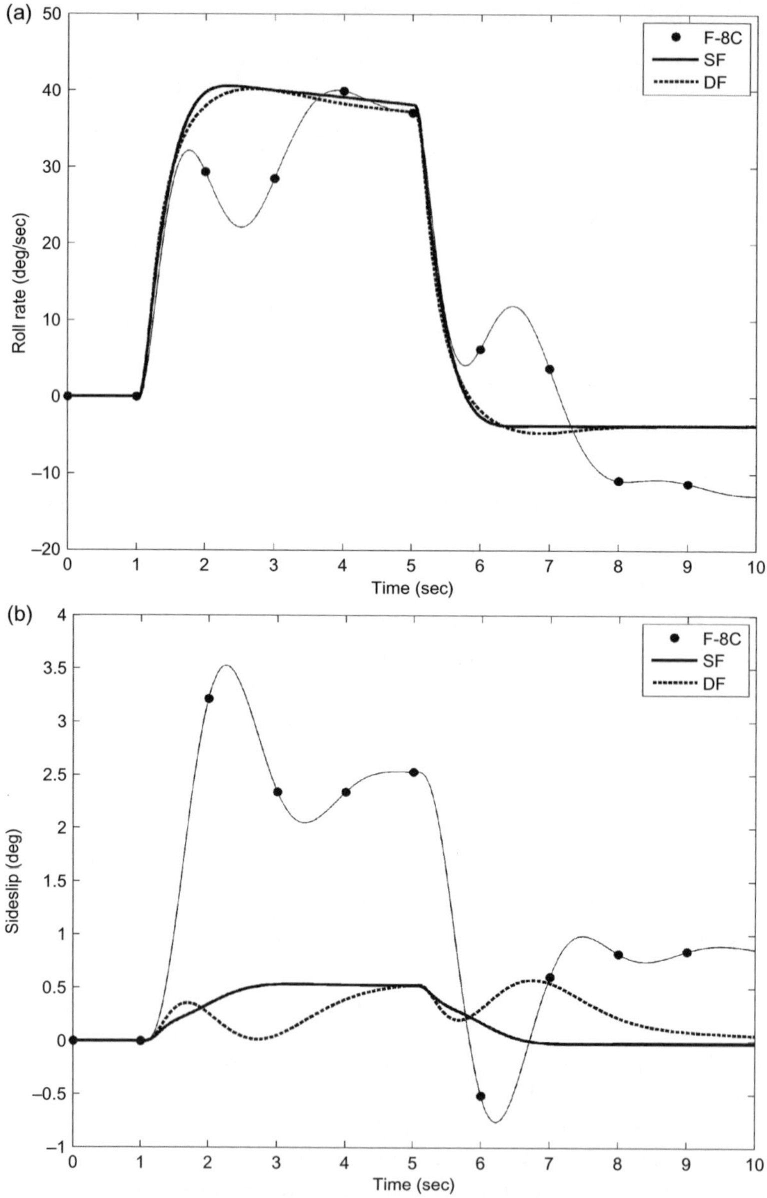

Figure 9.6 System response to aileron pulse command input ($p_{cmd} = 40\,deg\,/\,sec$) (a) Roll rate response; (b) Sideslip response; (c) Yaw rate and turn rate response; (d) Actuator position response; (e) Actuator rate response

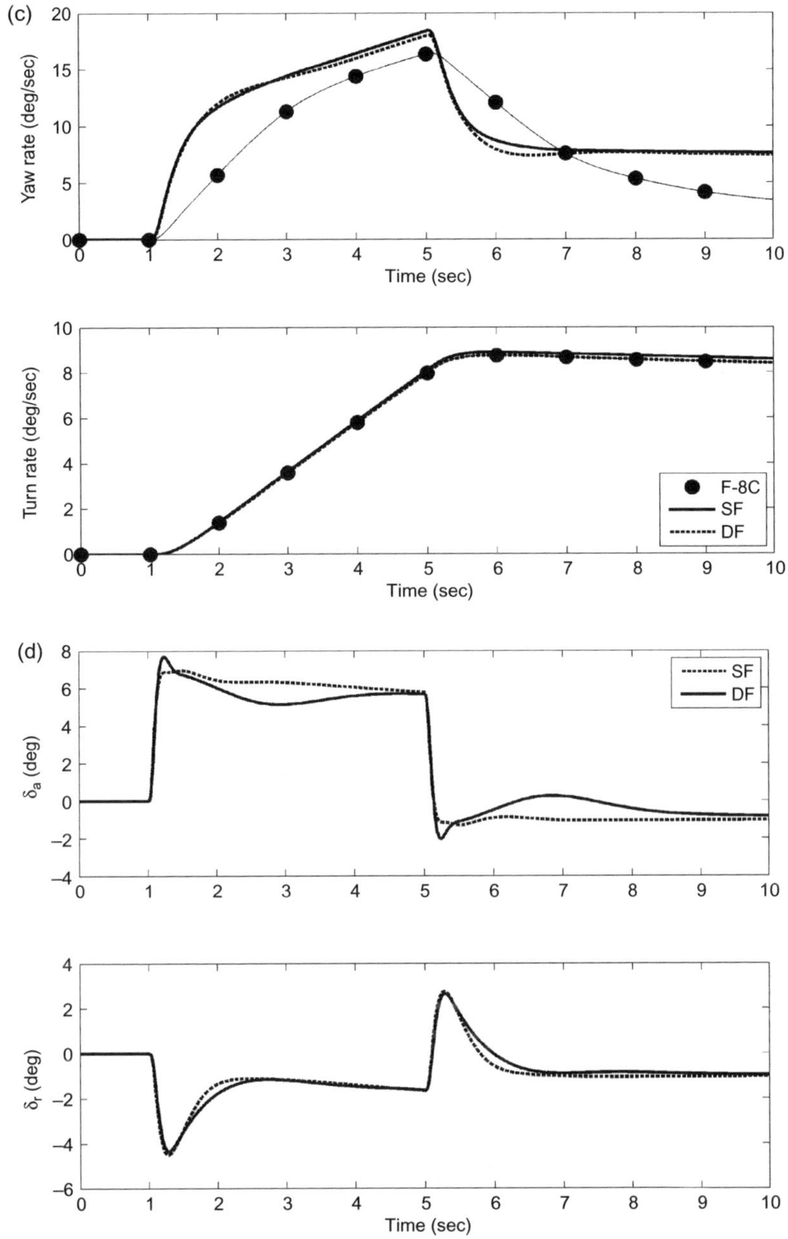

Figure 9.6 Continues

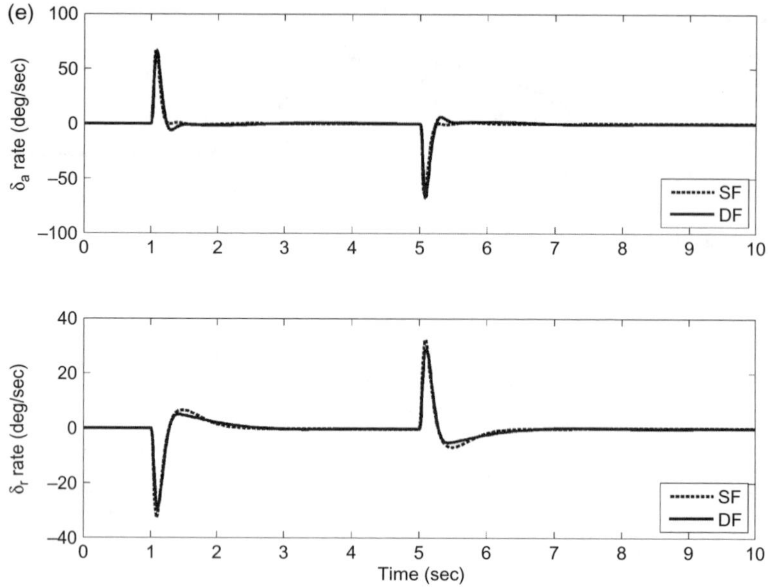

Figure 9.6 Concluded

the turn co-ordination performance, using the method proposed in section 9.5 (Figure 9.5). The low bank angle response of the DF design indicates heavy augmentation of the effective dihedral (c_{l_β}). This aspect will be further discussed in section 9.7. Again fine-tuning of this response is possible as indicated earlier.

9.6.2 Handling qualities performance

From Table 9.2, it is seen that the F-8C aircraft is marginally level 2 in both roll mode and Dutch roll mode damping. The feedback designs (Table 9.10) assign all rigid body modes to be inside the level 1 boundary. Turn coordination performance for large roll inputs has been summarised in Table 9.12. It is seen that the basic F-8C has level 2 handling qualities. For SF and DF designs, the sideslip excursions are very well within the level 1 handling qualities requirements.

Table 9.12 Turn coordination performances for large roll command (p_{max} = +40 deg/sec)

Configuration	k	β_k* (deg)	n_{ymax}† (g's)
F-8C	0.34	10.4	−0.350
SF	0.44	1.33	−0.088
DF	0.43	1.02	−0.080

*Requirement: (section 8.3), for positive roll, $\beta_k < +6$ deg, $\beta_k > -2$ deg.
†Requirement: (section 8.3), $|n_{y_{max}}| \le 0.1$g.

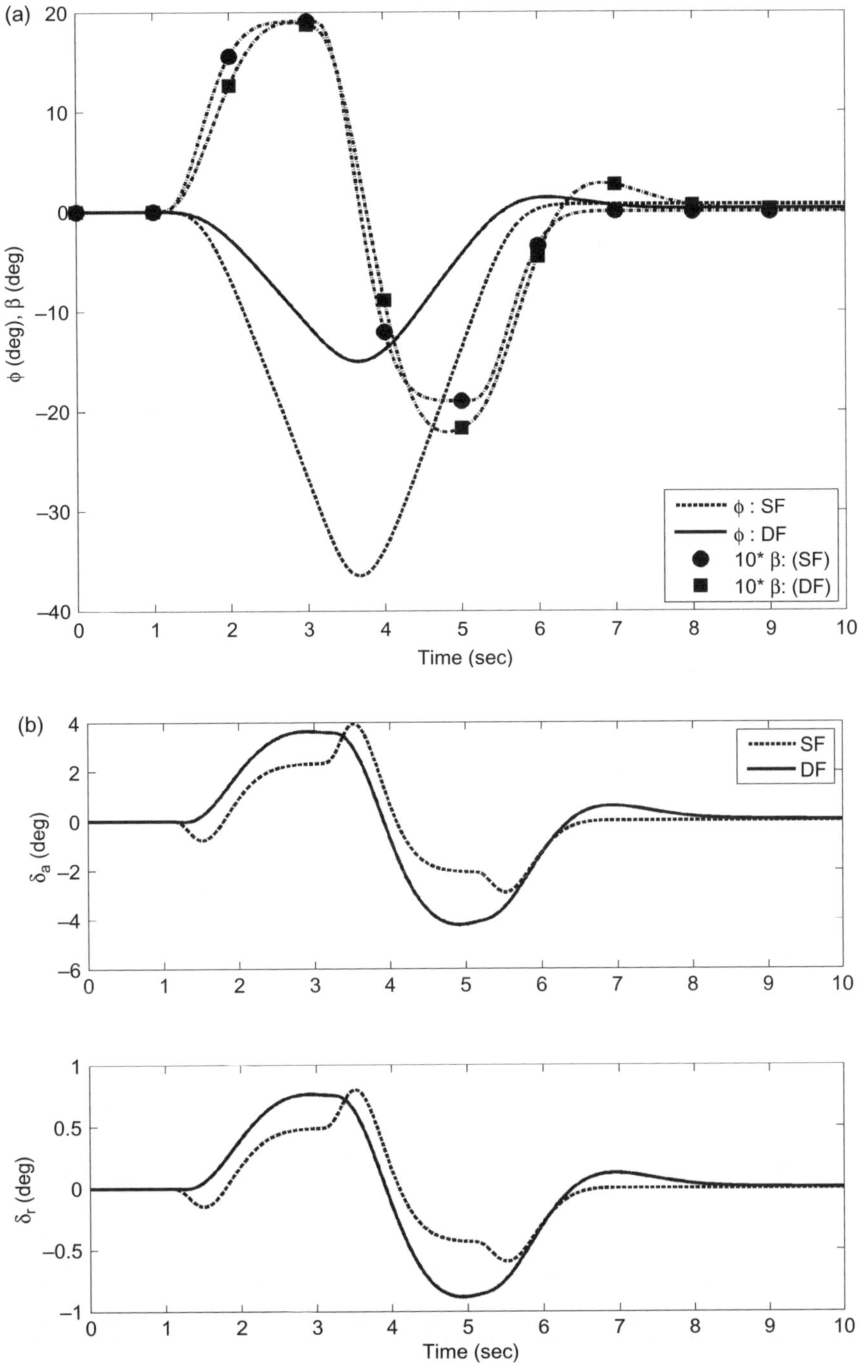

Figure 9.7 Rudder doublet command input (a) Sideslip and bank angle response; (b) Actuator response

9.6.3 Departure resistance characteristics

Aircraft designed for high agility will be required to manoeuvre at high angle of attack. It is well known that at high angle of attack, due to several aerodynamic deficiencies, aircraft tend to depart from controlled flight in lateral–directional modes typically characterised as wing drop, wing rock, nose slice, etc. The loss of control power of conventional aerodynamic surfaces at high angle of attack also limits the level of departure resistance augmentation using feedback controllers. Criteria for assessing the departure resistance of aircraft have been developed for use with linear perturbation models (8.2). Table 9.13 summarises: (i) static directional stability (c_{n_β}), (ii) effective dihedral (c_{l_β}), (iii) c_{n_β} dyn and (iv) LCDP for all the configurations. From Table 9.13, it is seen that the basic F-8C aircraft has negative directional static stability and low departure resistance (LCDP). The large effective dihedral of the F-8C ensures satisfactory c_{n_β} dyn. The SF design has low feedback gains to both aileron and rudder channels. This results in low augmentation of the static directional stability. For the same reason the effective dihedral remains close to the F-8C, resulting in satisfactory c_{n_β} dyn. However, the high control interconnect gain improves substantially the LCDP parameter. The sideslip derivatives cannot be computed for the DF design since there is no explicit sideslip feedback. However, to derive an estimate of the implicit augmentation of these parameters, it is required to transform the closed-loop system matrix to an equivalent rigid body system matrix. This is accomplished by computing an *equivalent* SF feedback that assigns all the rigid body eigenvalues and eigenvectors of the DF design and an equivalent interconnect gain (g_x) that achieves response matching of sideslip response to a doublet aileron input. The *equivalent* SF gains are listed in Table 9.14. The high equivalent sideslip feedback gains in both the aileron and rudder channels indicate heavy augmentation of the sideslip derivatives. The bank angle response to rudder input in Figure 9.6a corroborates the fact that the effective dihedral has been reduced by DF design. The net effect of the equivalent sideslip feedback is (i) static directional stability has improved, (ii) c_{n_β} dyn has reduced compared

Table 9.13 Departure resistance characteristics

Configuration	**g_x***	**Derivative values (rad^{-1})**					
		C_{n_β}	**C_{l_β}**	**$C_{n\delta_a}$†**	**$C_{l\delta_a}$†**	**C_{n_β}dyn**	**LCDP**
F-8C	0.000	−0.0016	−0.1017	0.0289	0.0426	0.2671	0.0674
SF	−0.763	−0.0006	−0.1016	0.1076	0.0444	0.2679	0.2455
DF‡	−0.900	0.0287	−0.0403	0.1217	0.0448	0.1341	0.1382

*Control interconnect gain.
†Control derivatives with interconnect.
‡Derivative values computed using an *equivalent* SF design (Table 9.14).

Table 9.14 Equivalent SF gains for DF design

Configuration	Channel	p	r	β	φ
	Aileron	–0.1972	0.2692	1.4491	–0.0224
DF	Rudder	–0.0196	0.4394	1.4159	–0.0316
	g_x		–0.90		

with the F-8C aircraft due to reduction in effective dihedral and (iii) LCDP has not been augmented to the level of the SF design. However, c_{n_β}dyn value meets the industry's design aim of c_{n_β}dyn > 0.1.

9.6.4 Single-loop stability margins

From MIL-F-9490D requirements, gain and phase margins have to be established for both aileron and rudder channels using the criteria described in section 8.6.1. Figure 9.8 shows the loop transfer function plots for both the aileron and rudder channels. The plot also shows the exclusion boundary, which the loop transfer function should not cross. All the designs show that adequate gain (GM) and phase (PM) margins exist as shown in Table 9.15.

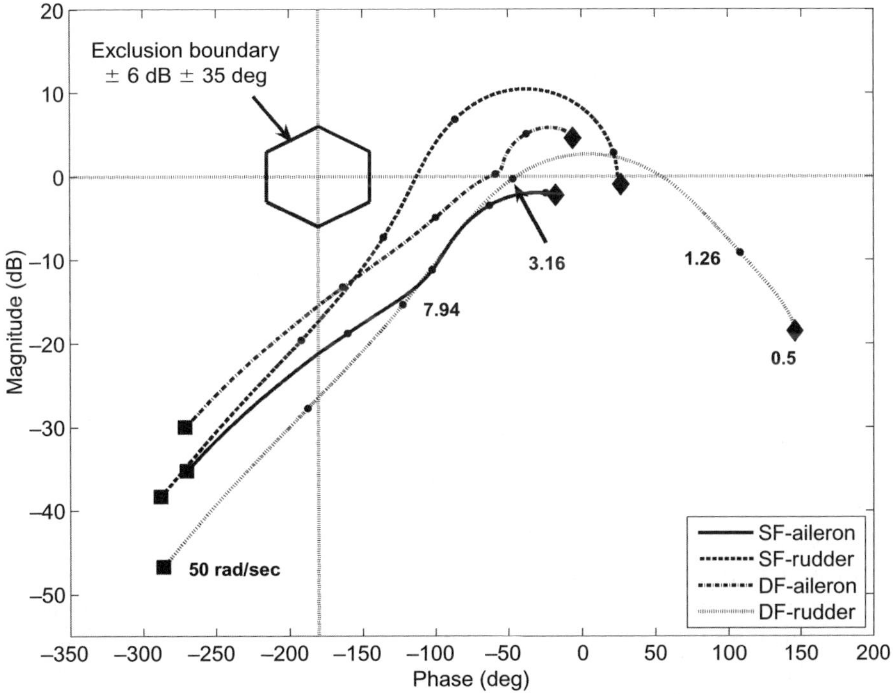

Figure 9.8 Loop transfer functions for stability margin estimates

Table 9.15 Single-loop gain and phase margins

Configuration	Aileron channel				Rudder channel			
	Frequency (rad/sec)	GM (dB)	Frequency (rad/sec)	PM (deg)	Frequency (rad/sec)	GM (dB)	Frequency (rad/sec)	PM (deg)
SF	24.20	21.40	0.05	100.8	17.09	17.25	0.75 4.60	−155 67
DF	23.54	15.38	3.56	117.4	18.19	26.35	2.16 3.12	−126 135

9.6.5 Multivariable stability margins

The multivariable stability margin is defined in Chapter 8 (Table 8.5). The worst-case margin testing procedure detailed in section 8.6.2 will be used to assess the stability of the closed-loop system. Figure 9.9 shows the multivariable root locus plot for the SF design. Figure 9.9a shows the root locus as the scalar gain (ρ) is varied from 0 to 4.5 dB. The hardware mode frequency and damping at the highest gain are $\omega_h = 18.68$ rad/sec and $\zeta_h = 0.63$. If a phase offset of 33 deg is added at the Dutch roll mode frequency ($\omega_c = 8.8$ rad/sec and $\zeta_c = 0.45$), the equivalent time delay is $\tau_d = 0.068$ sec ($\tau_d = 0.6/\omega_c$). Figure 9.9b shows the simultaneous application of both a gain of $\rho = 2.5$ dB and phase delay of $\tau_d = 0.068$ sec perturbation. For this case the hardware mode frequency is $\omega_h = 22.7$ rad/sec with a damping $\zeta_h = 0.8$. The Dutch roll mode frequency is $\omega_c = 7.2$ rad/sec and damping $\zeta_c = 0.8$. Figure 9.10 shows the corresponding results for the DF design. Figure 9.10a shows the root locus as the scalar gain (ρ) is varied from 0 to 4.5 dB. The hardware mode frequency and damping at the highest gain are $\omega_h = 17.96$ rad/sec and $\zeta_h = 0.45$. If a phase offset of 33 deg is added at the Dutch roll mode frequency ($\omega_c = 4.33$ rad/sec and $\zeta_c = 0.9$), the equivalent time delay is $\tau_d = 0.139$ sec ($\tau_d = 0.6/\omega_c$). Figure 9.10b shows the simultaneous application of both a gain of $\rho = 2.5$ dB and phase delay of $\tau_d = 0.068$ sec perturbation. For this case the hardware mode frequency is $\omega_h = 22.04$ rad/sec with a damping $\zeta_h = 0.67$. The Dutch roll mode frequency is $\omega_c = 2.59$ rad/sec and damping $\zeta_c = 0.79$. The aircraft industry practice is to ensure that the damping of the hardware mode $\zeta_h > 0.5$ for the nominal design. Except for the DF design case with a gain increase of 4.5 dB, the hardware mode damping is satisfactory even with additional gain and phase perturbations. The design of the phase advance filter plays a key role in controlling the hardware mode migration towards the imaginary axis. The multivariable root locus is a good graphical tool to optimally design these phase advance filters. In the present study, a phase advance filter $T_p(s) = \frac{2.0(s+9.5)}{s+19}$ is used in the aileron channel for both SF and DF designs. The phase advance filters increase the high frequency ($\omega > 4$ Hz) gain of the closed-loop system. Aircraft industry practice is to impose limits on this gain to avoid the rigid body dynamics coupling with structural modes. In this study the phase advance filter adds a high frequency gain of +6 dB.

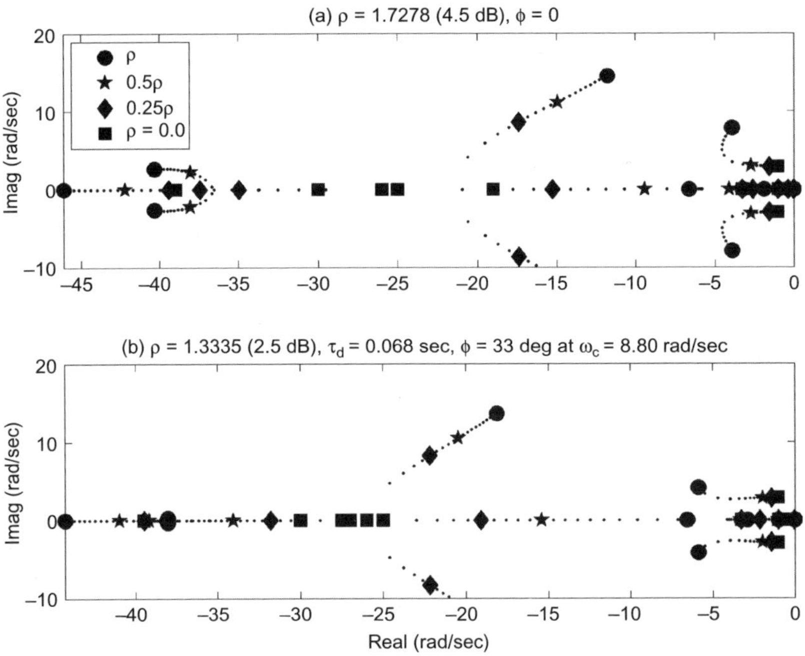

Figure 9.9 Multivariable stability margin–root locus (SF – design)

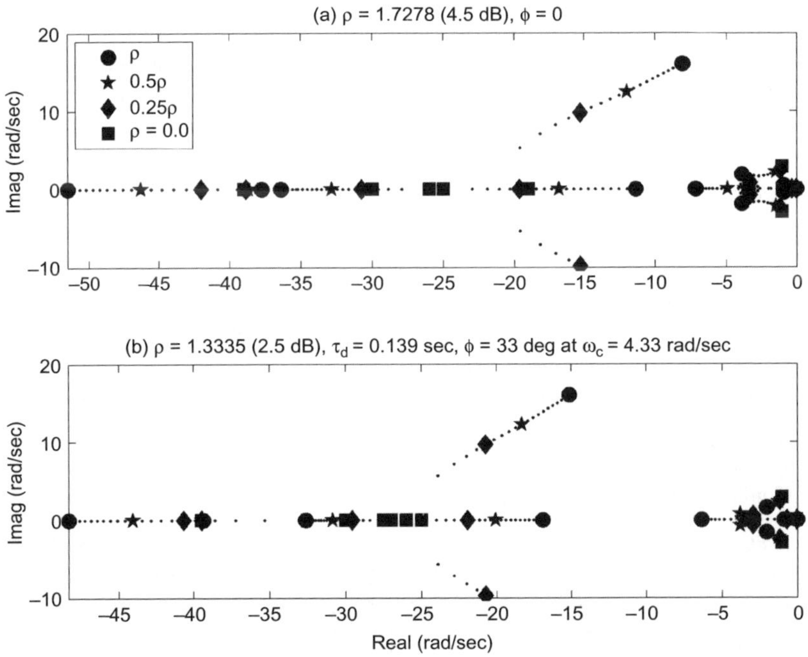

Figure 9.10 Multivariable stability margin–root locus (DF – design)

9.6.6 Gibson's PIO resistance criterion

The primary pilot control in the lateral–directional axes is the roll stick force (δ_s) in pounds, which essentially commands roll rate. The pilot's observed variable is bank angle. In high-gain tasks, the stability of the $\left(\frac{\varphi}{\delta_s}\right)$ loop, closed through the pilot's gain, becomes important in predicting control instability, which is usually referred to as PIO. Figure 9.12 gives the components of the command path (inset). The gain K_c is the command path gain from stick force to aileron displacement (deg/lb). The command path gain is usually reduced for small stick inputs to minimise PIO tendencies. The pilot control input is passed through a command filter to smooth out very abrupt pilot commands. The Gibson criterion requirements are given in section 8.3. A feedforward gain k_L across the command filter provides a lead in the command path to improve the Gibson's phase rate characteristics. The maximum roll rate attainable for full stick deflection is limited by the peak aileron actuator rate capability. For the present study, the F-8C, aileron actuator rate capability is assumed as 100 deg/sec (Appendix B). Budgeting a rate requirement of 10 deg/sec for turbulence (section 9.6.7) and sensor noise disturbance regulation, the net peak actuator rate available for maximum roll rate command is 90 deg/sec. For each design, the command path gain K_c is determined such that the full stick roll rate commanded, factoring the peak actuator rate demand due to feed-forward gain k_L, does not exceed this peak allowable actuator rate. Table 9.16a summarises the command path parameters. The command gain K_c (used in Table 9.16a) is reduced for small stick displacements, typically for ±10% travel (k_D), for precision control. The average phase rate characteristics are computed based on this gain.

Figure 9.11 shows the frequency response of stick force to bank angle. The F-8C aircraft frequency response crosses into the ratcheting boundary, primarily due to the low damping of the Dutch roll mode. The frequency responses of

Table 9.16 Command path parameters and phase rate characteristics

(a) Command path scaling

Full stick force (δ_s)	**Full stick roll rate command (δ_c)**	**Small stick input (±10%) gain (k_D)**
±9 lb	±50 deg/sec	0.5

(b) Average phase rate characteristics*

Configuration	**K_C# (deg/lb)**	**k_L#**	**φ_{avg} (deg/Hz)**	**f_c (Hz)**	**HQ**
F-8C	2.76	0	91.57	0.6442	Level 1
SF	2.32	0.3333	29.71	0.9607	Level 1*
DF	3.12	0.1666	37.46	0.9729	Level 1*

*Chapter 8, section 8, Table 8.1.
Figure 9.12, Command filter, pc = 4.

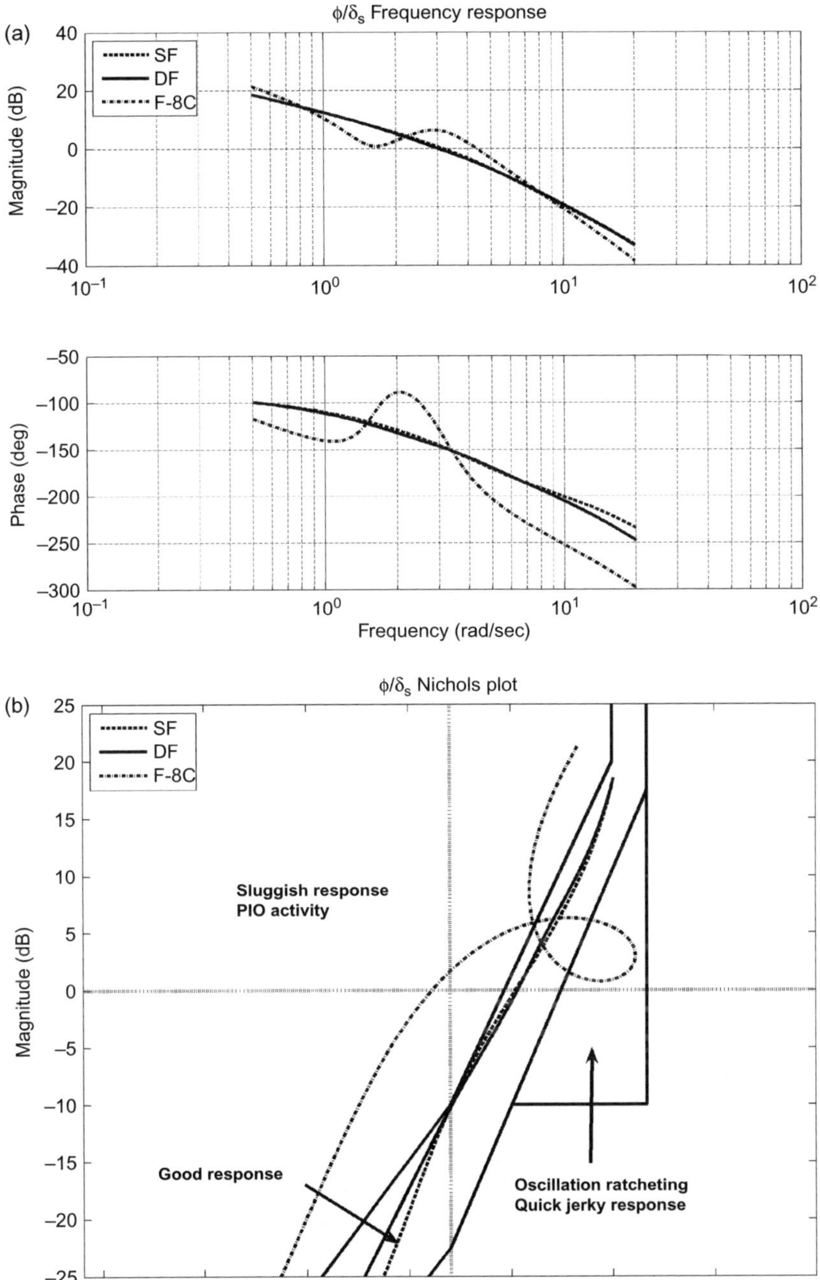

Figure 9.11 Lateral PIO resistance criterion (a) Roll stick force to bank angle frequency response characteristics; (b) Roll stick force to bank angle PIO criterion

both SF and DF designs fall inside the *good response* boundary of the Gibson template. Table 9.16b lists the phase rate characteristics of all the configurations. Both SF and DF designs have level 1* handling qualities. Even though the F-8C aircraft is level 1 in phase rate, the frequency response does not lie inside the *good response* boundary of the Gibson template.

9.6.7 Turbulence response

Appendix A describes a Dryden turbulence model as well as a method to compute the aircraft turbulence response. The Dryden model is appended to the *truth model* (Appendix B) to generate the RMS response of the aircraft. Table 9.17 gives a summary of the responses to a *severe* sideslip gust input for the SF and DF designs. The control margin of the actuator rate response defined in section 8.3 requires that, using DF design, an aileron actuator rate budget of 16 deg/sec ($3\dot{\delta}_a$) must be allocated for disturbance inputs.

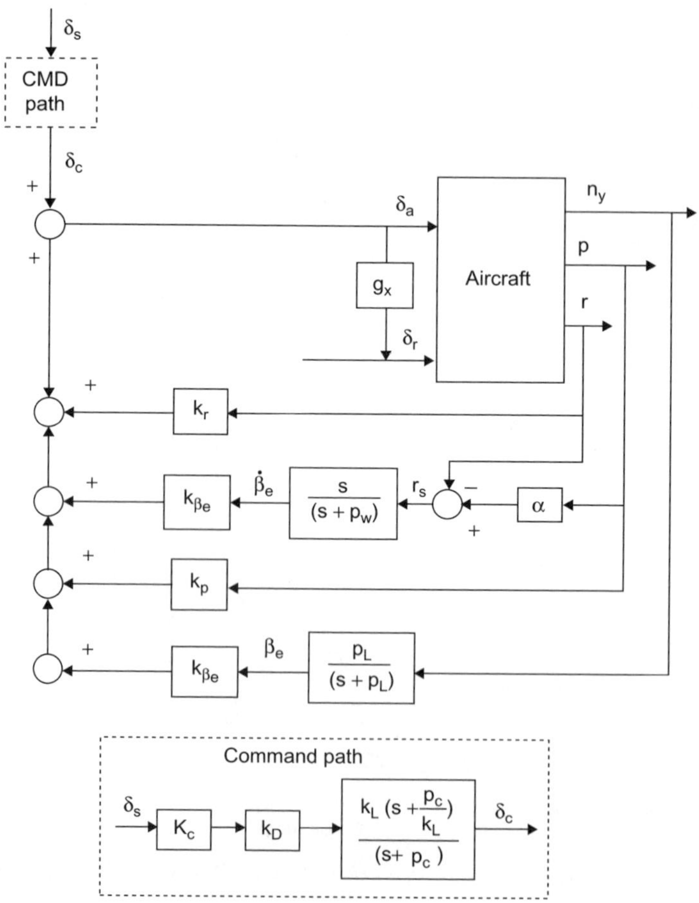

Figure 9.12 Schematic of dynamic output feedback (DF) controller (aileron channel)

Table 9.17 RMS response to severe turbulence (β-gust)

Configuration	p	r	β	φ	δ_a	$\dot{\delta}_a$	δ_r	$\dot{\delta}_r$
F-8C	6.26	0.21	1.27	3.49	–	–	–	–
SF	2.60	0.52	1.09	1.81	0.94	4.21	0.84	2.78
DF	2.51	1.07	1.27	1.06	1.21	5.40	0.55	1.53

RMS β_g intensity = 2.85 deg.
All angular rotation and rate response are in degrees and degree per second, respectively.

In summary, detailed handling qualities and control system performance analysis in this section reveals that both the SF and DF controller structures are able to improve the F-8C aircraft handling qualities (turn coordination, departure resistance) to level 1 while meeting the feedback control characteristics of robustness (stability margins), disturbance rejection (turbulence response) and control activity (actuator rate limits).

9.7 Roll/yaw damper design

In this section a design process, usually adopted in the aircraft industry, using classical control techniques is presented and its relation to the eigenstructure assignment approach used so far will be examined. The roll/yaw damper (RYD) controller structure is depicted in Figure 9.13.

The resulting DF controller has the form

$$\begin{bmatrix} \delta_a \\ \delta_r \end{bmatrix} = \begin{bmatrix} k_p & 0 & 0 \\ 0 & k_y & k_r \end{bmatrix} \begin{bmatrix} p \\ \beta_e \\ \dot{\beta}_e \end{bmatrix} \tag{9.20}$$

The feedback sensors are body axis roll rate, body axis yaw rate and lateral acceleration. The angle of attack signal (α) is needed to generate the stability axis yaw rate (r_s). The roll rate feedback gain to the aileron channel is used to augment the roll mode damping (L_p). The β_e and $\dot{\beta}_e$ feedback gains to the rudder channel augment the Dutch roll mode natural frequency and damping. In a roll command manoeuvre (p_c, Figure 9.13), the reference command to rudder channel (r_c, Figure 9.13) is zero. Thus, the rudder channel feedback loop has the effect of minimising both the sideslip and sideslip rate excursions for a roll input. This enforces the turn co-ordination. An ARI gain (g_x) is used to minimise the Dutch roll mode contamination in the roll rate response and also reduce sideslip excursions (Appendix A). The washout filter in (9.4) ensures that, in steady-state turns, the yaw rate feedback does not oppose the achievement of the desired turn rate. The design process, based on classical control techniques, is iterative in nature, and graphical design aid such as root locus is used. The controller has four gain parameters (k_p, k_r, k_y and g_x) for performance optimisation.

The following loop-at-a-time RYD design is usually followed:

Step 1. Close the β_e to δ_r loop, and select k_y gain to augment the Dutch roll mode natural frequency. This gain augments the effective $c_{n_{\beta\,dynamic}}$ of the aircraft.

Step 2. Close the $\dot{\beta}_e$ to δ_r loop, and select k_r gain to improve the Dutch roll damping.

Step 3. Choose the ARI gain (g_x) to locate the complex conjugate zero of the p to δ_a transfer function (ω_φ, ζ_φ – Appendix A) as close to the Dutch roll mode as possible. It should be noted that the interconnect gain (g_x) does not alter the Dutch roll mode. Optimum selection of ω_φ and ζ_φ reduces the contamination of Dutch roll mode in the roll rate response and also sideslip excursions for a roll stick input.

Step 4. Close the p/δ_a loop and select k_p gain to improve the roll rate response. As k_p is increased, the Dutch roll mode approaches the zero (ω_φ, ζ_φ). The relative location of the zero with respect to the Dutch roll mode in the complex plane influences the adverse or proverse sideslip characteristics. Higher gain (k_p) implies good disturbance rejection characteristics. However, it also results in (i) roll ratcheting due to low roll mode time constant (roll ratcheting is avoided by having a command path filter to give the pilot a desirable crisp roll rate response as discussed in section 9.6.6), (ii) reduction of the stability margin, (iii) reduction of the damping of the hardware modes that is usually compensated by having phase advance filters and (iv) increase of aileron actuator rate demands.

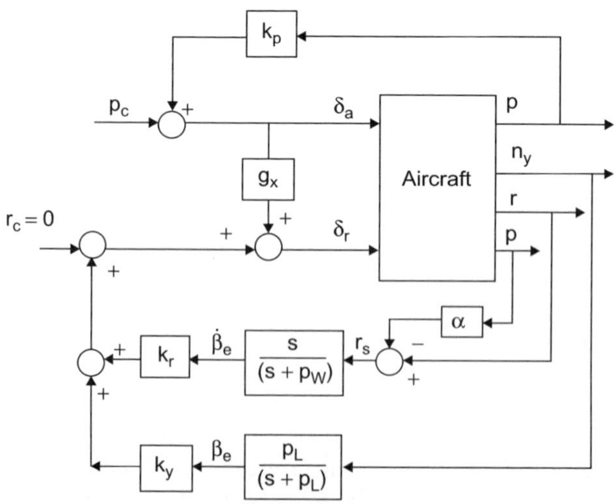

Figure 9.13 Schematic of roll/yaw damper

The RYD design is illustrated using the high angle of attack flight condition (FC2) of the F-8C aircraft as detailed earlier. In order to get a good starting solution to the above iterative design process, eigenstructure assignment using output feedback with p, β_e and $\dot{\beta}_e$ as feedback variables is employed. Using Theorem 4.1, roll mode and Dutch roll modes can be assigned with three eigenvector elements arbitrarily chosen. The resulting feedback controller has the form

$$\begin{bmatrix} \delta_a \\ \delta_r \end{bmatrix} = \begin{bmatrix} k_p & k_a & k_b \\ k_c & k_y & k_r \end{bmatrix} \begin{bmatrix} p \\ \beta_e \\ \dot{\beta}_e \end{bmatrix} \tag{9.21}$$

The feedback matrix has six gain elements including the entire cross feed gains. In order to get a solution with only the three non-zero gains (k_p, k_r and k_y) of (9.20), the following objective function, including the turn co-ordination criterion (Table 9.5), is minimised using the constrained optimisation approach of section 9.5:

$$J = \min \left\{ \left(\frac{\beta_{rms}}{p_{rms}} \right)_{\delta_a \text{ doublet}} + w_a k_a + w_b k_b + w_c k_c \right\} \tag{9.22}$$

The gain weights in (9.22) are used to force the respective cross feed gains to near zero. Table 9.18 lists the resulting starting solution.

Table 9.18 Starting solution for roll/yaw damper design

(a) Feedback gains

	p	β_e	$\dot{\beta}_e$
Aileron	–0.1193	–0.0018	0
Rudder	0	–0.1367	–0.2438

(b) Design parameters (based on *design model*)

λ_L	λ_W	g_x	ω_d	ζ_d	ω_φ	ζ_φ
–3.33	–0.5649	–0.5013	3.4	0.69	2.15	0.604

(c) Eigenvalues (based on *design model*)

Mode	**Eigenvalue**
Lag filter	–4.0
Dutch roll	–2.36 ± j2.45
Roll/washout filter	–0.56 ± j0.49
Spiral	–0.109

Since the Dutch roll natural frequency and damping of the starting values are adequate, the values of k_r and k_y are retained for the RYD design. The first two steps of the RYD design consist of closing the (β_e to δ_r) and ($\dot{\beta}_e$ to δ_r) loops with these gains. The resulting root loci are shown in Figures 9.14 and 9.15. The closed-loop eigenvalues for these loop closures are given in Table 9.19. It is seen that there is a large separation between the Dutch roll mode and the p/δ_a zero for ARI gain $g_x = -0.5013$ (Table 9.18). In order to reduce this separation, the locus of the zero as a function of g_x is generated as shown in Figure 9.16. This corresponds to step 4 of the RYD design. It is seen that if the damping of the zero (ζ_φ) is to be closer to the Dutch roll mode damping ($\zeta_d = 0.57$), its natural frequency (ω_φ) cannot be close to the Dutch roll mode ($\omega_d = 3.36$). On the other hand, if the natural frequency of the zero is located close to the Dutch roll mode natural frequency ($g_x = -1.4$), simulation studies revealed that the feedback system has large proverse sideslip characteristics. The relative location of the zero with respect to the pole in the complex plane influences the nature of the sideslip response.

Thus, using the RYD controller structure, an ideal zero location in the proximity of the Dutch roll mode cannot be selected. Further the zero location is independent of the gains k_y and k_r as the following approximate relationships of the zero in terms of the aircraft derivatives (A.5) indicate:

$$k_\delta = \left(\frac{N_\delta}{L_\delta}\right)$$

$$2\zeta_\varphi\omega_\varphi = 2\zeta_d\omega_d + k_\delta L_r, \qquad \omega_\varphi^2 = \omega_d^2 - k_\delta(L_r Y_\beta + L_\beta) \tag{9.23}$$

From (9.23) it is seen that the RYD design does not augment the roll derivatives L_r and L_β. Thus, the only parameter that is available for modifying the zero is k_δ. The augmented adverse ratio k_δ is a function of only the interconnect gain (g_x – (A.7)). In the SF and DF controller designs, both L_r and L_β are augmented, giving additional flexibility in optimally locating the zero as shown in Figure 9.4. Thus, in the present design, the value of g_x obtained from the starting solution is the optimal value for the RYD controller.

Table 9.19 Eigenvalues after step 2 of roll/yaw damper design

Mode	**Eigenvalue**
Lag filter	–3.55
Dutch roll	$-1.91 \pm j2.76$ ($\omega_d = 3.36$, $\zeta_d = 0.57$)
Roll/washout filter	$-0.24 \pm j0.49$
Spiral	–0.25
p/δ_a zero	$\omega_\varphi = 2.15$, $\zeta_\varphi = 0.604$

$$K = \begin{bmatrix} 0 & 0 & 0 \\ 0 & -0.1367 & -0.2438 \end{bmatrix}$$

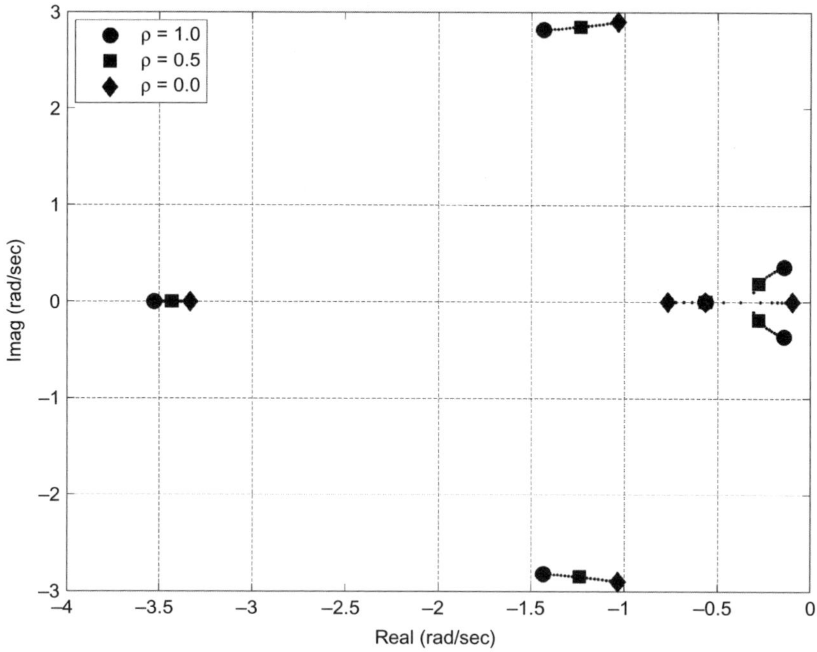

Figure 9.14 Gain root locus of β_e to δ_r loop closure (gain = $\rho * k_y$)

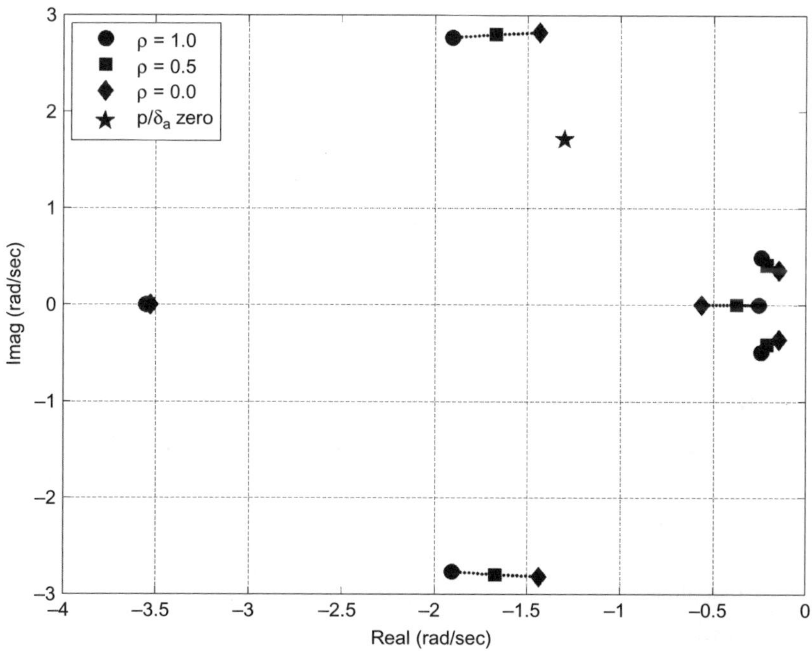

Figure 9.15 Gain root locus of $\dot{\beta}_e$ to δ_r loop closure (gain = $\rho * k_r$)

The final step in the RYD design is to close the p/δ_a loop with the gain k_p as high as possible with the limitations alluded to earlier. In this case the roll rate gain is increased to $k_p = -0.6$ (from the starting value of $k_p = -0.1193$ in Table 9.18). Table 9.20 shows the modification in the eigenvalues of the closed-loop system.

Figure 9.17 shows the p/δ_a loop root locus. The roll gain effectively eliminates the coalescing of the roll mode and washout filter roots, and examination of the closed-loop eigenvectors results in the mode identifications shown in Table 9.20.

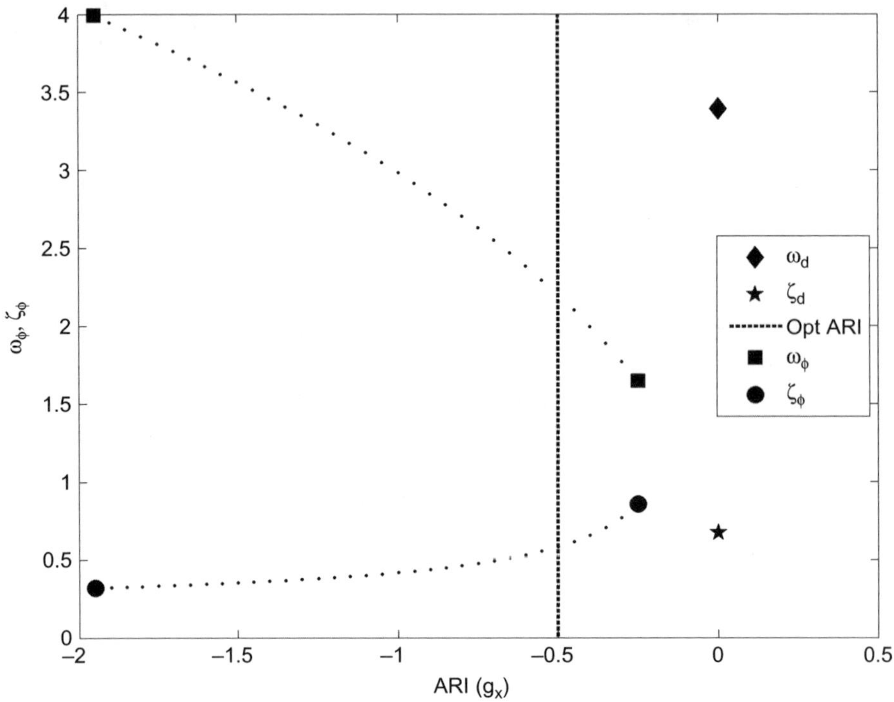

Figure 9.16 Locus of p/δ_a *transfer function zero*

Table 9.20 Closed-loop eigenvalues of roll/yaw damper design

Mode	**Eigenvalue**
Roll mode	–11.10
Dutch roll	$-1.72 \pm j1.39$ ($\omega_d = 2.21$, $\zeta_d = 0.78$)
Lag filter	–2.16
Washout filter	–0.7422
Spiral	–0.0266

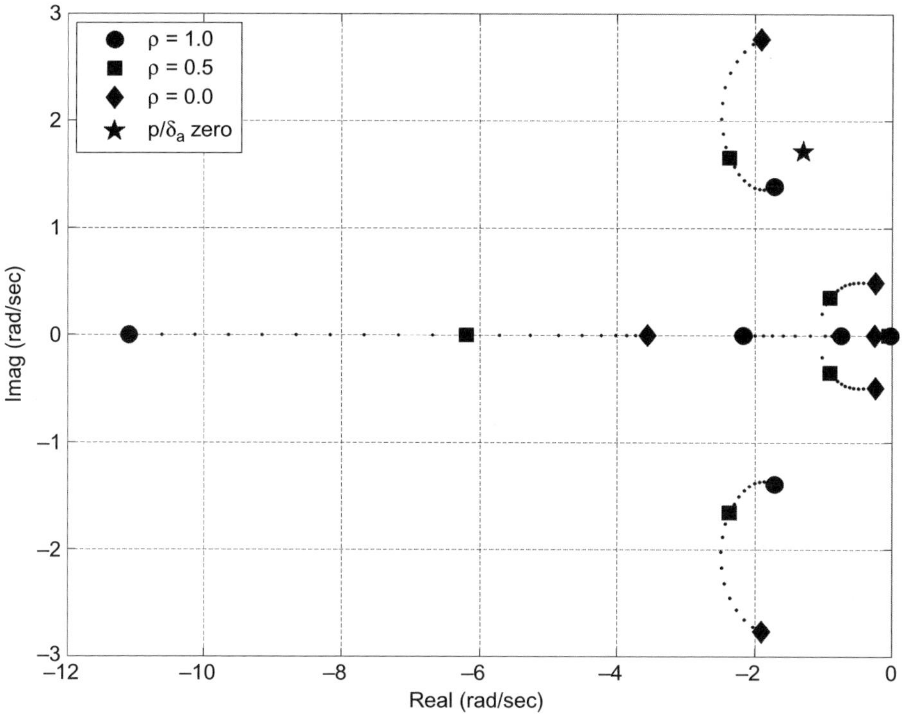

Figure 9.17 Gain root locus of p to δ_a *loop closure (gain =* $\rho * k_p$*)*

Figure 9.18 shows the loop transfer functions of both the aileron and rudder loops. The rudder loop has adequate stability margins. The aileron loop just meets the exclusion boundary requirement. Further the eigenvalues of the *truth model* also show that the dominant hardware mode has a damping $\zeta = 0.34$. Typical design guideline is $\zeta = 0.5$. A slight redesign of the phase advance filter will be needed to correct this deficiency or the gain k_p needs to be reduced.

Figures 9.19 and 9.20 compare the sideslip response of SF, DF and RYD designs for a pulse and doublet roll stick command input. For a pulse input (Figure 9.19), the sideslip response for the RYD design is proverse with an initial non-minimum phase response. More importantly, the sideslip response builds up after the pulse input is terminated. This is due to the sideslip response being contaminated with the spiral mode. However, the RYD control structure does not have explicit freedom to control the spiral mode eigenvector shape. The SF design has the ideal sideslip response characteristics, while the DF design response exhibits slight spiral mode contamination.

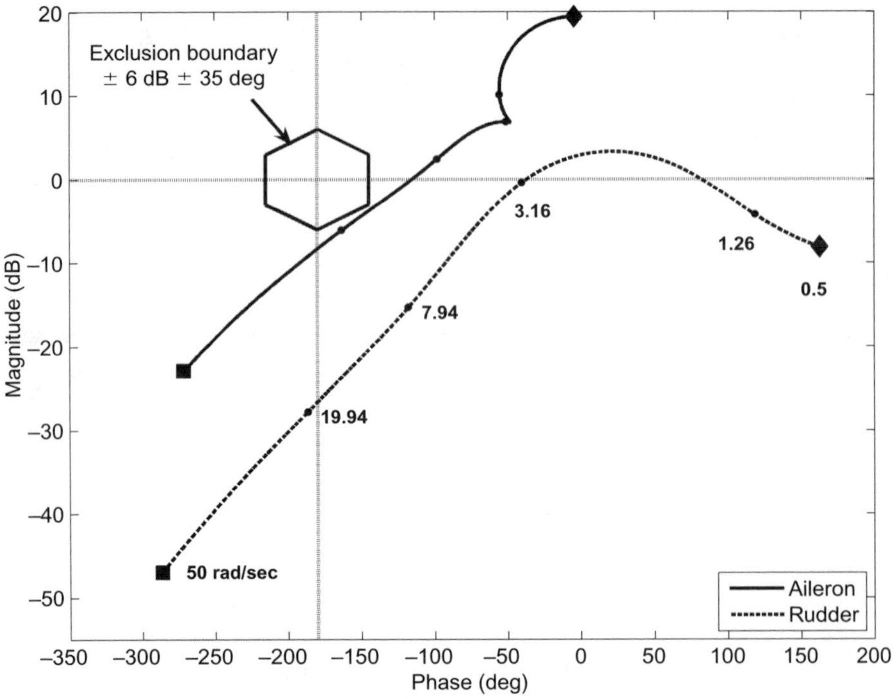

Figure 9.18 Loop transfer functions of RYD design

The discussions in this section reveal that the RYD controller structure is derived based on a sound flight mechanics analysis of the feedback control requirements. The sequence of closing the control loops (steps 1–4) results in minimal change in the desired characteristics achieved in the previous loop closures. Each loop closure is structured to effect enhancement of a selected performance requirement. The F-8C design study in this section reveals two major shortcomings of the RYD design: (i) inadequate flexibility in locating the optimum p/δ_a zero using ARI gain and (ii) lack of freedom to control the contamination of the spiral mode in the sideslip response for roll command inputs. This results in degraded turn co-ordination as compared with the SF and DF designs. Eigenstructure optimisation, with a reduced feedback variable set, provides a good starting point to the RYD design iterations ((9.21) and (9.22)). This shows how close the eigenstructure assignment methodology, as postulated in this chapter, is to the aircraft industry's design philosophy. The feedback control law in (9.21) brings additional flexibility in the selection of p/δ_a zero by allowing augmentation of the roll derivatives in (9.23). This eliminates one of the RYD design limitations referred to earlier. However, independent control of the spiral mode eigenvector shape would still require the DF feedback structure.

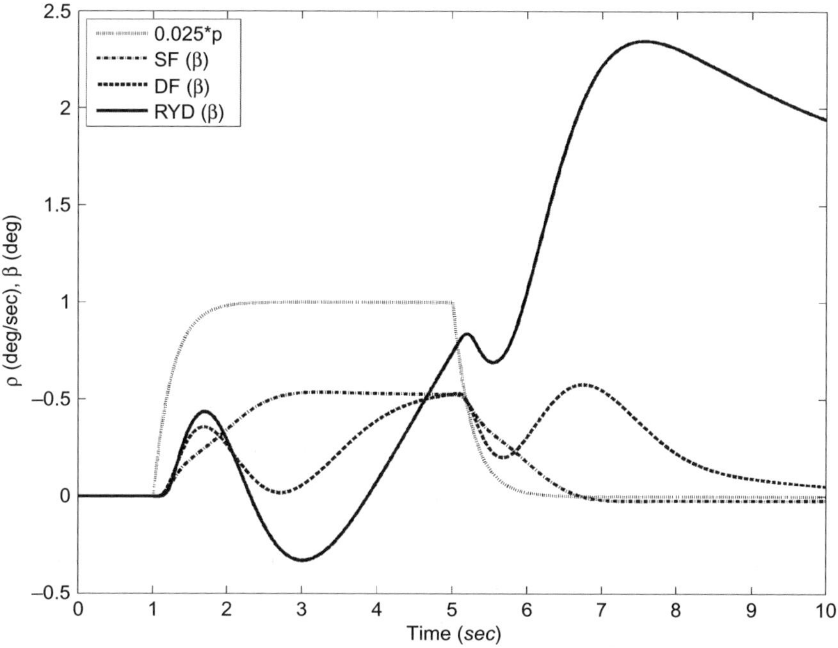

Figure 9.19 Sideslip response to a roll pulse command (p_{cmd} = 40 deg/sec)

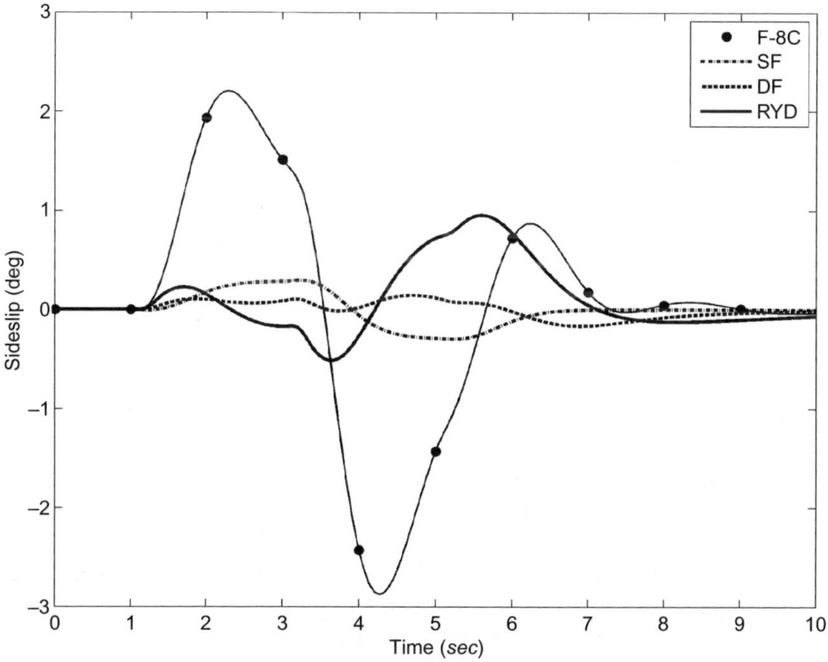

Figure 9.20 Sideslip response to a roll doublet command (p_{cmd} = 40 deg/sec)

9.8 Summary

In this chapter the application of eigenstructure assignment technique to improve the lateral–directional dynamics of an aircraft has been illustrated. It is shown that using the aircraft industry's preferred feedback sensor set of roll rate, yaw rate and lateral acceleration, all the performance goals detailed in Chapter 8 can be realised by employing a DF control structure. The SF design using roll rate, yaw rate, sideslip and bank angle sensors is used as a benchmark reference of achievable performance. The simplicity of selection of eigenstructure for decoupling of roll and spiral modes from sideslip response is highlighted. The eigenstructure modification process proposed has the following properties:

(1) Eigenstructure assignment is invariant with respect to ARI gain.
(2) The system response characteristics to aileron input remain invariant with respect to scaling of the special Dutch roll eigenvector structure specified.

These properties allow the designer to independently (i) fine-tune Dutch roll mode contamination in roll rate response and minimise sideslip excursions, to aileron input, using the ARI, without affecting the roll and spiral mode eigenstructures and (ii) fine-tune the bank angle response to rudder input by scaling the Dutch roll eigenvector without altering the system response to aileron input.

The study of the RYD control structure, usually preferred in the industry, reveals some limitations on design flexibility. It is shown that these can be overcome using eigenstructure synthesis approach. This study also reveals the close relationship between classical control approach and eigenstructure modification process. This makes it easier for the designer to migrate into the multivariable control design process for aircraft control law synthesis.

The detailed design study in this chapter covers practically all the performance goals usually addressed in the aircraft industry during an aircraft design project. Thus, the study has established a starting *design process* methodology for adapting eigenstructure synthesis to derive practical control laws. The good correlation between eigenstructure modification and handling qualities metrics makes it a viable design tool for flight vehicle systems having multiple control effectors.

References

1. SRINATHKUMAR, S.: 'Modal Control Theory and Application to Aircraft Lateral Handling Qualities Design', 1978, NASA TP 1234
2. ANDRY, A.N., SHAPIRO, E.Y., and CHUNG, J.C.: 'Eigenstructure assignment for linear systems', *IEEE Trans. Aerospace Electron. Syst.*, 1983, **AES-19**, (5), pp. 711–729
3. SOBEL, K.M., SHAPIRO, E.Y., and ANDRY, A.N.: 'Eigenstructure assignment', *Int. J. Control*, 1994, **59**, (1), pp. 13–37
4. MAGNI, J.-F., BENNANI, S., and TERLOUW, J. (eds.).: *Robust Flight Control – A Design Challenge*. Lecture Notes in Control and Information Sciences, vol. 224 (Springer-Verlag, Berlin, 1997)

5. LITTLEBOY, D.M.: 'Numerical techniques for eigenstructure assignment by output feedback in aircraft applications', Ph.D. thesis, University of Reading, UK, December 1994
6. FALEIRO, L.F.: 'The application of eigenstructure assignment to the design of flight control systems', Ph.D. thesis, Loughborough University of Technology, UK, 1998
7. DAVIDSON, J.B., and ANDRISANI, II, D.: 'Lateral–Directional Eigenvector Flying Qualities Guidelines for High Performance Aircraft', 1996, NASA TM 110306
8. MURPHY, P.C., and DAVIDSON, J.B.: 'A Control Law Design Method Facilitating Control Power, Robustness, Agility and Flying Qualities Tradeoffs: CRAFT', 1998, NASA TP-1998-208463
9. GARRARD, W.L., LOW, E., and PROUTY, S.: 'Design of attitude and rate command systems for helicopters using eigenstructure assignment', *J. Guidance Control Dyn.*, 1989, **12**, (6), pp. 783–791
10. CLARK, T., ENSOR J., and GRIFFIN S.J.: 'Desirable eigenstructure for good short-term helicopter handling qualities: the attitude command response case', *Proc. Inst. Mech. Eng. Part G. J. Aerospace Eng.*, 2003, **217**, pp. 43–55
11. FARINEAU, J.: 'Lateral electric flight control laws of a civil aircraft based upon eigenstructure assignment technique', *Proceedings of the AIAA Guidance, Navigation and Control Conference*, Boston, MA, 1989, AIAA-89-3594
12. WICHMAN, K.D., PAHLE, J.W., BAHM, C., DAVIDSON, J.D., BACON, B.J., MURPHY, P.C., *et al.*: 'High-Alpha Handling Qualities Research on the NASA F/A-18 High Alpha Research Vehicle', 1996, NASA TM 4773
13. DAVIDSON, J.B., MURPHY, P.C., LALLMAN, F.J., HOFFLER, K.D., and BACON, B.J.: 'High-Alpha Research Vehicle Lateral–Directional Control Law: Description, Analyses and Simulation Results', 1998, NASA-TP-1998-208465
14. WHITMORE, S.A., DAVIS, R.J., and FIFE, J.M.: 'In-Flight Demonstration of a Real-Time Flush Airdata Sensing (RT-FADS) System', 1995, NASA TM 104314
15. PAHLE, J.W., POWERS, B., REGENIE, V., CHACON, V., DEGROOTE, S., and MURNYAK, S.: 'Research Flight-Control System Development for the F-18 High Alpha Research Vehicle', 1991, NASA TM 104232

Chapter 10
Aircraft longitudinal handling qualities design

10.1 Introduction

The design of stability augmentation systems for the pitch axis of an aircraft has traditionally been accomplished as a single-input design, since the coupling of the longitudinal axis dynamics with the lateral–directional axes is assumed, at least at low angle of attack (AoA), to be negligible. With the advent of high-performance fly-by-wire aircraft, multiple control surfaces are being used to incorporate advanced pitch control modes such as decoupled direct lift, pitch pointing or vertical translation [1] and also improve handling qualities at high AoA flight conditions using a combination of aerodynamic and thrust vectoring controls [2]. This results in a multi-input design, and eigenstructure control concepts also become relevant in the design process. For a single-input controller, however, the design reduces to an eigenvalue assignment problem with no freedom in selection of eigenvector shapes.

Analytical study, based on eigenstructure assignment, for the design of pitch pointing control using both a moment and direct lift control surface has been reported [3]. Implicit model following (IMF) controller design, using eigenstructure assignment, has been demonstrated in an in-flight simulation experiment using the VISTA (F-16) aircraft [4]. Control law design based on eigenstructure assignment has also been attempted in the YF-22 aircraft program [5]. However, the final controller structure, which geared the pitching moment control and thrust vectoring nozzles, resulted in a single-input pole placement design.

In this chapter, the applicability of eigenvalue and eigenstructure assignment procedures to design the pitch axis controller of an aircraft will be discussed in detail. As in the case of the lateral–directional controller design discussed in Chapter 9, the emphasis is to evolve a *design process* that can be adapted to meet aircraft industry's design practices.

10.2 Flight mechanics analyses of control problem

10.2.1 Short-period model and time response

The longitudinal motion of a stable aircraft is composed of two distinct oscillatory modes characterising the short-term (short-period mode) and long-term (phugoid mode) motions of the aircraft. The aircraft handling qualities design issues revolve around the dominant short-period dynamics. Assuming phugoid and short-period modes are decoupled, study of the aircraft dynamics using the short-period mode approximation is found to be adequate. Thus, the state variable model for the short-period dynamics can be approximated as

$$\begin{bmatrix} \dot{\alpha} \\ \dot{q} \end{bmatrix} = \begin{bmatrix} Z_\alpha & 1 \\ M_\alpha & M_q \end{bmatrix} \begin{bmatrix} \alpha \\ q \end{bmatrix} + \begin{bmatrix} Z_{\delta_e} \\ M_{\delta_e} \end{bmatrix} \delta_e \tag{10.1}$$

where AoA (α) and pitch rate (q) are the state variables and elevator (δ_e) is the control variable. The matrix elements in (10.1) are the respective stability and control derivatives. The pitch rate transfer function is given by

$$\left[\frac{q(s)}{\delta_e(s)}\right] = \frac{K_\theta(s+z_\theta)}{s^2 + 2\zeta_{sp}\omega_{sp}s + \omega_{sp}^2}$$

$$\omega_{sp}^2 = (Z_\alpha M_q - M_\alpha), \qquad 2\zeta_{sp} = -(M_q + Z_\alpha), \qquad z_\theta = \left(\frac{1}{T_{\theta_2}}\right) \tag{10.2}$$

$$\text{Let } k = \left[\frac{Z_{\delta_e}}{M_{\delta_e}}\right]; \quad \text{then } z_\theta = (kM_\alpha - Z_\alpha); \quad K_\theta = M_{\delta_e}$$

The short-period natural frequency (ω_{sp}), damping (ζ_{sp}) and the numerator zero (z_θ) uniquely characterise the pitch attitude (θ) response of the aircraft. One of the design objectives is to optimise the time response of AoA, pitch attitude and flight path angle ($\gamma = \theta - \alpha$).

Gibson [6] elegantly identifies critical flight mechanics parameters that influence the pitch axis time response of the aircraft. These parameters facilitate in understanding the control law design constraints. Figures 10.1 and 10.2 graphically illustrate the interrelationship between these key parameters. Figure 10.1 shows the pitch attitude response to a pulse elevator input. The pitch attitude dropback (db) is defined as the incremental change in attitude response after the input is removed. From Figure 10.1, for an aircraft with no high-order stability augmentation, the following analytical relationships can be established:

$$t_\gamma = \left(\frac{2\zeta_{sp}}{\omega_{sp}}\right), \qquad dbq = \left(\frac{db}{q_{ss}}\right), \qquad dbq = (T_{\theta_2} - t_\gamma) \tag{10.3}$$

where q_{ss} is the steady-state pitch rate. The sign of dropback/overshoot ratio (dbq) is determined by the value of T_{θ_2}. For the case where pitch axis response shaping is optimised using only elevator control (single input), T_{θ_2} cannot be modified using feedback control. Thus, independent control of dbq and t_γ is not possible. For example, a faster flight path response by augmenting ω_{sp} results in higher pitch rate overshoot and higher attitude dropback. A positive value of dbq is desirable (as shown in Figure 10.1), since negative value indicates sluggish attitude response ($t_\gamma > T_{\theta_2}$). Gibson has also established the relationship between the pitch rate overshoot (ratio of peak to steady-state pitch rate – q_m/q_{ss})

and dbq as shown in Figure 10.2. For precision tracking the 'satisfactory' boundary (Figure 10.2) limits the dbq values in the range of 0–0.25. This consequently limits the improvement in flight path delay (t_γ).

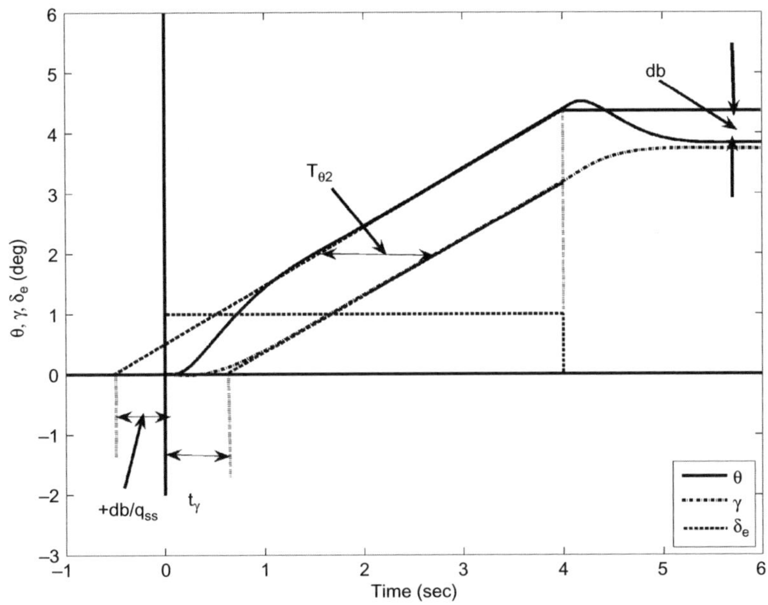

Figure 10.1 Pitch attitude and flight path response to elevator pulse input

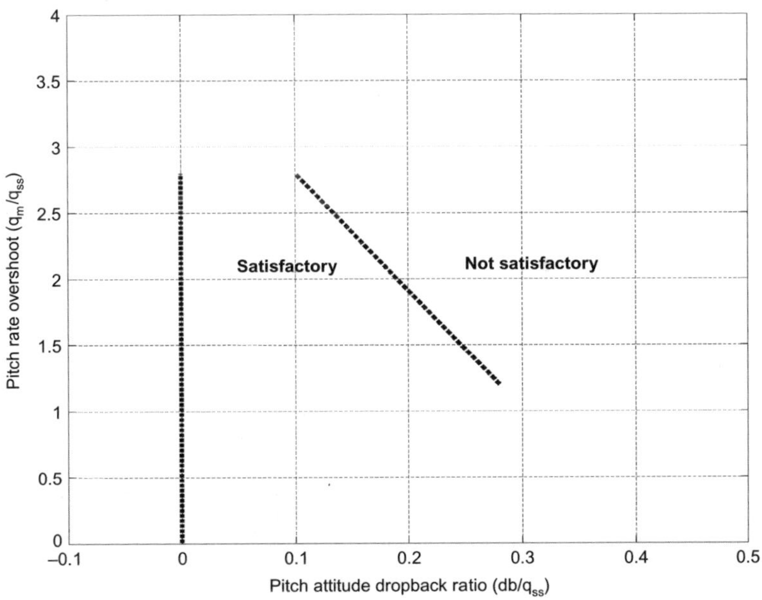

Figure 10.2 Boundaries of dropback ratio for precision tracking

10.2.2 Control interconnect to augment pitch rate zero

The ability to modify the pitch rate numerator zero (z_θ) yields some flexibility to independently control the pith attitude and flight path responses. If the aircraft is equipped with at least two control inputs in the pitch axis, such as (i) elevator and direct lift flap, (ii) elevator and canard or (iii) elevator and thrust vector nozzle, etc., it is possible to modify T_{θ_2} by a suitable control interconnect between the two controls. This is similar to the aileron/rudder interconnect concept discussed in Chapter 9.

Consider a two-input short-period model of the form (10.1) with control distribution matrix given by

$$\hat{B} = BG$$

where

$$B = \begin{bmatrix} Z_{\delta_e} & Z_{\delta_f} \\ M_{\delta_e} & M_{\delta_f} \end{bmatrix}, \quad G = \begin{bmatrix} 1 & 0 \\ g_x & 1 \end{bmatrix} \tag{10.4}$$

Then the augmented elevator control derivative ratio k of (10.2) takes the form

$$\hat{k} = \left(\frac{Z_{\delta_e} + g_x Z_{\delta_f}}{M_{\delta_e} + g_x M_{\delta_f}} \right) \tag{10.5}$$

From (10.2) it follows that z_θ gets modified by the factor $\hat{k}$ in (10.5).

10.2.3 Estimation of angle of attack and angle of attack rate signals

The primary feedback sensors in the pitch axis are AoA, pitch rate (q) and normal acceleration (n_z). In the case of lateral–directional control, the advantage of using the sideslip rate estimate ($\dot{\beta}_e$) as a feedback variable is demonstrated in Chapter 9. Similarly, in the pitch axis control, use of AoA rate ($\dot{\alpha}$) signal as a feedback variable is found to be useful in speeding up the AoA response. An estimate of $\dot{\alpha}$ signal can be derived using the force equation (10.1) as

$$\dot{\alpha} = Z_\alpha \alpha + q \tag{10.6}$$

resulting in an estimate for $\dot{\alpha}$ as

$$\dot{\alpha}_e = \left(\frac{s}{s + p_w} \right) q, \qquad p_w = -Z_\alpha \tag{10.7}$$

In the event of a failure in the AoA sensor signal, an estimate of α can be derived using the lagged normal acceleration sensor signal as

$$\alpha_e = \left(\frac{p_L}{s + p_L}\right)\alpha_{n_Z}, \qquad \alpha_{n_Z} = \frac{n_Z}{Z_\alpha} \tag{10.8}$$

where p_L is the lag filter root.

10.3 Aircraft model for design studies

For illustrating the design process, an aircraft longitudinal rigid body state variable model [4] will be used.

The state variable model is given by

$$\begin{aligned}
\dot{\mathbf{x}} &= \mathbf{A}\mathbf{x} + \mathbf{B}\mathbf{u} \\
\mathbf{y} &= \mathbf{C}\mathbf{x} + \mathbf{D}\mathbf{u} \\
\mathbf{x} &= [\alpha \quad q \quad \theta \quad V]^T, \qquad \mathbf{u} = [\delta_e \quad \delta_f]^T, \qquad \mathbf{y} = n_z
\end{aligned} \tag{10.9}$$

The respective matrices are given by

$$\begin{aligned}
A &= \begin{bmatrix} -0.8461 & 0.9951 & 0 & -0.0002 \\ M_\alpha & -0.7940 & 0 & 0.0001 \\ 0 & 1 & 0 & 0 \\ -4.5149 & -0.1891 & -32.1716 & -0.0111 \end{bmatrix}, \quad B = \begin{bmatrix} -0.1289 & -0.1526 \\ -9.1994 & 0.4611 \\ 0 & 0 \\ 3.8904 & -9.0298 \end{bmatrix} \\
C &= [15.3135 \quad 0.0887 \quad 0 \quad 0.0029], \qquad D = [2.3330 \quad 2.7619] \\
M_\alpha &= \ 1.0252, \quad \text{Configuration BSS: stable static stability} \\
M_\alpha &= +2.0504, \quad \text{Configuration RSS: relaxed static stablity}
\end{aligned} \tag{10.10}$$

The basic aircraft configuration (BSS) in (10.10) corresponds to the stable static stability condition. In the present study, the aircraft is *intentionally* made to be statically unstable (shown as RSS configuration in 10.10). This typically represents an aft C.G. loading of the aircraft. This modification is done to study the control design issues that arise with an RSS aircraft configuration. In this case, the short-period mode appears as two aperiodic modes, with one mode being unstable. The two control surfaces are horizontal tail (δ_e) and flaperon (δ_f), respectively. The changes in control derivatives, especially M_{δ_e} and M_{δ_f}, due to the aft shift in the centre of gravity are ignored. All angular deflections and rates are in radian units; acceleration is in g units and velocity is in ft/sec.

The open-loop RSS aircraft model eigenvalues and eigenvectors are given in Table 10.1. The control system design is carried out using the full rigid body model (10.10).

Table 10.1 RSS configuration eigensystem

State variable component	**Eigenvalues**			
	Short period		**Phugoid**	
	$\lambda_1 = -2.25$	$\lambda_2 = +0.6139$	$\lambda_{ph} = -0.0076 \pm j0.0740$, $\omega_{ph} = 0.0744$ rad/sec, $\varsigma_{ph} = 0.0102$	
α	0.0905	0.0078	0	0
q	−0.1275	0.0112	0.0002	0
θ	0.0567	0.0183	−0.0001	−0.0023
V	0.9861	−0.9997	1.0	0

Numerator zero $(1/T_{\theta_2}) = 0.8755$.

The assumed actuator and flight control system (FCS) hardware characteristics are given in Table 10.2. The hardware dynamics are added to the aircraft rigid body dynamics given in (10.10) to construct a *truth model* to assess the performance of the design in the presence of these additional lags.

10.4 Control of relaxed static stability aircraft

Relaxing the static stability of an aircraft results in enhanced performance such as (i) reduced trim drag, (ii) increased lift/drag ratio, (iii) reduced tail plane size, etc. Attaining these performance benefits tacitly assumes that artificial aircraft stability is provided using multiple redundant fly-by-wire FCS. Thus, from aircraft safety point of view, the reliability of the FCS hardware becomes critical. This consideration also dictates the feedback controller structure. McRuer *et al.* [7] detail different feedback control structures, using different sensor complement, for stabilising an RSS aircraft and highlight the merits of the choices. In particular, the *superaugmented* flight controller, using pitch rate command/attitude hold control structure, has been analysed in detail. The merits of this controller using in-flight simulation studies have also been reported [8]. Many production aircrafts have been successfully flying using the *superaugmented* control concept. The other alternative feedback variable is AoA. The ability to directly augment the static stability derivative (M_α) using AoA makes this an attractive option. However, the reliability of the AoA sensor system does not compare well with the highly reliable pitch rate sensor. Use of normal acceleration sensor (n_z) as a surrogate signal for AoA has also been considered in the event of AoA sensor system failure. The AoA/n_z command system along with a pitch damper using pitch rate sensor is usually referred to as a *conventional* controller.

In sections 10.5 and 10.6, the design of flight control laws using both *conventional* and *superaugmented* controller structures will be addressed. The control problem will be formulated as an eigenvalue optimisation problem using only the horizontal tail as the control variable. As indicated earlier, for a single-input system, the eigenstructure design degenerates to a pole assignment problem. The optimum location of the assignable eigenvalues is the only design freedom available. In section 10.7 the design of multivariable flight control laws using both horizontal tail and flaperon controls will be studied by formulating an eigenstructure optimisation problem.

10.5 Conventional controller design

The AoA/n_z command controller structure is known to give tight flight path control and found to be especially useful during landing approach and final flare. Figure 10.3 gives a schematic of the controller structure. Inclusion of the washout and lag filter results in a sixth-order state variable *design* model. Feedback from pitch rate acts as a pitch damper. The AoA rate feedback facilitates speeding up of the AoA response time. Dominant feedback from AoA or n_z determines the response variable being commanded. The command path filter is used to fine-tune the longitudinal handling qualities properties described in Chapter 8.

10.5.1 Feedback design

The design objectives for feedback design are (i) a crisp AoA response, (ii) quick pitch rate response with overshoot limited by the bounds given in Figure 10.2, (iii) tight feedback loop (high gain) to improve disturbance rejection properties without violating the stability margin boundaries and (iv) optimisation of the flight path delay (t_γ) bounded by the constraint of (10.3). Figure 10.3 has a dynamic feedback structure similar to the lateral–directional controller structure discussed in Chapter 9. There are four gains available to assign the desired eigenvalues. As indicated earlier, only the short-period mode needs to be modified for short-term response shaping.

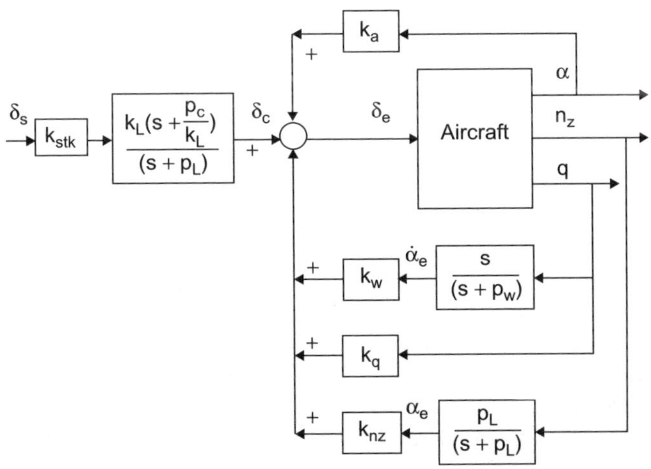

Figure 10.3 Schematic of conventional controller

Output feedback theory of Chapter 4 reveals that two gains k_a and k_q are adequate to assign the selected short-period mode by directly augmenting Z_α, M_α and M_q. In the event of an AoA failure, a reversionary control law using feedback gains k_q, k_w and k_{nz} is designed to assign the short-period mode and one additional mode such as speeding up the normal acceleration lag mode. All the four gains can be used for response optimisation. In such a case, the short-period mode and the filter roots corresponding to the dynamic filter (10.7) and (10.8) can also be assigned. It should be noted that the normal acceleration lag filter root (p_L) and the pitch rate washout filter root (p_w) appear as zeros in the q/δ_e transfer function. If the respective closed-loop roots are assigned as p_w and p_L, pole – zero cancellation occurs and the feedback gains k_w and k_{nz} are identically zero. The farther the closed-loop poles are assigned with respect to the filter roots, more effective are the corresponding feedback loops.

Feedback causes the migration of the unstable short-period root to the left half plane, and in this process the phugoid mode undergoes changes in natural frequency and damping. Depending on the controller structure, the phugoid mode may split into two aperiodic roots with one root becoming unstable. Any unstable low-frequency mode results in the aircraft speed becoming marginally divergent. This aspect will be further discussed as part of the simulation studies.

10.5.2 Command filter design

The criteria used to optimise the feedback design, as discussed in the previous section, generally will not meet some of the handling quality requirements detailed in Chapter 8. The command filter is fine-tuned to meet these requirements. The command filter can be one or more lead/lag filter sections to shape the θ/δ_e frequency response [6]. In the present study only one filter section, as shown in Figure 10.3, will be used for tuning the frequency response. The filter pole p_c and feedforward gain k_L are determined by formulating the following optimisation problem:

$$J = w1 * (t_\gamma) + w2 * \left(\frac{q_m}{q_{ss}} \right) + w3 * (dbq_r - dbq)^2, \quad \text{s.t. } 0 \le dbq_r \le 0.25 \tag{10.11}$$

The time domain parameters in (10.11) are graphically computed using the time response characteristics for an elevator pulse input as shown in Figure 10.1. The weighting factors w1, w2 and w3 are selected to arrive at a solution that satisfies the constraint in Figure 10.2. Selection of the reference parameter dbq_r determines the achievable flight path delay (t_γ) due to the constraint (10.3). In Figures 10.3 and 10.4 the control input to the command filter (δ_s) is the stick force in pounds. The stick path gain k_{stk} is scaled such that for the maximum aft stick deflection (Table 10.2; 5 in, 35 lb), the elevator command input (δ_c in degrees) produces a specified maximum normal acceleration $n_{z_{max}}$ dependent on the flight condition. The stick force gradient k_{stk} (deg/lb) is typically halved for small stick inputs (±10% of travel) to achieve precision control. This lower value is used for assessing Gibson's θ/δ_s frequency response based handling qualities criteria. The stick filter typically modifies the θ/δ_s frequency response characteristics in the 1–3 rad/sec range to improve dropback and phase rate characteristics. If the optimised command filter is a lag filter, it has the adverse effect of slowing the AoA response and increasing the flight path delay for pilot input. However, the disturbance rejection response time is not affected since it is solely dependent on the feedback design. If the command filter is a lead filter, it has the beneficial effect of quickening the AoA response and reducing the flight path delay.

10.6 Superaugmented controller design

The superaugmented controller is derived using only the reliable pitch rate sensor as the feedback variable. Figure 10.4 gives a schematic of the controller. The pilot commands pitch rate ($\delta_c = q_c$). The integrator in the error path (q_e) results in a pitch rate command/attitude hold control structure. The *optional* washed out pitch rate feedback ($\dot{\alpha}_e$) can be used to speed up the AoA response. The primary feedback loop design consists of tuning two gains k_q and k_θ. A commonly used control scheme for analysis of this primary loop is shown in Figure 10.5. The time constant $T_q = (k_q/k_\theta)$ plays a key role in shaping the closed-loop response. The location of T_q in relation to the integrator closed-loop pole, summarised in Reference 9, indicates that even though a conventional θ/q_c response can easily be obtained, the AoA and flight path response may be unsatisfactory. Only by choosing $1/T_q$ to be near to $1/T_{\theta_2}$, the AoA response to a step command can be kept from ramping off and instead reaches a desired steady-state value. This will enable precision control of flight path through pitch attitude cues. Thus, the design procedure consists of selecting k_q and T_q iteratively, using root locus methods to assign the desired short-period mode and optimise the AoA and pitch rate response. Then k_θ is computed from T_q.

For eigenvalue assignment problem formulation, the state variable model for the pitch rate command system of Figure 10.4, without washout filter feedback, takes the form

$$\begin{bmatrix} \dot{x} \\ \dot{z} \end{bmatrix} = \begin{bmatrix} A & 0 \\ c & 0 \end{bmatrix} \begin{bmatrix} x \\ z \end{bmatrix} + \begin{bmatrix} B \\ 0 \end{bmatrix} \delta_e + \begin{bmatrix} 0 \\ 1 \end{bmatrix} q_e \qquad (10.12)$$

where A and B are the rigid body dynamics, c is the output vector for pitch rate state variable and z is the integrator state. The feedback control law has the form

$$\delta_e = k_q q + k_\theta z \qquad (10.13)$$

The output feedback algorithm can be used to assign the desired short-period mode.

If the $\dot{\alpha}_e$ feedback is also used (Figure 10.4), a dynamic feedback structure results. The A and B matrices get augmented with the filter state dynamics. The feedback law takes the form

$$\delta_e = k_q q + k_\theta z + k_w \dot{\alpha}_e \tag{10.14}$$

Using control law (10.14), in addition to the short-period mode, the closed-loop washout filter root can also be assigned. Indeed if the assigned value is same as the washout filter value (pole – zero cancellation), $k_w = 0$ and we have the control law (10.13). Thus, the closed-loop location of the filter root determines the effect of $\dot{\alpha}_e$ feedback on AoA response shaping.

The command filter design follows the same procedure detailed in section 10.5.2. The parameters of the command filter are optimised to improve the handling qualities performance.

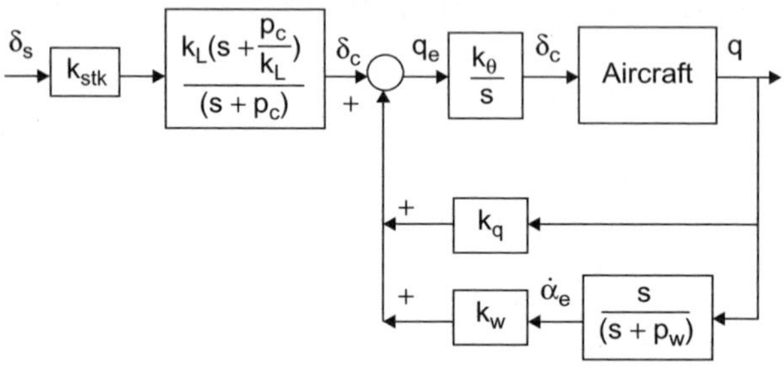

Figure 10.4 Schematic of superaugmented controller

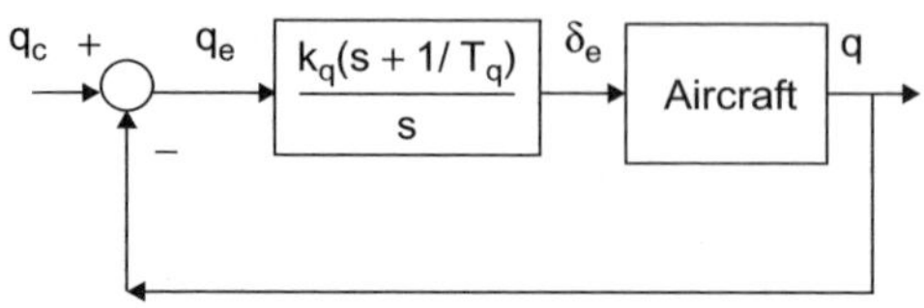

Figure 10.5 Simplified schematic of superaugmented controller

10.7 Single-input controller performance assessment

In sections 10.5 and 10.6, design of feedback control laws using three different controller structures has been investigated. In this section, their performance will be assessed using both handling quality metrics and desirable control system performance criteria. The following control laws will be evaluated:

1. Conventional controller using α and q feedback (CF)
2. Reversionary controller (AOA-fail) using q, α_e and $\dot{\alpha}_e$ feedback (RF)
3. Superaugmented controller using q and $\int q_e$ feedback (SF)

10.7.1 Control law performance analysis

As in the case of lateral–directional control law design (Chapter 9), the control law performance is assessed using the truth model that includes additional filters composed of FCS hardware and phase lead filters associated with the control law. Table 10.2 gives these filter definitions.

Table 10.3 gives the feedback gain magnitudes of the control laws. CF and RF control laws have the short-period mode assigned at ω_{sp} = 3 rad/sec and ζ_{sp} = 0.7. The SF control law has the short-period mode assigned at ω_{sp} = 4 rad/sec and ζ_{sp} = 0.9. Table 10.4 summarises the command filter and stick gradient constants. The stick gradient is determined such that for full aft stick force (35 lb; Table 10.2) pulse input, the steady-state normal acceleration attained is 7 g. The k_{stk} gradient in Table 10.4 is half of this value and is used to assess the Gibson's PIO resistance characteristics for small-amplitude stick deflection (±10%). Table 10.5 lists the closed-loop eigenvalues based on the *design* model.

Table 10.2 Filters associated with truth model

S. no.	**Definition**	**Remarks**		
(a) Aircraft FCS hardware assumptions				
1	$T_1(s) = \frac{25}{s+25}$	Horizontal tail primary servo (rate limit: 80 deg/sec) Deflection: TEU = 25 deg, TED =10 deg		
2	$T_2(s) = \frac{30}{s+30}$	Flaperon primary servo (rate limit: 80 deg/sec) Deflection: ±25 deg		
3	$T_3(s) = \frac{\omega_n^2}{s^2 + 2\zeta\omega_n + \omega_n^2}$ $\omega_n = 62.8$ rad/sec $\zeta = 0.7$	Secondary servo (for both channels)		
4	Longitudinal stick characteristics	Deflection (in)		Force gradient (lb/in)
		Aft	Fwd	7
		5	3	
(b) FCS filter assumption				
1	$T_4(s) = \frac{26}{s+26}$	Lumped effect of structural filters and computation delay (80-Hz sampling) Phase lag at 1 Hz = 13.6 deg (in each control channel)		
(c) Control law filter				
1	$T_5(s) = \frac{2.0(s+10)}{s+20}$	Phase advance filter (high-frequency gain = +6 dB) (in each control channel)		

Table 10.3 Feedback gains of controllers (δ_e channel)

Controller	α	q	$\int q_e$	α_e	$\dot{\alpha}_e$
CF	0.8889	0.2656	–	–	–
RF	–	1.1075	–	0.0016	–0.8342
SF	–	0.7181	2.0783	–	–

Table 10.4 Command path filter constants

Controller	k_{stk} (deg/lb)	p_c	k_L
CF	0.33	0.8178	2.68
RF	0.33	1.23	2.90
SF	0.33	2.87	0.44

Table 10.5 Closed-loop eigenvalues (design model)

Modes	Controller		
	CF	RF	SF
Short period	$-2.10 \pm j2.14$	$-2.10 \pm j2.14$	$-3.6 \pm j1.74$
Phugoid	$-0.0048 \pm j0.0522$	–	–
Low frequency-1	–	+0.0263	0.0
Low frequency-2	–	–0.0374	–0.99e – 4
Integrator	–	–	–1.05
Washout filter	–	–0.8413	–

The interaction of the unstable short-period mode with the phugoid mode is of special interest. This is illustrated in Figure 10.6. It should be noted that the final location of the roots in Figure 10.6a–c is different from the values given in Table 10.5. This is due to the interaction of the higher order modes of the 'truth' model that are captured in the root loci.

Based on Figure 10.6a–c, the following observations can be made. The CF control law retains the classical short-period and phugoid mode characteristics. The RF control law does not allow the unstable root to move to the left half plane due to a non-minimum phase zero in the n_z/δ_e transfer function. The resultant slow divergent mode is generally corrected by pilot control action similar to the spiral divergence in the lateral–directional axes. The SF control law has a stable and unstable eigenvalue pair very close to the origin. In both RF and SF control law cases, the classical phugoid mode splits into two aperiodic roots.

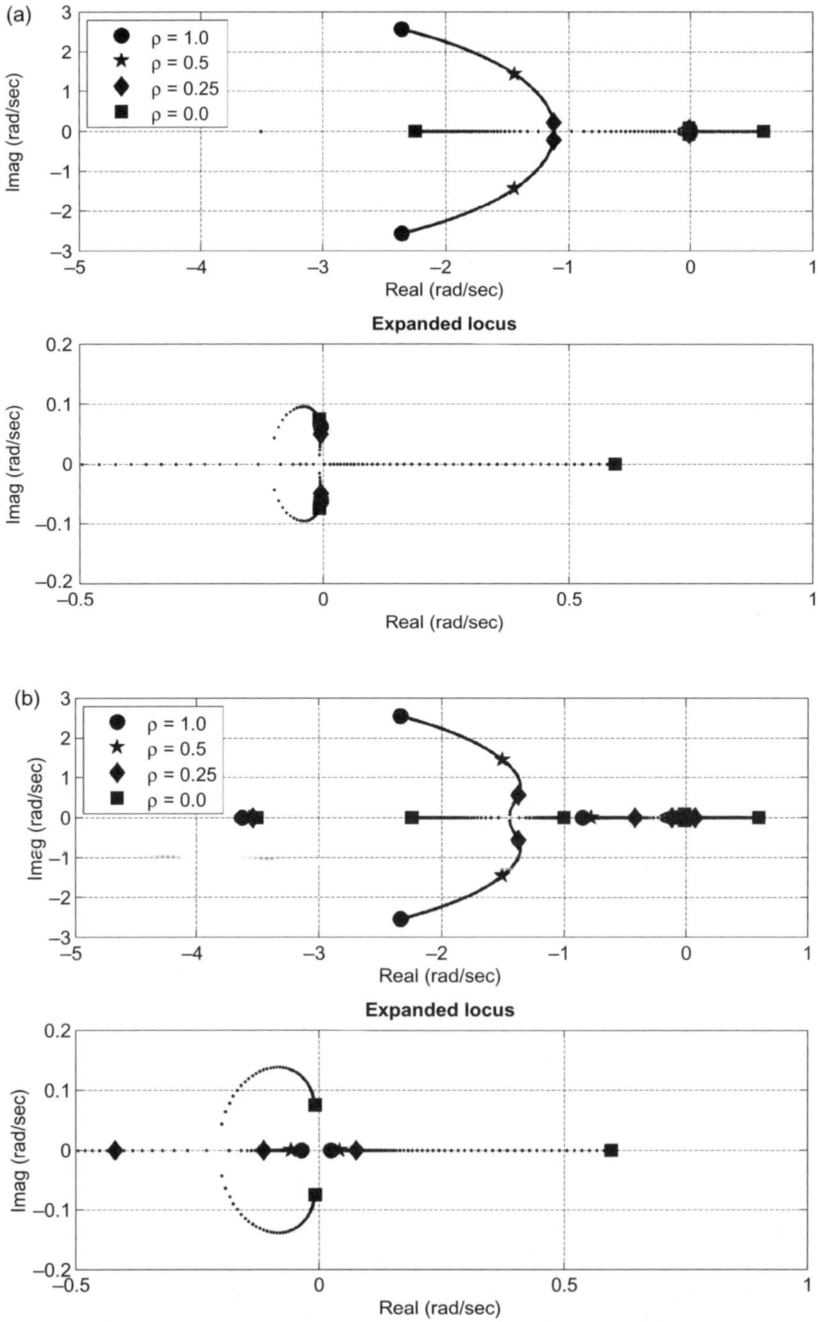

Figure 10.6 Root locus of unstable aircraft root migration (design model) (a) CF control law; (b) RF control law; (c) SF control law

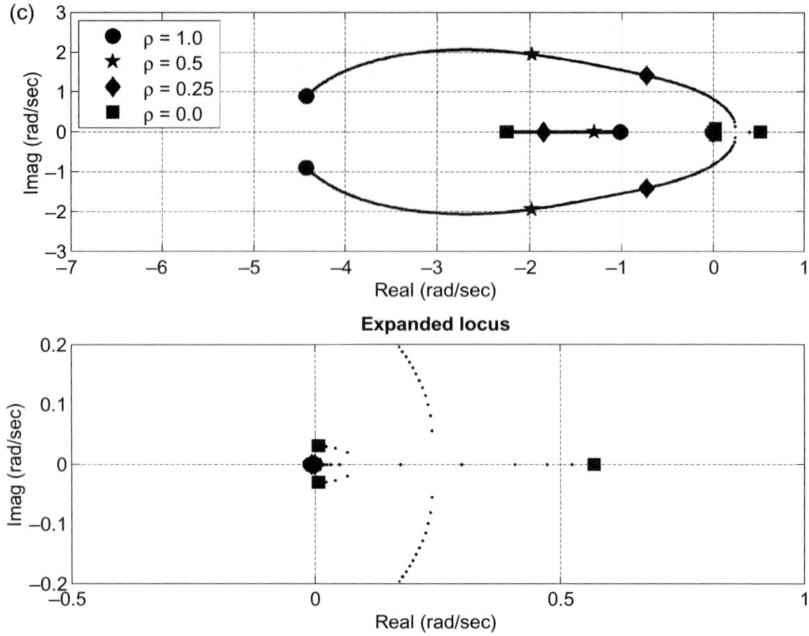

Figure 10.6 Concluded

10.7.2 Handling qualities characteristics

As discussed in Chapter 8, the performance of the control laws will be assessed based on the metrics defined therein. Figure 10.7a shows the LOES model parameters for the controllers. The dominant modes of Figure 10.7a are based on the respective 'truth' models. The difference in the dominant and LOES modes of CF and RF control laws is relatively lower as compared with the SF control law. Further SF control law has higher pitch time delay ($\tau_\theta = 0.152$). The low-frequency closed-loop integrator root of the SF control law causes (i) larger shift in the dominant and LOES modes and (ii) higher time delay.

The locations of the LOES modes in relation to the short-period mode thumbprint are shown in Figure 10.7b [10]. For the CF and RF designs even though the dominant modes were selected to be in the 'good' region of the thumbprint, the resulting LOES modes drop them into the 'acceptable' boundary. Increasing the assigned dominant mode natural frequency during eigenstructure synthesis can compensate this performance reduction. The SF design falls in the 'good' zone since the dominant mode is selected high, during the design, to compensate for the phase delay caused by the low-frequency integrator root ($\omega_{sp} = 4.34$ and $\zeta_{sp} = 0.945$ based on truth model). Based on pitch time delay, CF and RF designs meet level 1 requirements and SF design meets level 2 requirements.

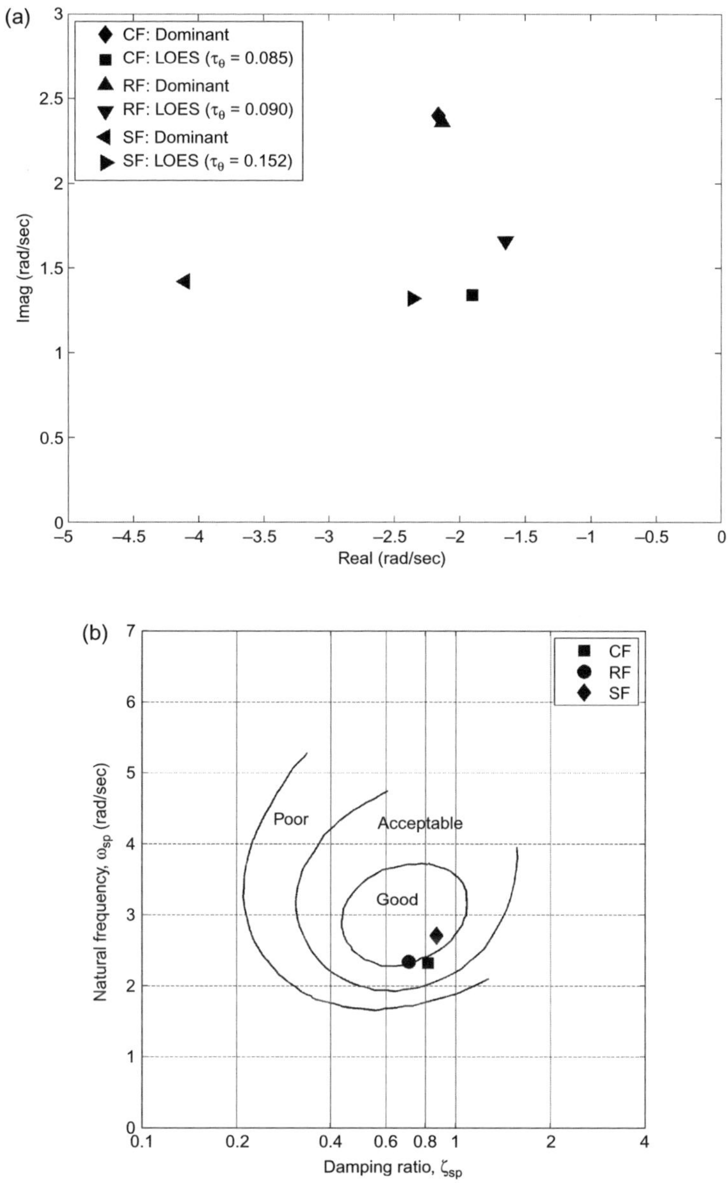

Figure 10.7 Lower order equivalent systems: (a) Dominant and LOES modes; (b) LOES short-period mode thumbprint

The control anticipation parameters (CAP), based on the LOES models, are given in Figure 10.8. CAP is computed using (8.6), with $n/\alpha = 17.14$, based on aircraft trim velocity of 192 m/sec and the open-loop aircraft value of $1/T_{\theta_2} = 0.8755$. The short period ω_{sp} is based on LOES model. All designs fall in level 1 region.

The bandwidth criterion results are given in Figure 10.9. All control laws are in the level 2 region. The SF control law has the highest phase delay. The stringent bandwidth requirement in level 1 region is a consequence of the requirement of superior disturbance rejection properties [9]. However, excessive bandwidth has other negative consequences such as (i) interaction with structural modes, (ii) higher pitch rate overshoot resulting in increased dropback and (iii) reduction in stability margin.

The bandwidth criterion also does not take into account the flight path dynamics that is essential during landing tasks. A unified bandwidth criterion [11], not yet included in the military handbook [9], is being proposed to overcome some of these deficiencies.

The Gibson's time domain precision tracking performance, based on attitude dropback, is shown in Figure 10.10. All the control laws have satisfactory dropback characteristics. It should be remembered that the command path filter controls the pitch attitude dropback parameter. Further for a single-input feedback system, with $1/T_{\theta_2}$ fixed at the open-loop value, direct trade-off between dropback (dbq) and flight path delay (t_γ) can be achieved by tuning the command filter.

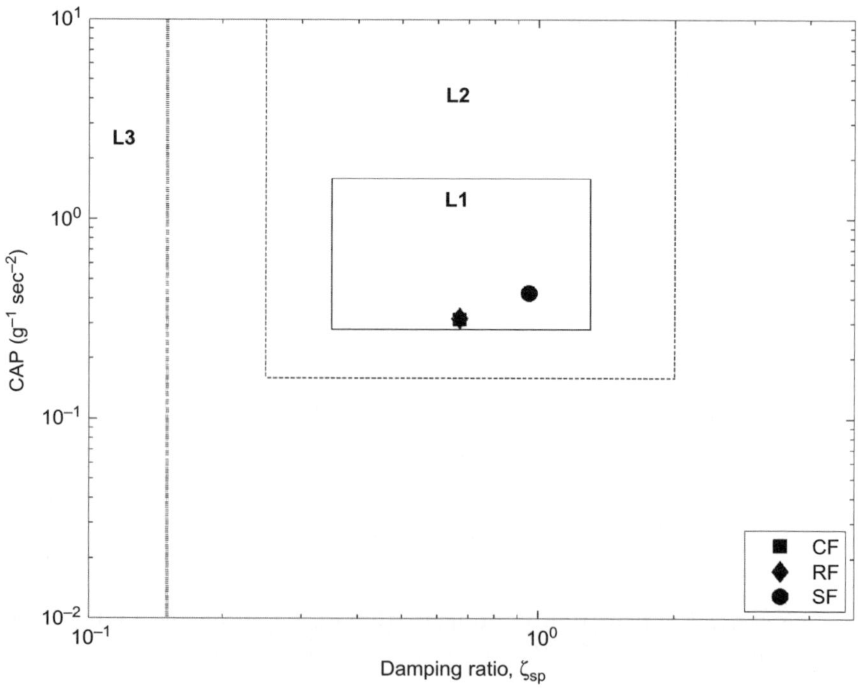

Figure 10.8 LOES short-period mode CAP chart

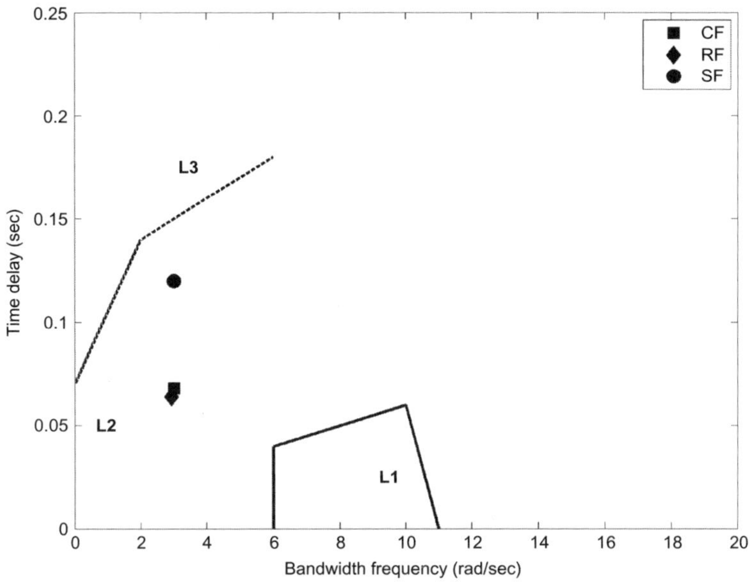

Figure 10.9 Bandwidth/phase delay characteristics

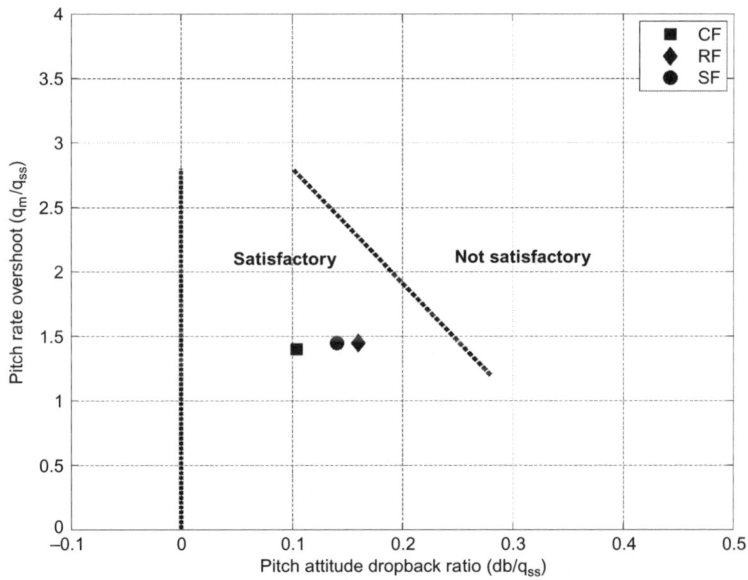

Figure 10.10 Precision tracking characteristics

The θ/δ_s frequency response of candidate designs is shown in Figure 10.11. The roll-off of the SF controller near −180 deg phase angle indicates poorer average phase rate characteristics compared with the conventional controller. The results of the Gibson's PIO-prone pitch loop gain boundary criterion are shown in Figure 10.12.

The SF controller marginally crosses into the PIO activity region. The results for pitch attitude frequency response for level 1 handling qualities are shown in Figure 10.13. All the controllers meet the level 1 design aim. Figure 10.14 shows the average phase rate characteristics, and the conventional controllers (CF, RF) are in the level 1* region while the SF design is on the border of level 1/level 2 boundary.

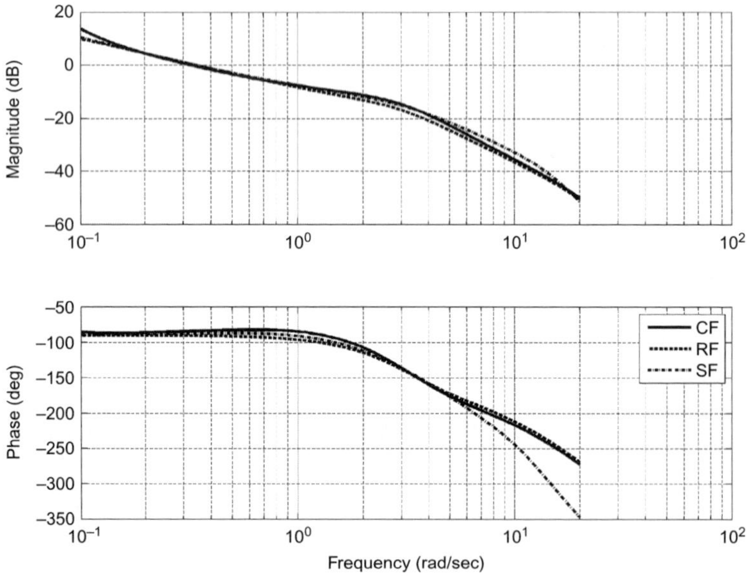

Figure 10.11 θ/δ_s *frequency response characteristics*

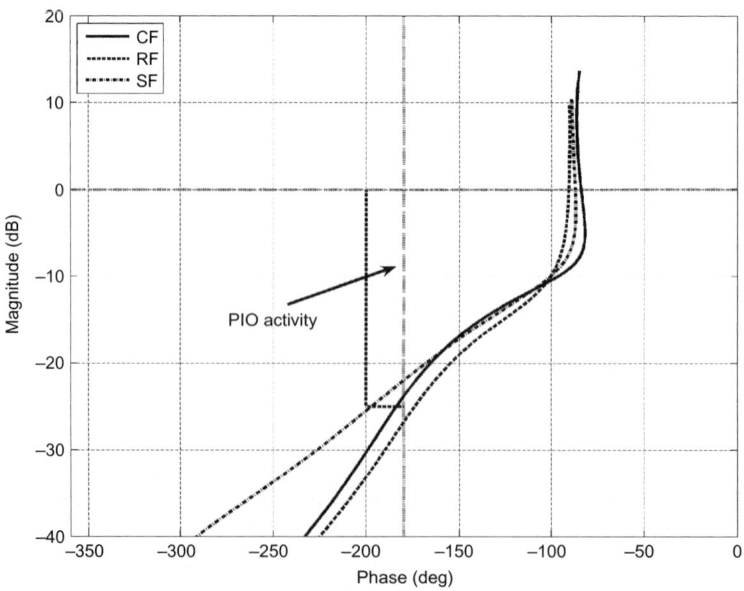

Figure 10.12 θ/δ_s *PIO activity gain boundary criterion*

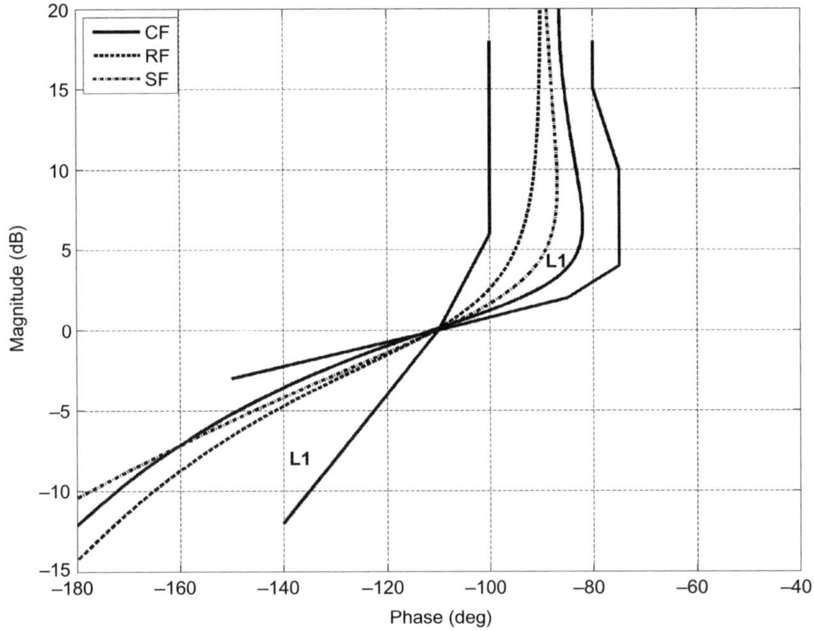

Figure 10.13 θ/δ_s *frequency response HQ boundary criterion*

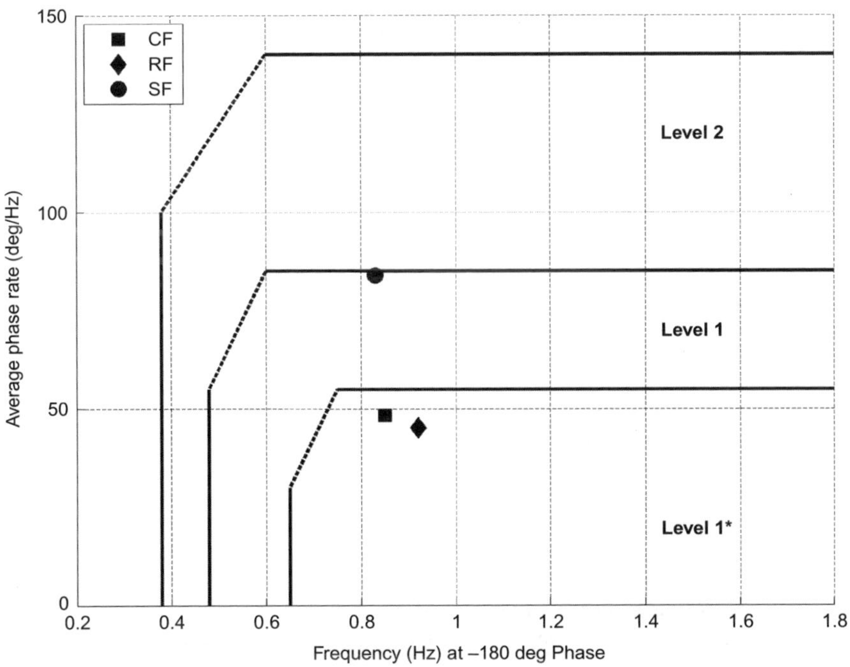

Figure 10.14 *Average phase rate criterion results*

In summary, all controllers meet the pitch axis handling qualities requirements discussed in Chapter 8. In general, the pitch rate command/attitude hold control scheme (SF) exhibits slightly inferior HQ performance compared with the conventional controller. This is primarily due to the integrator low-frequency root in the closed-loop system. Additional performance improvement can be attempted using the AoA rate feedback loop proposed in the controller structure, and/or optimisation of a two-section command path filter

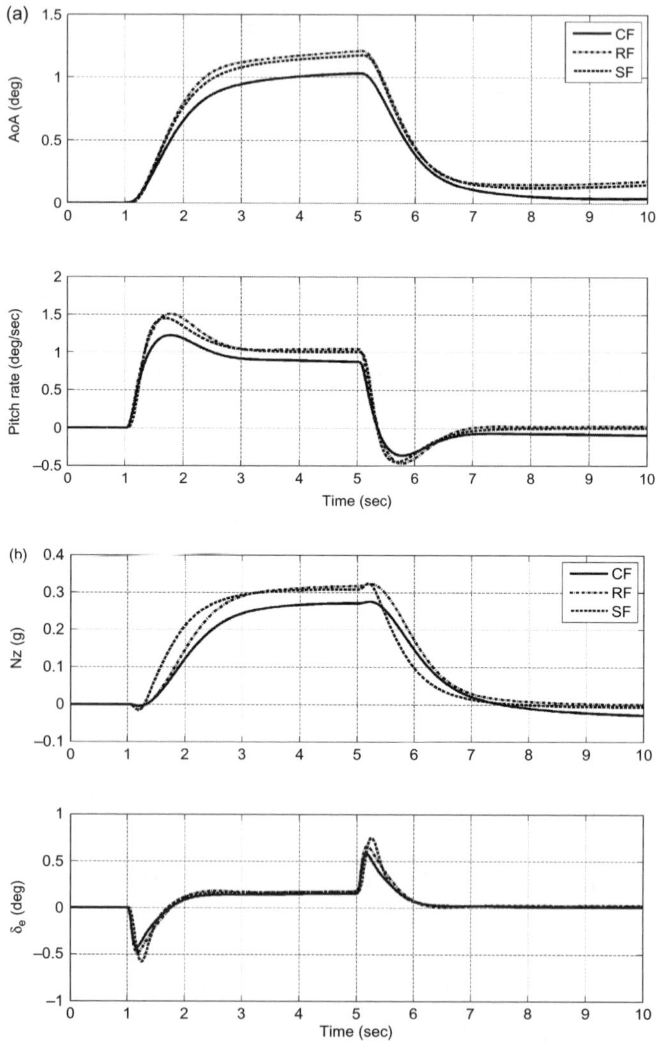

Figure 10.15 Time response characteristics of single-input controllers (δ_c pulse command: −1 deg) (a) Angle of attack and pitch rate response; (b) Normal acceleration and control deflection

can be attempted to shape the θ/δ_s frequency response near the frequency range of 1–5 rad/sec to improve average phase rate characteristics.

It should be noted that it is possible to formulate an ideal model with the desired θ/δ_s frequency response characteristics and synthesise the forward path controller parameters using an explicit model control scheme discussed in Chapter 7.

10.7.3 Time response performance

Figure 10.15 gives the time response characteristics of the single-input controllers CF, RF and SF. The conventional controller (CF) has relative poor response compared with RF and SF. This is caused by higher lag of the command filter (Table 10.5). The pitch rate response, in all cases, reaches its steady-state value in 2 sec, which is the maximum expected limit. All controllers have approximately the same response sensitivity for unit command input. The horizontal tail control deflections are also similar.

10.7.4 Stability margins

Figure 10.16 shows the robustness of the design using the stability margin templates discussed in Chapter 8. Adequate gain and phase margins exist for all designs. The SF controller, by its inherent controller structure, is known to exhibit relatively lower phase margins compared with the conventional controller structure. Another reason for the lower margins in the case of SF design is the higher closed-loop short-period frequency selected compared with the CF and RF designs.

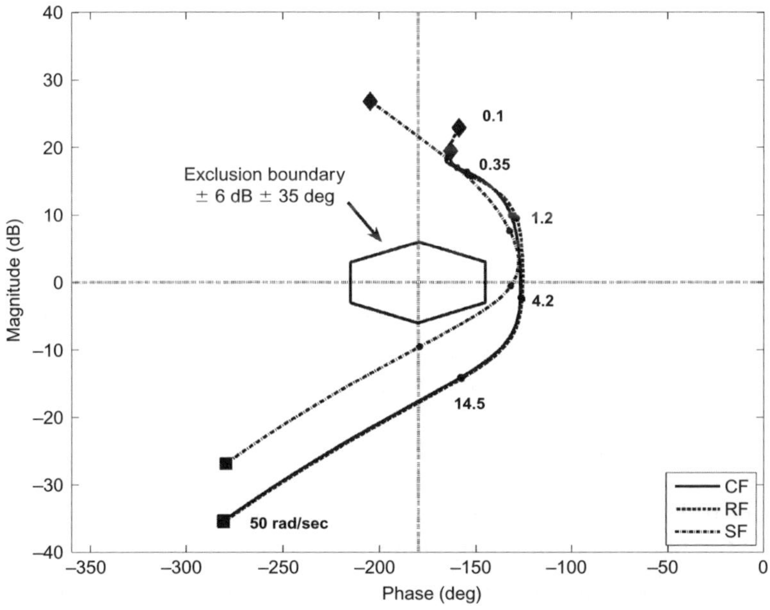

Figure 10.16 Stability margin results

10.8 Implicit model following control design

In this section a multivariable control design using the IMF concept (Chapter 7) will be studied. The pitch axis control design using both the horizontal tail and flaperon will be investigated. Use of two control inputs allows, in addition to eigenvalue placement, some control on eigenvector modification as well. Examination of the single-input design results in section 10.7 reveals that, since the pitch rate zero (z_θ) is invariant with respect to feedback, an improvement in dbq invariably results in higher flight path delay (t_γ). In the multivariable case it is possible to remedy this situation by assigning z_θ to a desirable location. The target dbq and flight path delay can thus be independently controlled.

A model following control problem will be formulated to synthesise the desired handling qualities properties. The control law will be derived as an IMF controller. This has advantages of feedback robustness properties. The first step is to construct the desired HQ model. For illustration a model based on Gibson's time domain criteria [6], discussed in section 10.2.1, will be derived.

For the two-input system considered, a desired short-period dynamics can be realised by selection of the modal parameters ω_{sp}, ζ_{sp} and T_{θ_2} (z_θ). The short-period state variable model derivatives that synthesise this dynamics are Z_α, M_α, M_q and k (10.2). The modal parameters define t_γ and dbq (10.3). A non-linear constrained optimisation problem to compute the model parameters for a defined triple ω_{sp}, ζ_{sp} and z_θ can be formulated with appropriate bounds set on the design parameters Z_α, M_α, M_q and k. Starting from the BSS aircraft model rigid body state variable model, the short-period model parameters are computed to minimise a performance index of the form

$$J = w1 * J_1 + w2 * J_2 + w3 * J_3$$
$$J_1 = (W_{sp} - \omega_{sp})^2, \quad J_2 = (Z_{sp} - \varsigma_{sp})^2, \quad J_3 = (Z_\theta - z_\theta)^2 \tag{10.15}$$

In the above performance index, W_{sp}, Z_{sp} and Z_θ are the desired target values of the model parameters, and w1, w2 and w3 are weighting factors on the performance indices J_1, J_2 and J_3.

As an example, an HQ model with the short-period mode is selected in the 'good' region of Figure 10.7b, as W_{sp} = 3 rad/sec and Z_{sp} = 0.7. For a desired DBQ = 0.2, $T_{\theta_2} = 0.6667$ or Z_θ = 1.5, the flight path delay is t_γ = 0.4667. To meet this model specification, the BSS configuration aircraft short-period model parameters Z_α, M_α, M_q and k are modified to minimise the objective function (10.15). The optimisation is carried out using the fourth-order rigid model and computing t_γ and dbq graphically as shown in Figure 10.1.

The general feedback scheme of the multi-input IMF controller is given in Figure 10.17. The following steps are involved in the design of the feedback controller to match the HQ model characteristics:

Step 1. Extract the eigenstructure of the short-period mode of the rigid body HQ model.

Step 2. Assign the extracted eigenstructure of the short-period mode to the BSS configuration aircraft using output feedback algorithm. It should be noted that using AoA and pitch rate feedback, the entire eigenstructure of the short period mode can be exactly assigned with four feedback gains.

Step 3. Using the 'truth' model of the plant, verify the achieved accuracy of the HQ parameters and the time response match between the model and plant states.

Step 4. Fine-tune, if necessary, some of the HQ parameters, especially dbq, using the command path filter shown in Figure 10.17.

Remark: For the reversionary control law (AoA fail), using q, n_z and $\dot{\alpha}_e$ as feedback variables, only an approximate solution to the model matching problem can be realised since direct augmentation of Z_α and M_α is not possible. An optimisation scheme can be adopted to get an acceptable fit between the model and plant state variable (α, q and θ) time response trajectories.

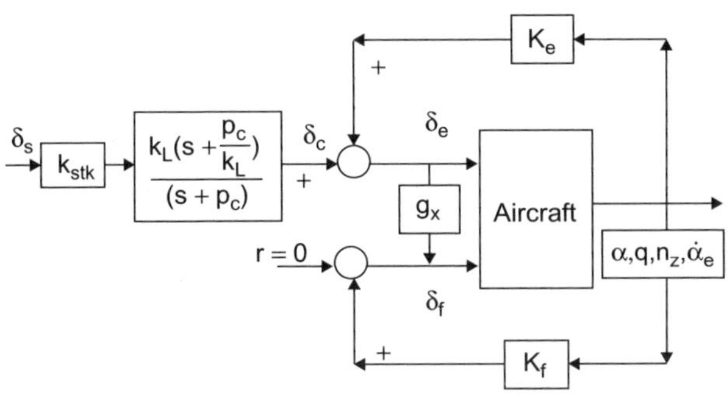

Figure 10.17 Multivariable feedback controller

The synthesis of *any* HQ model definition may not be feasible due to control system constraints such as actuator deflection and rate capability, stability robustness, etc. Thus, final definition of a *feasible* HQ model has to be arrived at after a few trial designs. The objective of venturing into a multi-input design is to improve, if possible, the performance of single-input designs discussed in section 10.7. One obvious objective is to independently assign required flight path delay and pitch dropback ratio (dbq) by selecting the appropriate z_θ. This assignment in turn also improves the AoA/n_z response time. The HQ1 model (Table 10.6) is the outcome of this requirement. The design based on this HQ1 model specification results in large flaperon control deflections (Figure 10.18). This is due to high augmentation of Z_α required to assign z_θ. This results in high feedback gain from AoA to flaperon control surface. The following analysis using literal form of the feedback control scheme clarifies this limitation.

The augmented short-period approximation of the closed-loop system matrix can be written as

$$\begin{bmatrix} \hat{Z}_\alpha & \hat{Z}_q \\ \hat{M}_\alpha & \hat{M}_q \end{bmatrix} = \begin{bmatrix} Z_\alpha & 1 \\ M_\alpha & M_q \end{bmatrix} + \begin{bmatrix} Z_{\delta_e} & Z_{\delta_f} \\ M_{\delta_e} & M_{\delta_f} \end{bmatrix} \begin{bmatrix} k_1 & k_2 \\ k_3 & k_4 \end{bmatrix} \tag{10.16}$$

where the control law is given by

$$\begin{bmatrix} \delta_e \\ \delta_f \end{bmatrix} = \begin{bmatrix} k_1 & k_2 \\ k_3 & k_4 \end{bmatrix} \begin{bmatrix} \alpha \\ q \end{bmatrix} \tag{10.17}$$

In particular, the augmented lift coefficient $\hat{Z}_\alpha$ in (10.16) that determines the closed-loop z_θ is given by

$$\hat{Z}_\alpha = Z_\alpha + Z_{\delta_e} k_1 + Z_{\delta_f} k_3 \tag{10.18}$$

The higher the value of the lift effectiveness of the direct lift control surface ($Z_{\delta f}$), the lower will be the AoA feedback gain k_3 to assign a desired $\hat{Z}_\alpha$. Lower gain implies lower δ_f excursions.

To arrive at an acceptable solution, the short-period mode is moved to the border of 'good' and 'acceptable' regions (Figure 10.7b), with $\omega_{sp} = 2.5$ and $\zeta_{sp} = 0.8$. Designs are carried out for a range of z_θ values to evaluate flaperon deflection requirements to attain 1 g incremental n_z (Figure 10.18). To achieve maximum normal acceleration of 6 g for full stick deflection, $z_\theta = 1.1$ design (HQ2 model, Table 10.6) results in a flaperon deflection of 22 deg. This is within the maximum flaperon deflection limit (±25 deg, Table 10.2). However, if the control surface is shared for both longitudinal and lateral control, the control authority for pitch axis gets reduced to ±12.5 deg. This would imply selecting $z_\theta = 1.0$ (HQ3 model, Table 10.6), to meet the full stick normal acceleration requirement of 6 g. The HQ1 model requirements (Figure 10.18, label A) severely limit the maximum normal acceleration attainable. For the present discussions, it is assumed that HQ2 model is a feasible design model definition.

The HQ parameters of all candidate models along with the BSS aircraft model are listed in Table 10.6. The BSS model has negative dbq, indicating a sluggish response. The HQ models have lower flight path delay and improved short-period natural frequency compared with the basic aircraft model. In Table 10.6, the graphically optimised dbq and t_γ of the HQ models, using the rigid body dynamics, do not add up to the analytically derived T_{θ_2} value of the short-period approximation. HQ2 model has a dropback ratio (dbq) higher than the maximum limit of 0.25. The command path filter (Figure 10.17) is tuned to reduce the dbq parameter to lie within the satisfactory boundary with a penalty of increasing the flight path delay.

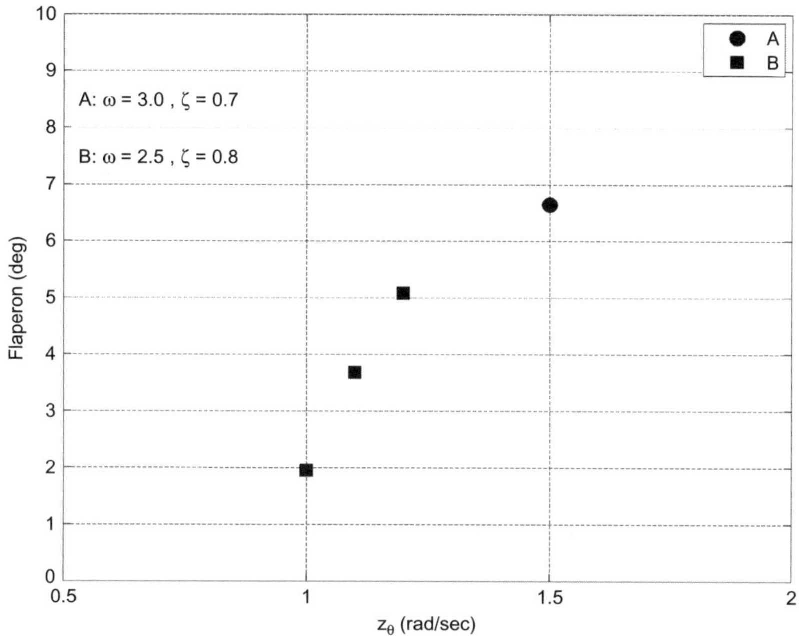

Figure 10.18 Steady-state flaperon deflection for step δ_e command (short-period model) ($\Delta nz = 1.0$ g)

The HQ1 and HQ2 short-period state variable models are given by

HQ1 model:

$$\begin{bmatrix}\dot{\alpha}\\ \dot{q}\end{bmatrix}=\begin{bmatrix}-1.5676 & 1\\ -4.8741 & -2.6311\end{bmatrix}\begin{bmatrix}\alpha\\ q\end{bmatrix}+\begin{bmatrix}-0.1289 & -0.1526\\ -9.1995 & 0.4611\end{bmatrix}\begin{bmatrix}\delta_e\\ \delta_f\end{bmatrix}$$

HQ2 model:

$$\begin{bmatrix}\dot{\alpha}\\ \dot{q}\end{bmatrix}=\begin{bmatrix}-1.1410 & 1\\ -2.9870 & -2.6311\end{bmatrix}\begin{bmatrix}\alpha\\ q\end{bmatrix}+\begin{bmatrix}-0.1289 & -0.1526\\ -9.1995 & 0.4611\end{bmatrix}\begin{bmatrix}\delta_e\\ \delta_f\end{bmatrix} \tag{10.19}$$

Table 10.6 Handling qualities model parameters

Model	ω_{sp} (rad/sec)	ζ_{sp}	T_{θ_2}(sec)	z_θ (rad/sec)	t_γ (sec)	dbq (sec)
BSS	1.3003	0.6317	1.2009	0.8327	1.1580	–0.1829
HQ1	3.0000	0.7000	0.6667	1.5000	0.5053	0.2228
HQ2	2.5000	0.8000	0.9091	1.1000	0.7088	0.2693
HQ3	2.5000	0.8000	1.0000	1.0000	0.7127	0.3667

The design parameters of IMF design (MF), based on HQ2 model, are given in Table 10.7.

Table 10.7 IMF controller parameters

Controller	**Target model**	**δ_e Channel**		**δ_f Channel**	
		α	q	α	q
(a) Feedback gains					
MF	HQ2	0.6180	0.2136	1.4042	–0.2136

Controller	**k_{stk} (deg/lb)**	**p_c**	**k_L**
(b) Command path filter constants			
MF	0.20	2.37	1.36

Modes	**Controller (MF)**	
	ω_n	ζ
(c) Closed-loop modal characteristics (based on truth model)		
Short period	2.66	0.771
Phugoid	5.57e – 2	7.69e – 2
z_θ	1.1098	

10.8.1 Time response performance

The match between the HQ2 model and the IMF controller is shown in Figure 10.19. The good match indicates that the closed-loop system has the desired properties of the HQ2 model. The AoA and n_z response to a pulse command for CF and MF controllers is shown in Figure 10.20. It is seen that the ramping (sluggish) response of the AoA response evident in the CF controller is absent in the MF controller. The normal acceleration response of the MF controller does not exhibit the lagging characteristics of the CF controller. The control deflection to attain 1 g incremental normal acceleration is shown in Figure 10.21. The horizontal tail control deflections are similar for both controllers. The expected flaperon deflection to achieve 1 g incremental n_z is also seen.

10.8.2 Handling qualities characteristics

The IMF control law (MF) does not have any additional dynamic elements, as it is only a partial state feedback controller. The LOES model optimisation is carried out with z_θ fixed at the closed-loop value, instead of at the plant z_θ. Table 10.8 summarises the CAP, bandwidth and phase rate characteristics of the MF controller.

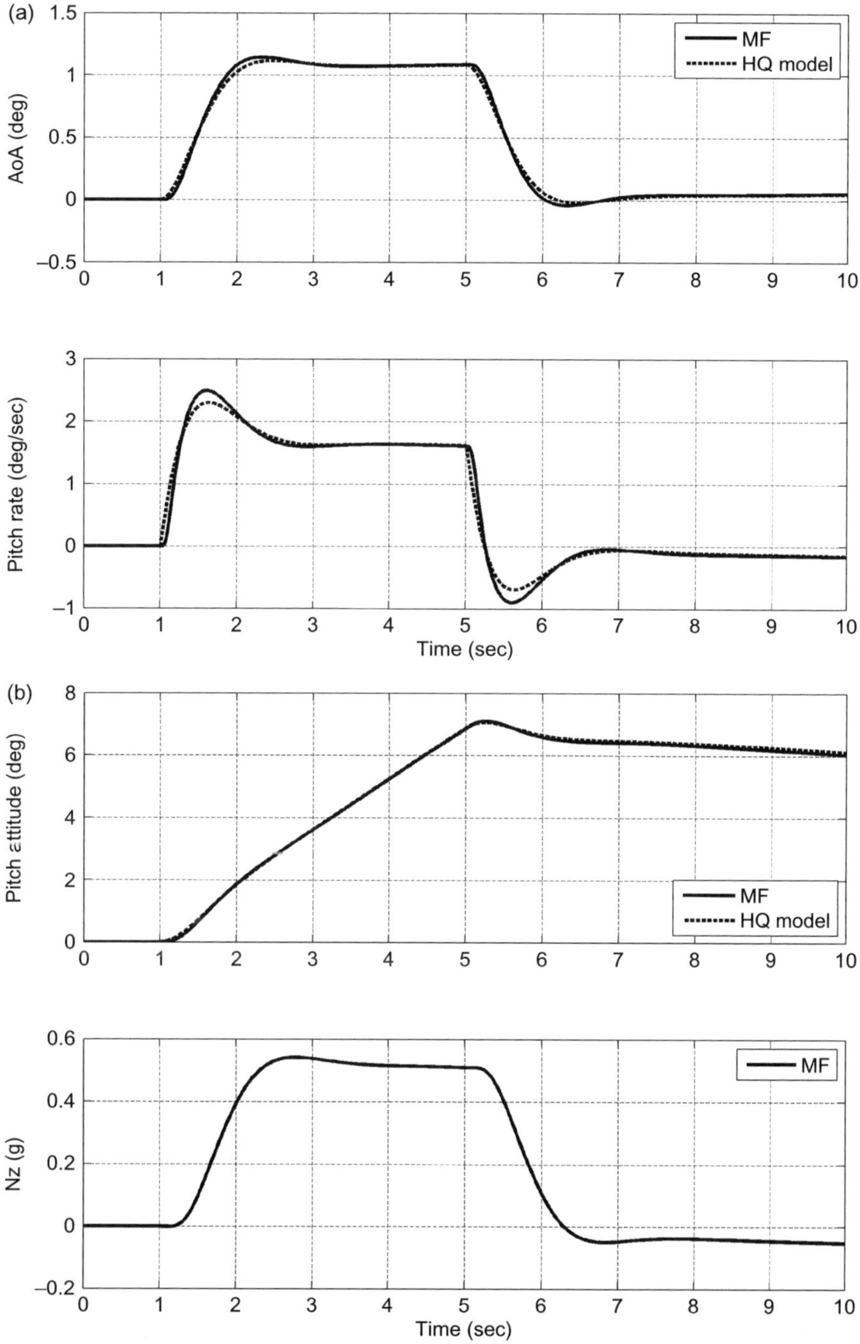

Figure 10.19 HQ2 model/IMF system time response match ($\delta_c = -2$ deg) (a) Angle of attack and pitch rate response; (b) Pitch attitude and normal acceleration response; (c) Control surface deflections

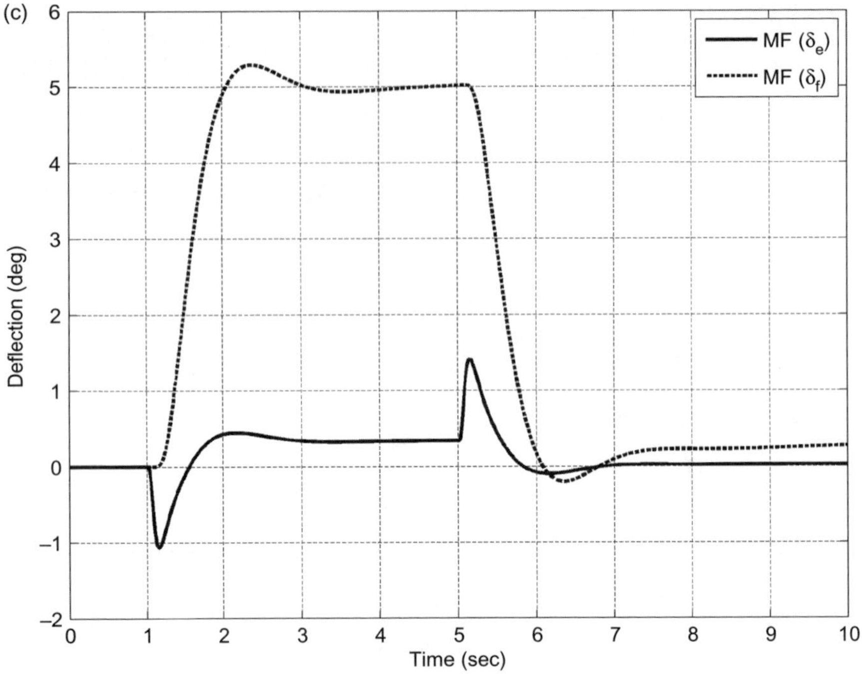

Figure 10.19 Concluded

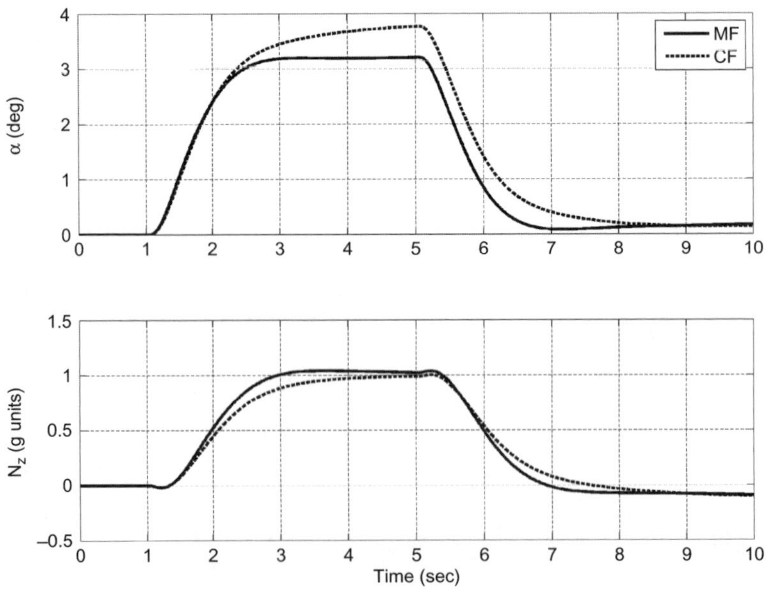

Figure 10.20 Comparison of CF and MF control laws AoA/n_z pulse response ($\delta_c = -2\,deg$)

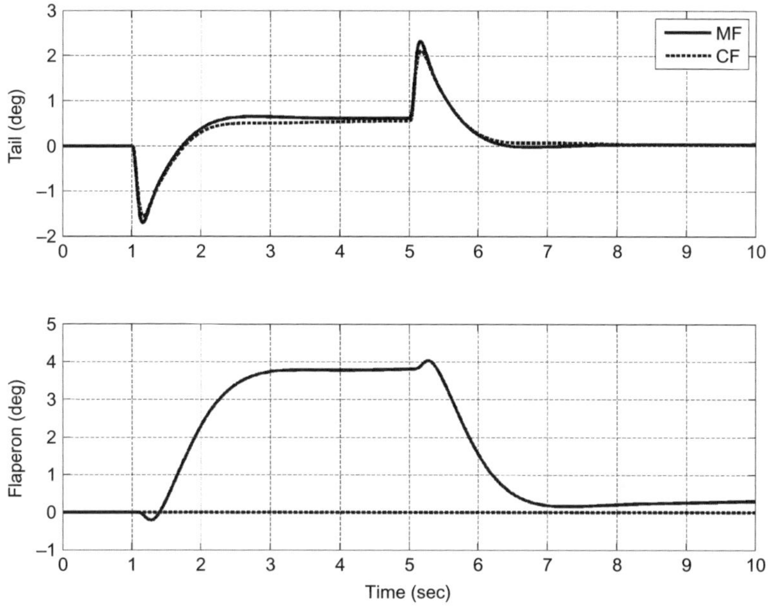

Figure 10.21 *Comparison of CF and MF control laws control surface deflections ($\delta_c = -2$ deg)*

Table 10.8 *Handling qualities properties of MF controller*

Dominant mode		LOES			CAP	HQ level
ω	ζ	ω	ζ	τ_θ		
(a) Control anticipation parameter (CAP) criterion						
2.66	0.77	2.49	0.816	0.077	0.3607	1

Bandwidth (rad/sec)	Phase delay (sec)	HQ level
(b) Bandwidth criterion		
3.0158	0.0608	2

Phase rate (deg/Hz)	Frequency (Hz)	HQ level
(c) Average phase rate criterion		
43.4773	1.0258	1*

As in the case of the single-input control (CF), the MF design has level 1 CAP, level 2 bandwidth and level 1* average phase rate characteristics. The LOES mode falls on the border of 'good' and 'acceptable' regions. The pitch attitude frequency response characteristics are given in Figure 10.22. The Gibson's frequency response criteria results are shown in Figures 10.23 and 10.24. The MF design satisfies the PIO criterion and has level 1 HQ properties. Figure 10.25 shows that the MF controller satisfies the Gibson's time domain dropback criterion.

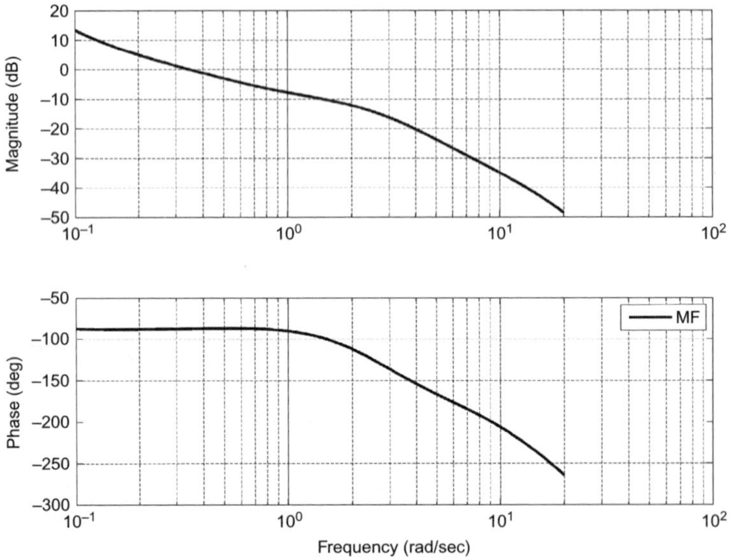

Figure 10.22 Pitch attitude frequency response (MF controller)

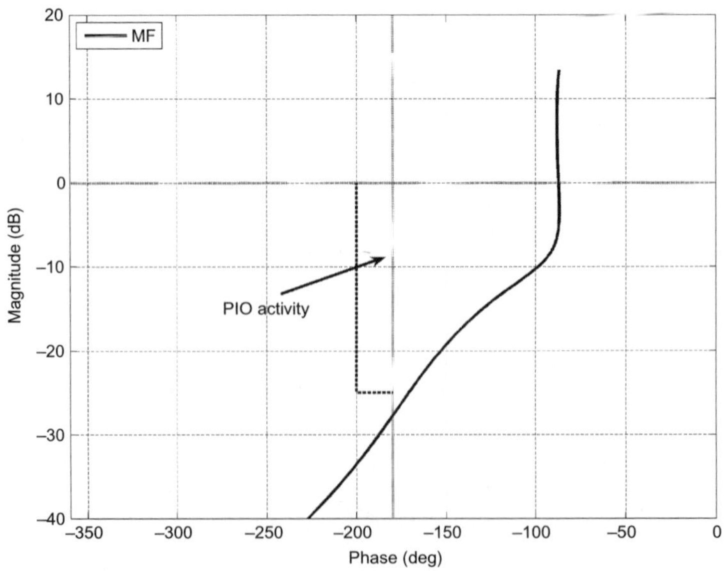

Figure 10.23 θ/δ_s PIO activity gain boundary criterion (MF controller)

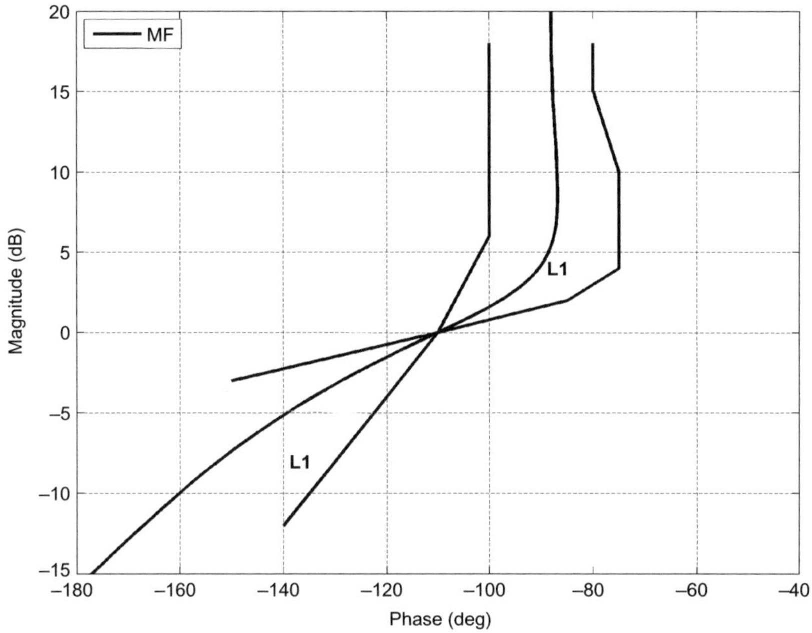

Figure 10.24 θ/δ_s *frequency response HQ level boundary criterion (MF controller)*

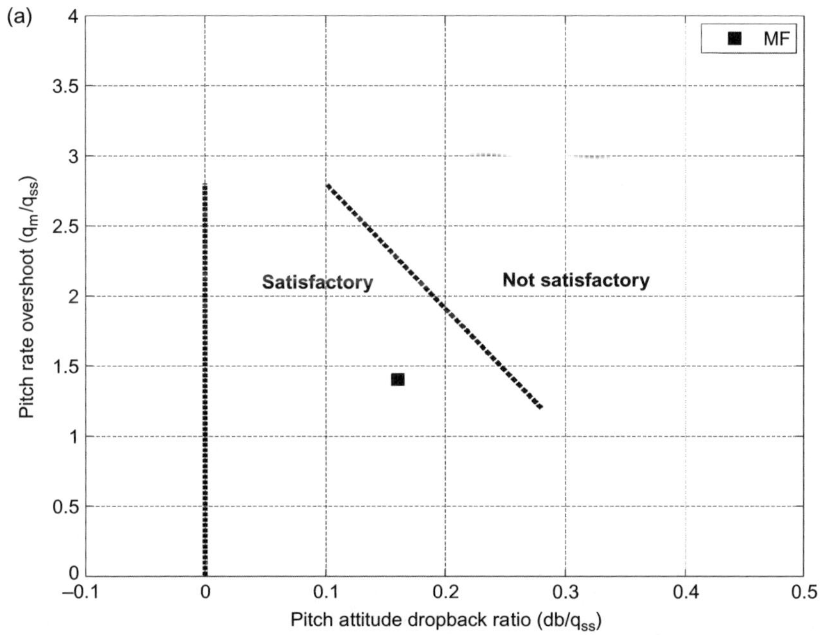

Figure 10.25 *MF controller HQ characteristics (a) Precision tracking characteristics; (b) LOES short-period mode CAP chart; (c) Bandwidth/phase delay characteristics*

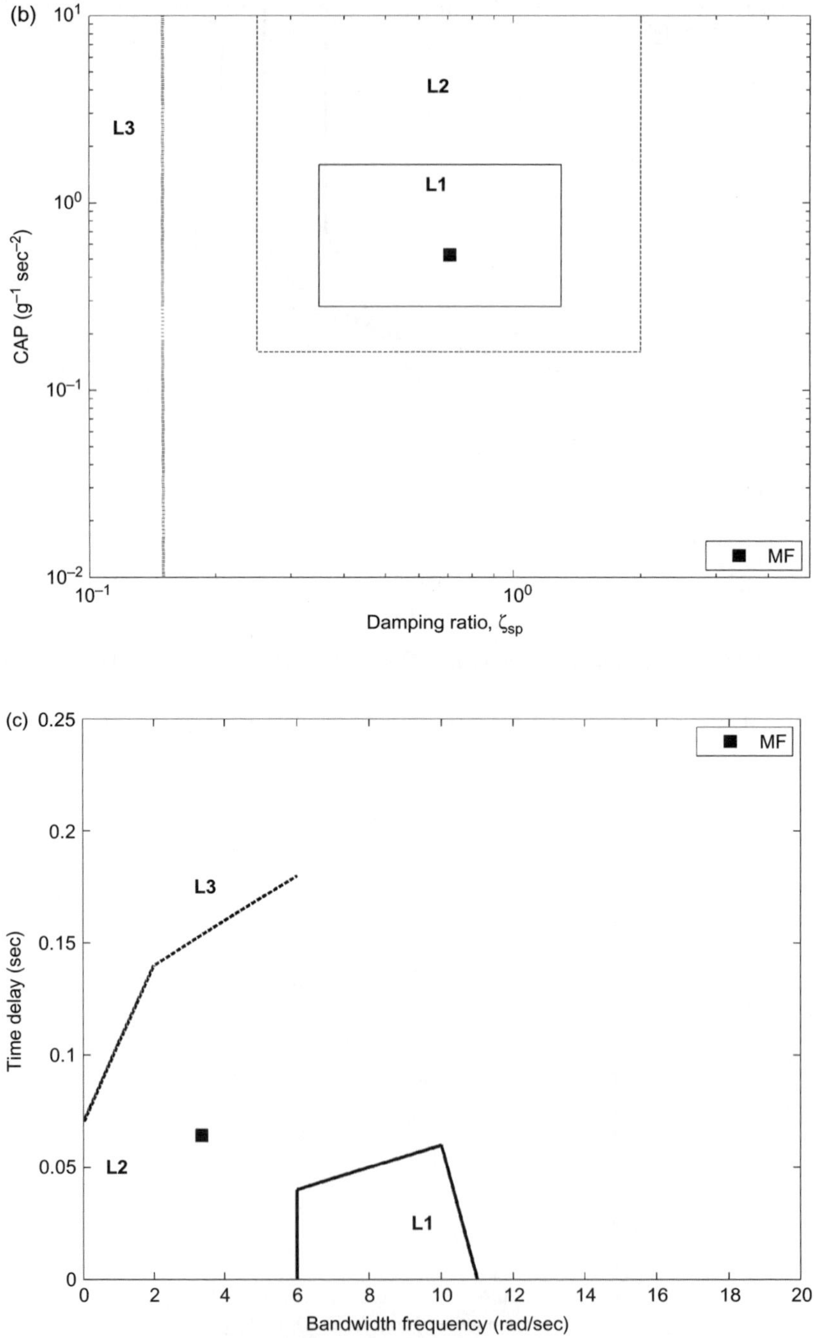

Figure 10.25 Concluded

Table 10.9 summarises the flight path delay and dropback ratio achieved by different designs. The MF design has the lowest flight path delay. Based on the normalised flight path delay with a designed dropback ratio of 0.2, the improvement in flight path delay using the MF controller varies from 4.95% (CF) to 11.3% (SF).

Table 10.9 Flight path delay/dropback ratio summary

Controller	CF	RF	SF	MF
Flight path delay (sec)	0.8981	0.8676	0.9181	0.8018
Dropback ratio (dbq)	0.1041	0.1577	0.1415	0.1606
Normalised flight path delay (sec), dbq = 0.2	0.8021	0.8253	0.8596	0.7624

In summary, the use of flaperon as a second control input marginally improves the performance of the single-input designs in terms of (i) flight path delay and (ii) AoA/n_z response time. The lift control effectiveness of the flaperon is not adequate to achieve a larger shift in z_θ to further improve the flight path delay. However, the design process outlined in this section will find use in applications that use direct lift devices with higher control effectiveness.

10.9 Pitch pointing mode controller design

If an aircraft is equipped with at least two control surfaces, advanced pitch axis control modes such as decoupling of the pitch attitude (θ) and flight path (γ) responses, termed as pitch pointing mode (PPM), can be incorporated. In the PPM mode the nose can be pointed in the pitch axis without changing the flight path with the result $\theta \cong \alpha$. The pitch pointing capability has been demonstrated in the AFTI-F16 flight test program [1]. Application of eigenstructure control for designing pitch pointing control laws has also been studied in Reference 3. In this study, a feedback control law in conjunction with a forward path command generator has been used to achieve the pitch attitude and flight path mode decoupling. From a practical design/implementation perspective, changing the feedback control law to just engage this special mode is not an attractive proposition. Instead, if the mode is mechanised entirely as an explicit model following (EMF) controller, while retaining the normal operational feedback controller, engaging/disengaging of the PPM can be seamlessly achieved. This approach is used in the following discussions to design an EMF controller that generates steering control signals to the horizontal tail and flaperon control surfaces to achieve the desired decoupling between θ and γ.

In the EMF scheme, the first step is to derive the required PPM model. Towards this end starting from the BSS model (stable configuration), a feedback control law, using both the horizontal tail and flaperon control, is derived that minimises the following objective function:

$$J = \min\left\{\frac{\mathrm{rms}(\gamma(t))}{\mathrm{rms}(\theta(t))}\right\} \tag{10.20}$$

where the responses are computed for a pulse δ_e input. The optimisation parameters are the short-period natural frequency and damping and two entries in the short-period mode eigenvector. Since decoupling is sought between θ and γ, by synthesising a decoupled eigenvector structure, the optimisation is best carried out using $[\gamma \quad q \quad \theta \quad V]^T$ as the state vector. Substituting γ for α as a state variable can be effected using a simple co-ordinate transformation. Output feedback eigenstructure assignment algorithm, using AoA and q as feedback variables (four gains), is used to generate the feedback solutions during the optimisation iteration process. The required PPM model with the following short-period dynamics results:

$$\begin{bmatrix}\dot{\alpha}\\ \dot{q}\end{bmatrix} = \begin{bmatrix}-1.5898 & 1\\ -4.8509 & -2.6088\end{bmatrix}\begin{bmatrix}\alpha\\ q\end{bmatrix} + \begin{bmatrix}0.1289 & -0.1526\\ -9.1994 & 0.4611\end{bmatrix}\begin{bmatrix}\delta_e\\ \delta_f\end{bmatrix} \tag{10.21}$$

The optimum feedback control law that constructs the above PPM model is given by

$$\begin{bmatrix}\delta_e\\ \delta_f\end{bmatrix} = \begin{bmatrix}-0.0508 & 0.1676\\ -5.2337 & -0.1583\end{bmatrix}\begin{bmatrix}\alpha\\ q\end{bmatrix} \tag{10.22}$$

The model has a short-period mode with $\omega_{sp} = 1.75$ and $\zeta_{sp} = 0.7$.

Figure 10.26 gives a schematic of the EMF control scheme. The model states are $x_m = [\alpha_m \quad q_m \quad \theta_m]^T$. The model has two outputs $y_m = C_{m_{opt}} x_m$. The output matrix $C_{m_{opt}}$ is optimised to minimise the mismatch between plant and model states for a pulse command input using the tunable CGT controller scheme discussed in Chapter 7. A single-input conventional feedback controller (PF) for the stable BSS model is designed to meet the handling qualities requirements of the aircraft using the procedure already outlined in section 10.5. The gain matrices and modal characteristics are given in Table 10.10.

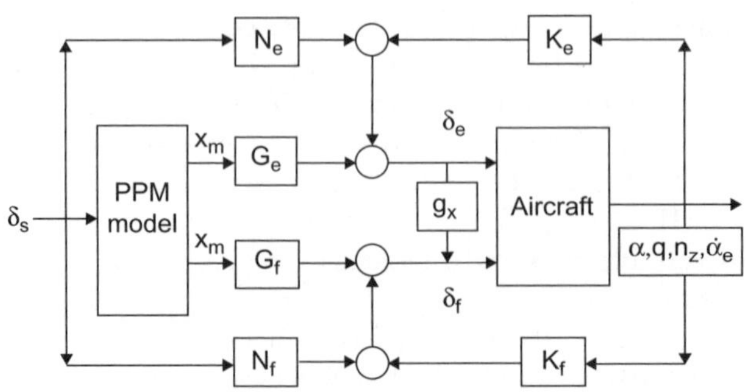

Figure 10.26 Schematic of pitch pointing mode (PPM) controller

With the feedback loop closed using the PF control law, the explicit model gain matrices G_e, G_f, N_e and N_f are optimised using the tunable command generator design described in Chapter 7. The model following controller can match two selected outputs of the model with those of the aircraft since the aircraft has two independent controls. However, using the tunable CGT controller design, a linear combination of the model states α_m, q_m and θ_m (resulting in two outputs) is optimised. The design procedure is outlined in Example 7.3. The optimised model output matrix is used to compute the explicit model gain matrices. The EMF steering control law has the form (Figure 10.26)

$$\begin{aligned} \delta_e &= G_e x_m + N_e \delta_s \\ \delta_f &= G_f x_m + N_f \delta_s \end{aligned} \tag{10.23}$$

The respective matrices are given by

$$\begin{aligned} C_{m_{opt}} &= \begin{bmatrix} -0.5156 & 0.0023 & 0.5215 \\ 1.5412 & 0.0060 & -0.5735 \end{bmatrix} \\ G_e &= [-2.1207 \quad -0.1037 \quad 1.5407], \qquad N_e = 0.9563 \\ G_f &= [-23.7527 \quad -0.1571 \quad 18.9981], \qquad N_f = -0.3706 \end{aligned} \tag{10.24}$$

Table 10.10 Feedback controller (PF) parameters

Controller	**δ_e Channel only (K_e), $K_f = 0$**		
	α	q	g_x
(a) Feedback gains			
PF	0.5659	0.2702	0

Controller	**k_{stk} (deg/lb)**	**p_c**	**k_L**
(b) Command path filter* constants			
PF	0.20	1.13	2.82

*Not shown in Figure 10.26.

Modes	**Controller (PF)**	
	ω_n	ζ_n
(c) Closed-loop modal characteristics (based on truth model)		
Short period	3.18	0.682
Phugoid	6.27e – 2	7.53e – 2
z_θ (plant)	0.8327	

Figure 10.27 shows the decoupling of the pitch attitude and flight path angle achieved by the EMF controller. Figure 10.28 demonstrates the model matching

characteristics of the controller, and finally Figure 10.29 indicates the control effort needed by the EMF controller to achieve the desired θ/γ decoupling.

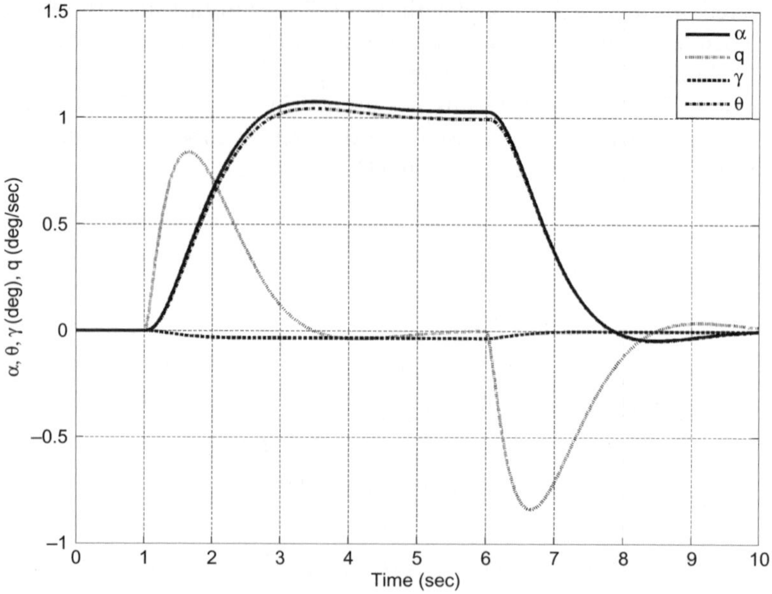

Figure 10.27 Pitch pointing mode pulse response ($\delta_s = -0.333$ deg)

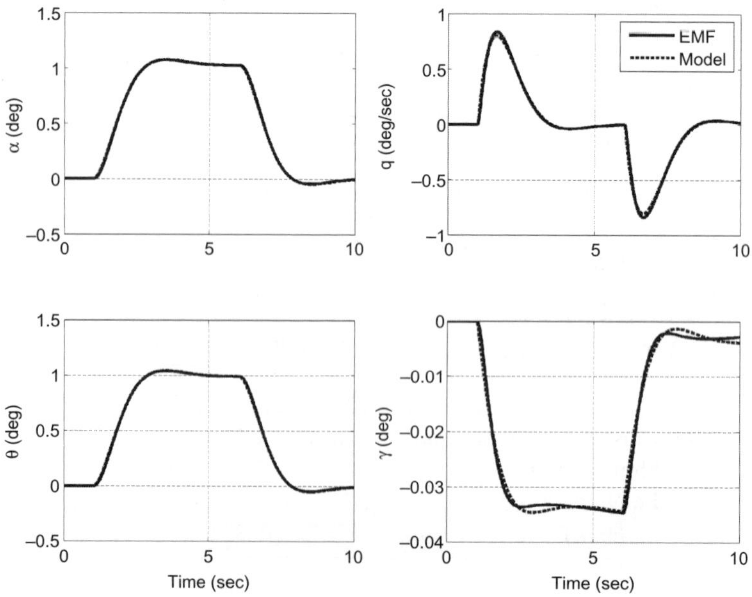

Figure 10.28 PPM model matching of the EMF controller ($\delta_s = -0.333$ deg)

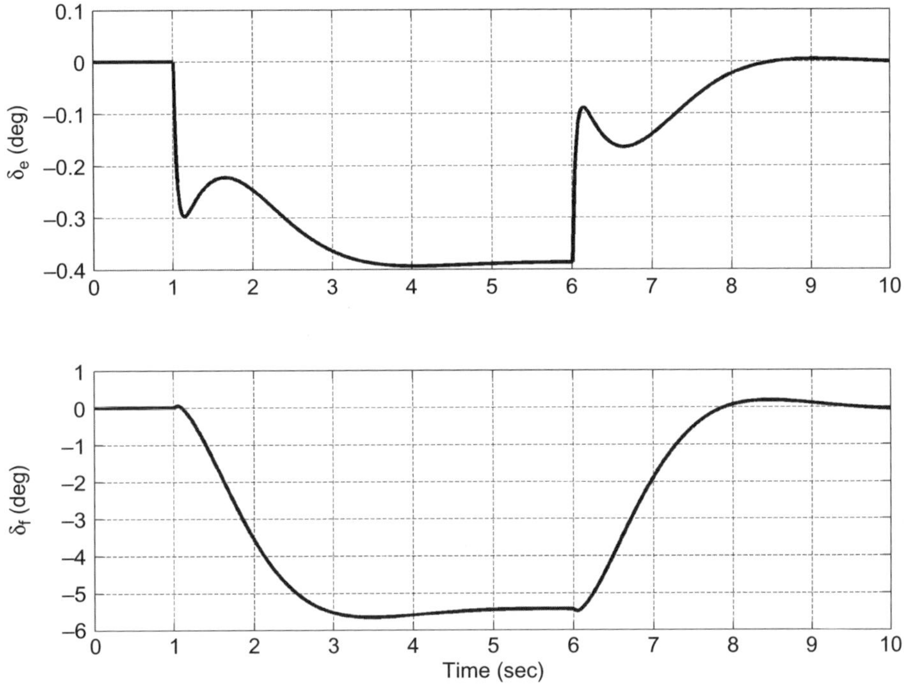

Figure 10.29 Control surface deflections ($\delta_s = -0.333$ deg)

The results show that excellent decoupling of the pitch attitude and flight path angle has been achieved and the EMF controller produces steering control signals to the plant to accurately follow the PPM model response. However, as in the case of the IMF controller design of section 10.8, the flaperon control deflections needed to achieve the performance are high. In the present design, to generate 1 deg of pitch attitude change requires close to 6 deg of flaperon deflection. Thus, with the full control deflection of 25 deg of flaperon, only 4 deg of pitch pointing can be achieved. The present design has a decoupling ratio of $\gamma/\theta = 0.035$ (3.5%). By relaxing the decoupling requirement by appropriately modifying the objective function (10.20), such as defining a higher decoupling γ/θ target ratio, the flaperon deflection can be reduced. Again, as in the IMF handling qualities design, the lift control effectiveness of the direct lift surface dictates the achievable pitch pointing range capability.

10.10 Summary

In this chapter, the pitch axis controller design of an aircraft is studied. If the aircraft has only one control surface in the pitch axis, eigenstructure synthesis is not possible and the design reduces to a simple eigenvalue assignment problem. Since the aircraft dynamic performance is dictated by the short-period mode, the natural frequency and damping of the mode can be modified by feedback to improve the response characteristics. The Gibson's time and frequency domain criteria are usually satisfied by proper design of the command filter. If the aircraft is equipped with more than one control surface in the pitch axis, some control can be exercised in shaping the eigenvectors. Again since the main interest is in shaping the short-period mode response, the complex conjugate pair of eigenvectors associated with the short-period mode needs to be synthesised. As discussed in Chapter 9, the direct synthesis of real mode and its associated eigenvector, for a specified requirement, is relatively easy, and synthesis of complex eigenvectors is best achieved by formulating an appropriate optimisation problem. In this chapter, it is shown that the multivariable pitch axis design can be conveniently formulated as a model following control problem since the design objectives can be easily translated into an appropriate dynamic model. Examples of both IMF and EMF control design formulations are included to highlight the utility of eigenstructure assignment algorithms. Thus, for the pitch axis design, using multiple inputs, optimisation techniques play a major role both in constructing ideal models and in the subsequent design of the controllers to match the performance of these ideal models.

References

1. ISHMAEL, S.D., and MCMONAGLE, D.R.: 'AFTI/F-16 Flight Test Results and Lessons', 1983, NASA TM 84920
2. ASTROFF, A.J., HOFFLER, K.D., and PROFFITT, M.S.: 'High-Alpha Research Vehicle (HARV) Longitudinal Controller: Design, Analyses and Simulation Results', 1994, NASA TP 3446
3. SOBEL, K.M., and SHAPIRO, E.Y.: 'A design methodology for pitch pointing flight control systems', *J. Guidance Control Dyn.*, 1985, **8**, (2), pp. 181–187
4. CRASSIDIS, A.L., and MARKOFSKI, A.R.: 'VISTA pole and zero placement using eigenstructure assignment', *Proceedings of the AIAA Atmospheric Flight Mechanics Conference*, 2002, California, AIAA-2002-4698
5. HARRIS, J.J., and BLACK, G.T.: 'F-22 control law development and flying qualities', *Proceedings of the AIAA Atmospheric Flight Mechanics Conference*, 1996, California, AIAA-96-3379-CP, pp. 155–168
6. GIBSON, J.C.: 'Handling qualities for unstable aircraft', *Proceedings of the ICAS Congress*, 1986, London, pp. 433–445, paper ICAS-86-5.3.4
7. MCRUER, D., JOHNSTON, D., and MYERS, T.: 'A perspective on superaugmented flight control: advantages and problems', *J. Guidance Control Dyn.*, 1986, **9**, (5), pp. 530–540
8. CHALK., C.R.: 'Flying qualities of pitch rate command/attitude hold control systems for landing', *J. Guidance Control Dyn.*, 1986, **9**, (5), pp. 541–545
9. ANONYMOUS: 'Flying Qualities of Piloted Aircraft', 1997, MIL-HDBK-1797

10. PADFIELD, G.D.: 'Capturing requirements for tilt rotor HQ – case studies in virtual engineering', *Aeronautical J.*, 2008, **112**, (1134), pp. 433–448
11. MITCHELL, D.G., HOH, R.H., APONSO, B.L., and CLIDE, D.H.: 'Proposed Incorporation of Mission-Oriented Flying Qualities into MIL-STD-1797A', 1994, DTIC Accession No. ADA 294211

Chapter 11

Rotorcraft handling qualities design

11.1 Introduction

Design of full authority helicopter stability augmentation systems (SAS) for modern helicopters is becoming a reality. Several research helicopters, with fly-by-wire capability, are being operated, primarily to evaluate new control algorithms for improving the handling characteristics of helicopters [1–3]. An operational helicopter flight control system using multivariable control techniques has been successfully tested in flight [4]. Unlike an aircraft, from a control point of view, the helicopter is a complex vehicle exhibiting substantial pitch, roll and yaw inter-axis coupling. The presence of the rotor adds additional coupling dynamics interacting with the rigid body motion. The modelling of the helicopter for control design is thus associated with significant model uncertainties. Thus, the robustness of the controller to modelling uncertainty becomes a prime performance requirement. However, the major design challenge is the reduction of inter-axis coupling.

In aircraft control, since the coupling between the longitudinal and lateral–directional dynamics can be neglected, it is possible to derive simple controller structures for each axis based on flight mechanics analysis as illustrated in Chapters 9 and 10. In addition, availability of normal and lateral acceleration as primary feedback sensors aids in synthesising feedback signals that approximate angle of attack and sideslip signals. In case of helicopters since rigid body acceleration signals are corrupted by rotor and vibration dynamics, their use is not generally preferred. Thus, multivariable control techniques have been preferred for full authority fly-by-wire control law design. Designs based on H_∞ optimisation and non-linear dynamic inversion have also been assessed in experimental flight test programs [3,4]. The dynamic inversion is a form of explicit model following (EMF) control concept. The primary advantage claimed in this non-linear dynamic inversion method is the avoidance of the classical gain scheduled controller design methodology. The design approach is to carry the aircraft model on board the vehicle computer and compute the non-linear control laws in real time [5]. The H_∞ optimisation uses a two-degree-of-freedom controller structure with feedback and feedforward elements. The resulting design can be mapped to a classical state feedback (SF) and full-order observer controller structure [6].

Analytical studies applying eigenstructure assignment theory to design helicopter SAS have also been reported [7,8, and references therein]. In Reference 7, using all the four helicopter controls and full state variable feedback, it is demonstrated how a four-input, four-output decoupled system can be designed to meet the handling qualities specifications. A forward path precompensator is used for response shaping. In Reference 8, a full SF controller with four controls is designed for stability augmentation and mode decoupling. An additional outer-loop controller achieves the *response type* requirement. To implement these full SF controllers, a full-order observer is recommended to reconstruct the states using available helicopter sensors.

All of the design methods discussed above result in high dynamic-order filter (eighth order or higher) to be implemented in real time. However, to realise controllers that can be implemented in *production class* flying vehicles, the design process should yield solutions with minimal controller complexity. A design philosophy addressing this objective is proposed in Reference 9. This approach will be the basis for the discussions to follow in this chapter, and further improvements and extensions to this concept will be highlighted. The study illustrates a *design process* for evolving practical *low-order stable compensator-based* robust feedback controllers for improving the helicopter handling qualities in the pitch, roll and yaw axes.

11.2 Helicopter handling qualities requirements

The helicopter handling qualities specifications are defined in the document ADS-33E-PRF [10]. The relevant specifications applicable to the design study in this chapter are briefly summarised in Chapter 8. The pilot has to operate the helicopter in a wide range of mission task elements (MTE) under varied usable cue environment (UCE). The gist of the handling qualities specifications is directed at providing a vehicle where the pilot can accomplish these tasks efficiently. It is also clear that without substantial augmentation of the basic helicopter dynamics, these handling qualities goals cannot be attained. The handling qualities design process can be broadly classified into (i) a *feedback* controller design to reduce the severe inter-axis coupling and (ii) a *command path* controller that shapes the response to pilot control inputs. This response shaping effectively reduces the pilot workload when the helicopter is operated in a degraded UCE.

11.3 BO-105 helicopter model

The BO-105 helicopter has been chosen for the study in this chapter. The state variable model is described in Appendix C. Two candidate models designated as AFD and DLR are used for analysis. The helicopter rigid body model has state variables p, q, r, u, v, w, θ and φ. The control inputs are δ_c (collective), δ_a (longitudinal cyclic), δ_e (lateral cyclic) and δ_p (pedal). The eigenvalues of the rigid and rotor-augmented models are given in Tables 11.1 and 11.2, respectively. The phugoid mode is unstable (level 2), resulting in high pilot workload to maintain a stable platform. Augmenting the rotor model results in the coalescing of the pitch and roll modes with those of the corresponding flapping modes to form a complex conjugate pair of eigenvalues (except in the pitch mode of DLR model).

Table 11.1 Eigenvalues of BO-105 helicopter (rigid model)

Mode	**Model**	
	AFD	**DLR**
Roll	−8.3227	−8.5639
Pitch #1	−6.0397	−4.3595
Dutch roll	−0.5631 ± j2.5393	−0.3021 ± j2.4569
Phugoid	0.1088 ± j0.2796	0.0561 ± j0.3149
Pitch #2	−0.4921	−0.6456
Spiral	−0.0253	−0.0210

Table 11.2 Eigenvalues of BO-105 helicopter (rigid + rotor)

Mode	**Model**	
	AFD	**DLR**
Roll	−8.3827 ± j8.7153*	−8.5762 ± j8.4359*
Latitudinal flapping		
Longitudinal flapping	−9.2565 ± j2.8061†	−12.1248
Pitch #1		−5.9676
Dutch roll	−0.6748 ± j2.5968	−0.3444 ± j2.4852
Phugoid	0.1089 ± j0.2820	0.0542 ± j0.3178
Pitch #2	−0.4962	−0.6539
Spiral	−0.0253	−0.0210

*Roll mode and flapping mode coalesce.
†Pitch mode and flapping mode coalesce.

Figure 11.1 shows the dispersion in on-axis frequency response characteristics among the two candidate models. Figure 11.2 shows the effect of rotor dynamics on the frequency response characteristics (DLR model). In Figure 11.1, the dominant presence of the phugoid mode in the roll rate response to lateral cyclic input is indicative of the strong longitudinal to lateral cross coupling.

For constructing the helicopter state variable model from flight test data, there are a large number of parameters to be identified using a rigid body state variable model structure, with n-states (n = 8) and m-controls (m = 4). Thus, engineering judgment is needed to fix the model structure and the number of parameters to be identified. The dominant cross axis coupling in the helicopter can generally be identified using both off-axis stability and control derivatives. However, from a robust feedback design point of view, an ideal model structure should not include off-axis control derivatives, especially those that are difficult to identify. This would result in the inter-axis coupling effect being captured through stability derivatives alone. The feedback design has the property to minimise the sensitivity to uncertainty in these stability derivatives. On the other hand, if the cross axis effect is identified through off-axis control derivatives, high control interconnect gains have to be employed to alleviate the cross axes response. Control interconnect, being an open-loop programming scheme, is quite sensitive to modelling uncertainties.

Table 11.3 lists the control moment derivatives for the cyclic and pedal controls for both AFD and DLR models. The data reveal that there is significant dispersion in the estimates of the off-axis derivatives L_{δ_a}, M_{δ_e}, M_{δ_p}, N_{δ_c} and N_{δ_a}. This indicates the difficulty in identifying these derivatives from flight data. It is also clear that the DLR model structure uniformly de-emphasises capturing the cross axis response through off-axis control derivatives. In light of the discussions above, the DLR model is preferred for control design study in this chapter.

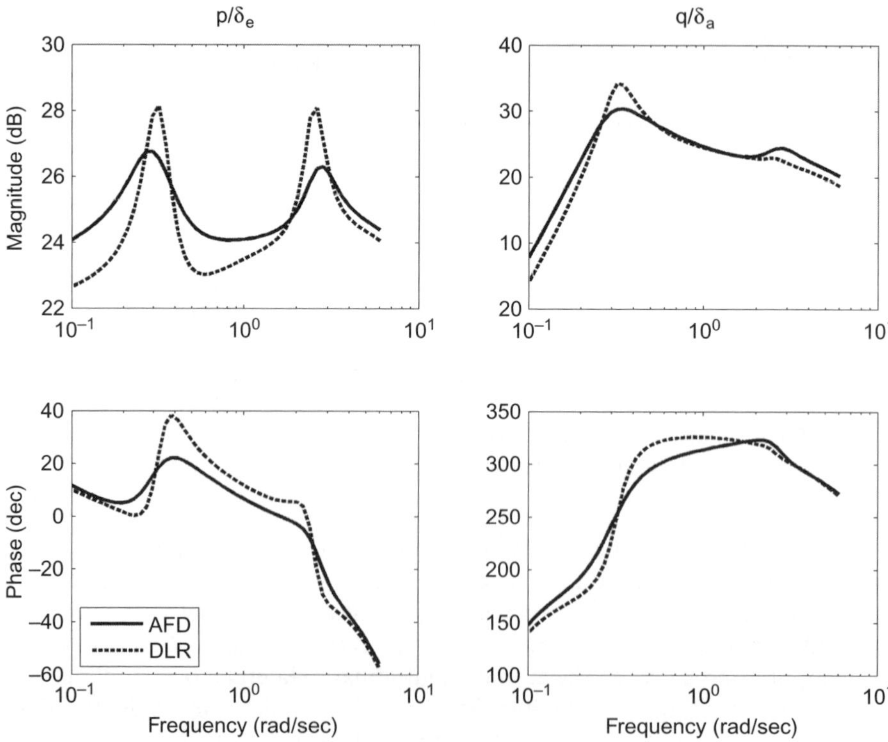

Figure 11.1 Frequency response of rigid body models

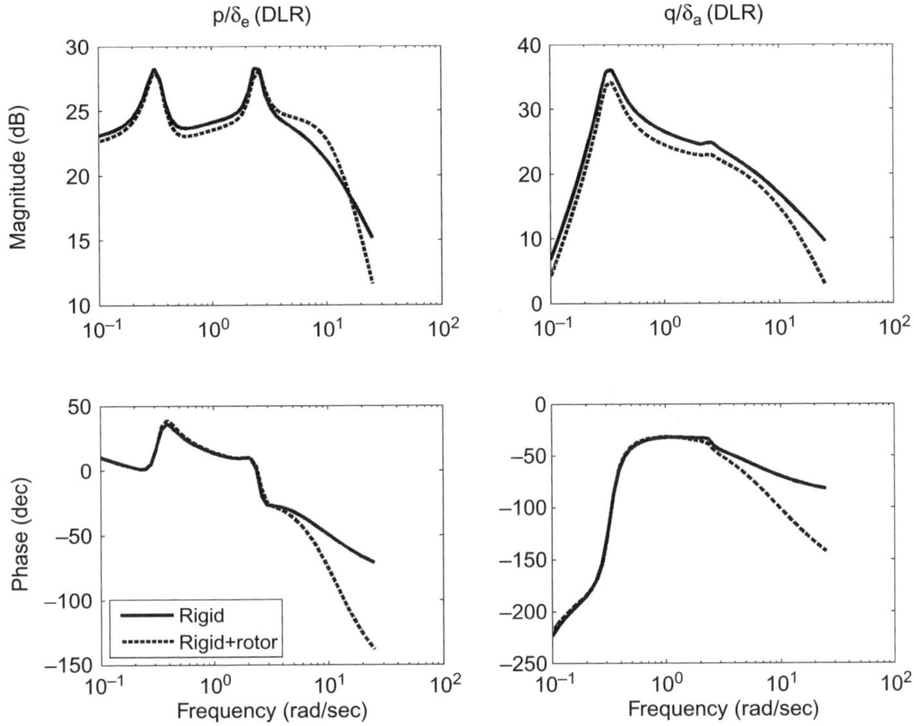

Figure 11.2 Effect of rotor dynamics on frequency response

Table 11.3 Control moment derivatives (deg/sec²)

Model	L_{δ_c}	L_{δ_a}	L_{δ_e}	L_{δ_p}
(a) Rolling moment derivatives				
AFD	16.61	59.75	146.51	−7.74
DLR	9.16	19.64	151.43	−8.02
Model	M_{δ_c}	M_{δ_a}	M_{δ_e}	M_{δ_p}
(b) Pitching moment derivatives				
AFD	20.91	80.21	0	3.72
DLR	16.33	76.12	−7.37	0
Model	N_{δ_c}	N_{δ_a}	N_{δ_e}	N_{δ_p}
(c) Yawing moment derivatives				
AFD	−14.6	−61.39	27.01	16.33
DLR	0	0	21.28	14.03

11.4 Feedback controller design

As indicated earlier, the helicopter controller has two components: (i) a feedback controller to reduce inter-axis coupling and (ii) a command path controller to alleviate pilot workload under degraded UCE. The feedback controller design process follows closely the method outlined in Chapter 9. The command path controller uses the tunable EMF algorithm described in Chapter 7. Figure 11.14 gives a schematic of the command path and feedback controllers. The schematic identifies the various control gain matrix elements that will be designed using the methods discussed in subsequent sections.

11.4.1 Feedback sensors and control law structure

In case of the helicopter, only the inertial rate (p, q and r) and attitude (θ and φ) sensors are available for feedback. As in the case of the aircraft, the attitude sensors cannot be used for high-gain feedback controllers due to sensor redundancy issues. The normal and lateral acceleration sensors are corrupted by their response to helicopter rotor dynamics and structural vibrations and thus are not preferred. The flow angle sensors (α, β) have similar limitations and are not used for control. With only the rate sensors available for feedback, a dynamic feedback controller structure results. However, attitude sensors can still be used in the low-bandwidth autopilot control loops to derive attitude command attitude hold (ACAH) functions. In Chapter 9, for the low-order lateral–directional aircraft control, a simple dynamic controller structure, using sideslip and sideslip rate estimates, was used to design a feedback system to match the SF solution. In the present case, such a simple approach is not possible because of the highly coupled longitudinal and lateral dynamics. Thus, a more general, functional observer based, dynamic controller structure discussed in Chapter 6 has to be used to match a benchmark SF solution.

11.4.2 Pitch–roll cross coupling

The BO-105 is a high-agility helicopter exhibiting severe pitch to roll inter-axis coupling. This is due to the stiff rotor systems and large hinge offsets needed for high agility. The main design objective of feedback is to minimise this coupling. The pitch–roll coupling arises from two sources as shown in the following simplified dynamics of the helicopter:

$$\begin{aligned} \dot{q} &= M_q q + M_p p + M_{\delta_a} \delta_a + M_{\delta_e} \delta_e \\ \dot{p} &= L_q q + L_p p + L_{\delta_a} \delta_a + L_{\delta_e} \delta_e \end{aligned} \tag{11.1}$$

The off-axis stability derivatives (M_p, L_q) and off-axis control derivatives (L_{δ_a}, M_{δ_e}) contribute to the cross coupling in the bandwidth (ω_{BW}) to neutral stability frequency (ω_{NS}) range. The eigenvector modification using eigenstructure assignment will alleviate the stability derivative cross coupling. Control interconnects have to be used to alleviate the off-axis control derivative coupling.

11.4.3 State feedback control law design

In the present study three control inputs δ_a (longitudinal cyclic), δ_e (lateral cyclic) and δ_p (pedal) will be used to derive a *benchmark* SF law using the rigid body states (p, q, r, u, v, w, θ and φ). The 8th-order rigid body model will be used as the design model, and the 14th-order truth model will be used for performance optimisation. The filters associated with computation and sensor filter lags (Appendix C) will not be included in the design process. The effect of these additional lags has to be compensated by lead filters in control channels as illustrated in the aircraft design example of Chapter 9. For eigenstructure synthesis a state variable design model results with n = 8 and m = 3. From Theorem 2.1, all rigid body modes can be arbitrarily assigned, and two eigenvector entries, in each mode, can be arbitrarily selected. Thus, 16 eigenvector parameters are available for eigenstructure modification. The eigenvalues are chosen (in the assignable range) to meet level 1 dynamic stability requirements [10] and also to achieve required eigenvector decoupling characteristics as discussed in Chapter 9.

The design process involves the following steps:

Step 1. Control interconnect selection.

For reducing the inter-axis coupling, four interconnect gains: (i) g_{EA} (δ_E to δ_A), (ii) g_{AE} (δ_A to δ_E), (iii) g_{AP} (δ_A to δ_P) and (iv) g_{EP} (δ_E to δ_P) are used. This modifies the original input distribution matrix (B) as follows:

$$G_X = \begin{bmatrix} 1 & 0 & 0 & 0 \\ 0 & 1 & g_{EA} & 0 \\ 0 & g_{AE} & 1 & 0 \\ 0 & g_{AP} & g_{EP} & 1 \end{bmatrix}, \qquad B_X = BG_X \tag{11.2}$$

The input distribution matrix B_X is used for SF and observer designs.

Step 2. Real mode eigenvector synthesis.

As discussed in the aircraft example of Chapter 9, using the mode-coupling matrix (2.21), the real mode and its free eigenvector elements are chosen to have dominant response in their expected state variables while decoupled from the other variables. In particular, the longitudinal (lateral) modes need to be decoupled from the lateral (longitudinal) state variables. For example, using the special eigenvector structure synthesis of section 3.4, the pitch mode #2 has an excellent vertical speed (w) state variable dominated eigenstructure for $\lambda = -0.815$ (an eigenvalue of matrix F). Using a similar approach, pitch #1, roll and spiral modes and their eigenvectors are synthesised to get the desired mode shapes.

Step 3. Complex modes and control interconnect optimisation.

The control interconnect gain matrix G_X together with phugoid and Dutch roll eigenvectors is optimised to minimise the following objective function:

$$J = \min_{\omega_{min} \leq \omega \leq \omega_{max}} \left[\left\{ \text{avg} \left| \frac{q(s)}{p(s)} \right|^2_{s=j\omega} + \text{avg} \left| \frac{p(s)}{q(s)} \right|^2_{s=j\omega} + \text{avg} \left| \frac{r(s)}{q(s)} \right|^2_{s=j\omega} \right\} \right] \tag{11.3}$$

For performance assessment, the ADS [10] require that each of the average functions in (11.3) be computed over the corresponding bandwidth (ω_{BW}) to neutral stability (ω_{NS}) frequency range. However, simulation studies revealed that, for optimising the performance, minimising the objective function in (11.3) in the range $\omega_{min} = 0.1$ rad/sec and $\omega_{max} = 15$ rad/sec (rotor coupled rigid body model frequency range) resulted in improved solutions. This ensures that the phugoid and Dutch roll mode coupling in pitch and roll rates are also minimised.

Additional constraints such as gain magnitude constraints (Table 9.5), robustness boundary constraints (section 8.6 (8.11)), etc. are added to iteratively arrive at an acceptable solution meeting control system design requirements.

11.4.4 Functional observer design

From the observer theory of Chapter 6, for an n-state, m-input and r-output *strictly proper* system

$$\begin{aligned} \dot{\mathbf{x}} &= \mathbf{A}\mathbf{x} + \mathbf{B}\mathbf{u} \\ \mathbf{y} &= \mathbf{C}\mathbf{x} \end{aligned} \tag{11.4}$$

a pth-order *known input* observer of the form

$$\dot{\delta} = F\delta + \mathbf{G}\mathbf{u} + \mathbf{H}\mathbf{y} \tag{11.5}$$

where $\delta \in R^p$, F is a diagonal matrix of p observer eigenvalues, and G and H are matrices of compatible dimensions, exists such that a state variable feedback of the form

$$u = \mathbf{K}\mathbf{x} \tag{11.6}$$

is estimated using the observer-augmented system

$$\begin{bmatrix} \dot{\mathbf{x}} \\ \dot{\delta} \end{bmatrix} = \begin{bmatrix} A & 0 \\ HC & F \end{bmatrix} \begin{bmatrix} \mathbf{x} \\ \delta \end{bmatrix} + \begin{bmatrix} B \\ G \end{bmatrix} \mathbf{u_a} \tag{11.7}$$

and an SF law of the form

$$\mathbf{u_a} = [NC \quad M] \begin{bmatrix} \mathbf{x} \\ \delta \end{bmatrix} \tag{11.8}$$

The order of the observer (p) has a lower bound given by

$$p = \min\{(n - r), \max(m, n - 2r + 1)\} \tag{11.9}$$

From (11.9), it is seen that a third-order observer, using inertial rate feedback sensors, is sufficient to estimate the SF law. However, the observer design freedom is restricted to only selecting observer eigenvalues (Chapter 6). In order to increase this design freedom, for performance optimisation, the observer order is increased to $p = 4$. With this choice, four eigenvalues and ten additional design parameters are available for design optimisation.

In classical state estimation theory, the main concern is to minimise the observer initialisation errors rapidly. Towards this goal, it is general practice to have high-bandwidth observer roots. This results in large transients during initialisation. However, when the observer is used in place of a state variable feedback, the observer initialisation is not a major concern. The observer can be looked upon as a dynamic compensator. Thus, the observer roots are optimally located to meet other feedback system performance criteria. Some of the candidate objectives for optimising the observer parameters are:

1. Minimising the inter-axis coupling of the observer-augmented system using (11.3).
2. Modal robustness of the observer-augmented system: From (11.7) and (11.8), the observer-augmented closed-loop system matrix is given by

$$A_c = \begin{bmatrix} A + BNC & BM \\ HC + GNC & F + GM \end{bmatrix} \tag{11.10}$$

   The modal robustness objective function is given by

$$J_2 = \min\{\kappa_2(X)\} \tag{11.11}$$

   where $\kappa_2(X)$ is the condition number of modal matrix X of A_c.
3. Minimising observer sensitivity to process noise disturbance: The known input observer constructs the SF control law estimate using both the system input and sensor output signals. The input channel signals inject system noise disturbance to the observer, through the gain matrix G. Minimising the magnitudes of G matrix gain elements will reduce this effect.
4. Location of observer roots: The observer roots are selected to ensure that they do not coincide with the other helicopter modes resulting in modal ill-conditioning. The multivariable gain locus helps in optimally locating the observer roots to achieve good closed-loop robustness characteristics.

11.5 Feedback controller performance analysis

The performance of the SF and observer feedback (OBS) designs is assessed using the truth model. The control laws are optimised based on the DLR reference models (design and truth). Tables 11.4a and b summarise the control laws.

Table 11.4a State feedback control law

Mode	Eigenvalue	Mode	Eigenvalue
(A) Specified rigid body model eigenvalues			
Pitch #1	−6.0	Roll	−8.0
Pitch #2	−0.80	Dutch roll	−1.5 ± j2.6
Phugoid	−0.1 ± j0.1	Spiral	−0.05

	p	q	r	w	u	v	θ	φ
(B) State feedback gains (deg/deg)								
δ_A	0.0059	−0.0340	0.0079	−0.0377	−0.0208	−0.0793	−0.0121	−0.0010
δ_E	0.0058	−0.0222	−0.0165	−0.0471	0.0388	0.1767	−0.0016	0.0034
δ_P	0.0588	−0.0153	−0.0984	−0.0373	0.0151	0.0730	−0.0106	0.0548

g_{EA}	g_{AE}	g_{AP}	g_{EP}
(C) Control interconnect gains (deg/deg) (11.2)			
0.1031	−0.1590	0.0889	0.5416

Table 11.4b Observer-based control law (OBS) (F = diagonal (−0.2, −1.0, −3.0, −5.0))

G			H		
0.5969	−1.8430	0.1487	0.0442	−0.0158	0.0393
−1.4639	−0.6012	1.0605	0.0613	0.0230	0.2882
−0.8846	0.1825	0.4778	0.0813	0.0537	0.2041
0.1133	0.4670	0.0290	−0.0827	0.0570	−0.0387

M				N		
−1.0258	1.6809	−0.9596	−0.5753	0.0175	−0.0035	−0.0674
−2.4521	−0.4235	0.1156	−0.1505	−0.0320	−0.0089	0.0156
−0.4971	−1.6421	0.9949	−0.5459	0.0311	−0.0280	−0.0179

11.5.1 Eigenvector decoupling characteristics

The BO-105 helicopter exhibits significant coupled response in roll and yaw rates for a longitudinal cyclic command input. This longitudinal to lateral coupling reduction is the major design objective. Many metrics can be used to quantitatively measure this coupling. Initially we will use the modal matrix norm to measure the coupling. Let the modes be ordered with longitudinal followed by lateral modes as

$$\Lambda = \mathrm{Diag}[\lambda_{P\#1}, \lambda_{P\#2}, \lambda_{PH}, \lambda_R, \lambda_{DR}, \lambda_S] \tag{11.12}$$

and the state variable eigenvector components be ordered as [w, q, u, θ, p, r, v, φ].

Then the modal matrix X (matrix of eigenvectors) has the structure

$$X = \left[\begin{array}{c|c} X_{Lon} & X_{Lat/Lon} \\ \hline X_{Lon/Lat} & X_{Lat} \end{array}\right] \tag{11.13}$$

The modal cross coupling between the longitudinal (lateral) and lateral (longitudinal) dynamics is determined by $X_{Lon/Lat}$ ($X_{Lat/Lon}$) submatrix. If these submatrices were to be null, the longitudinal and lateral dynamics are fully decoupled. Relative coupling can be measured using the norm of the submatrices. Table 11.5 illustrates the reduction in mode coupling by feedback using the norm measure. The table also shows how strong the longitudinal to lateral coupling ($X_{Lon/Lat}$) is for the BO-105 helicopter. Figure 11.3a and b shows the mode decoupling achieved by state variable feedback (*design* model). It is seen that feedback has substantially decoupled the longitudinal modes (Figure 11.3a) from the response of the lateral state variables. In particular, feedback has achieved significant decoupling of the bank angle state variable (φ) from pitch mode #2. It is also seen that the pitch mode #2 ($\lambda = -0.8$) is fully decoupled from all state variables except the vertical velocity (w).

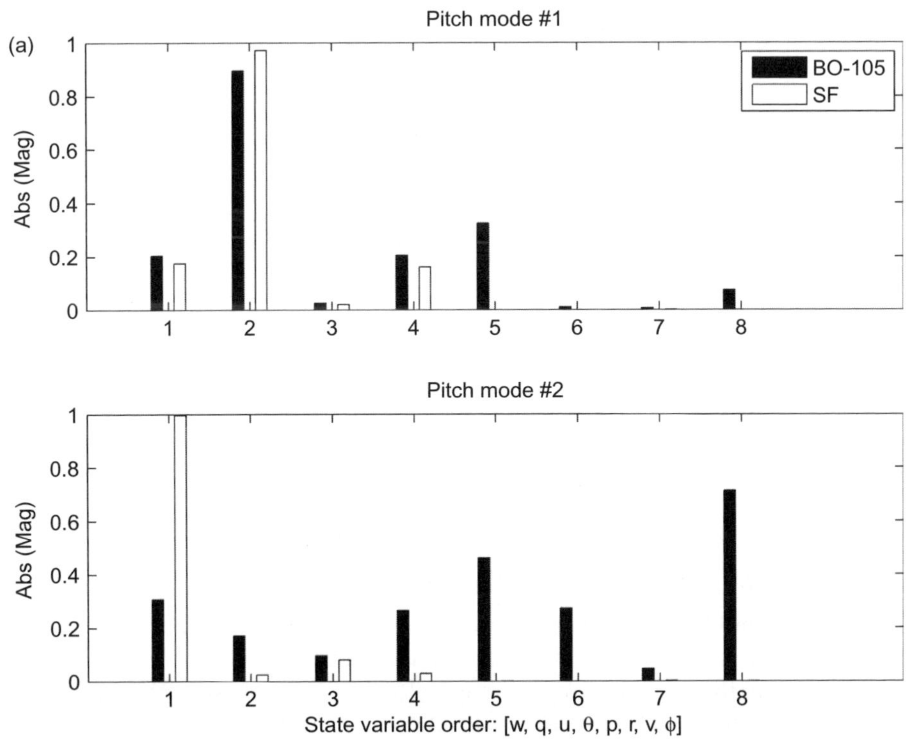

Figure 11.3 Eigenvector coupling characteristics (a) Pitch mode coupling (real modes); (b) Pitch mode coupling (complex mode)

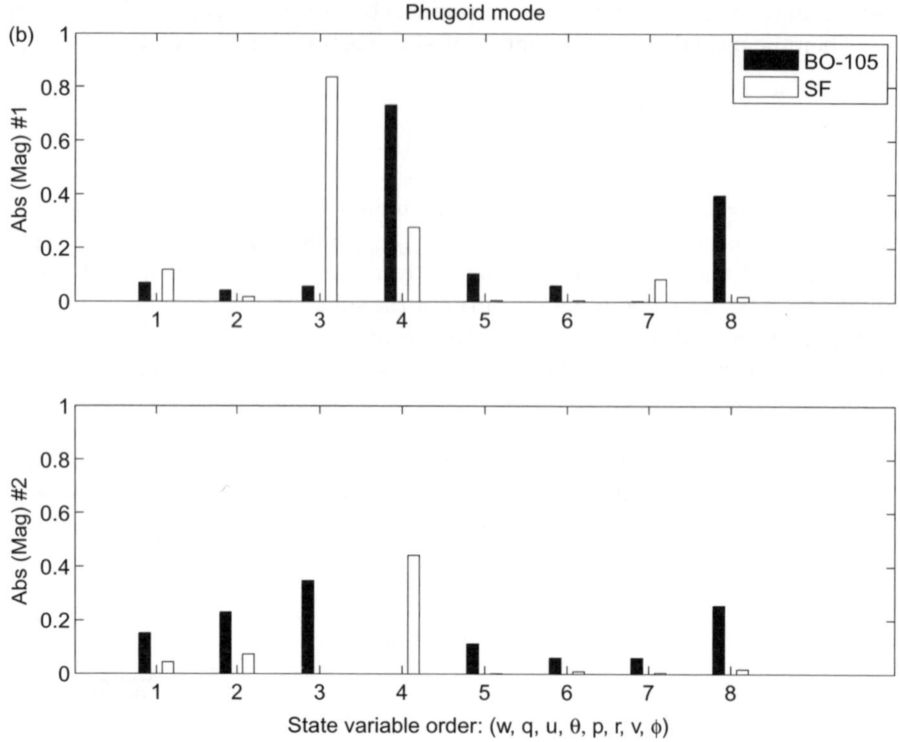

Figure 11.3 Concluded

Table 11.5 Modal matrix norm

Configuration	$\|X_{Lon}\|$	$\|X_{Lon/Lat}\|$	$\|X_{Lat/Lon}\|$	$\|X_{Lat}\|$
BO-105	1.0055	1.0065	0.1947	1.1072
SF	1.0799	0.0867	0.1405	1.1344

11.5.2 Dynamic stability and bandwidth

Figure 11.4a and b depicts the dynamic stability characteristics. For the BO-105, phugoid mode is unstable (level 1/2), and Dutch roll mode is level 3. The control laws substantially augment the damping to level 1. The phugoid mode damping of $\zeta = 0.35$ is required for divided attention operation [10]. Figure 11.5 illustrates the bandwidth of the system for target acquisition and tracking task.

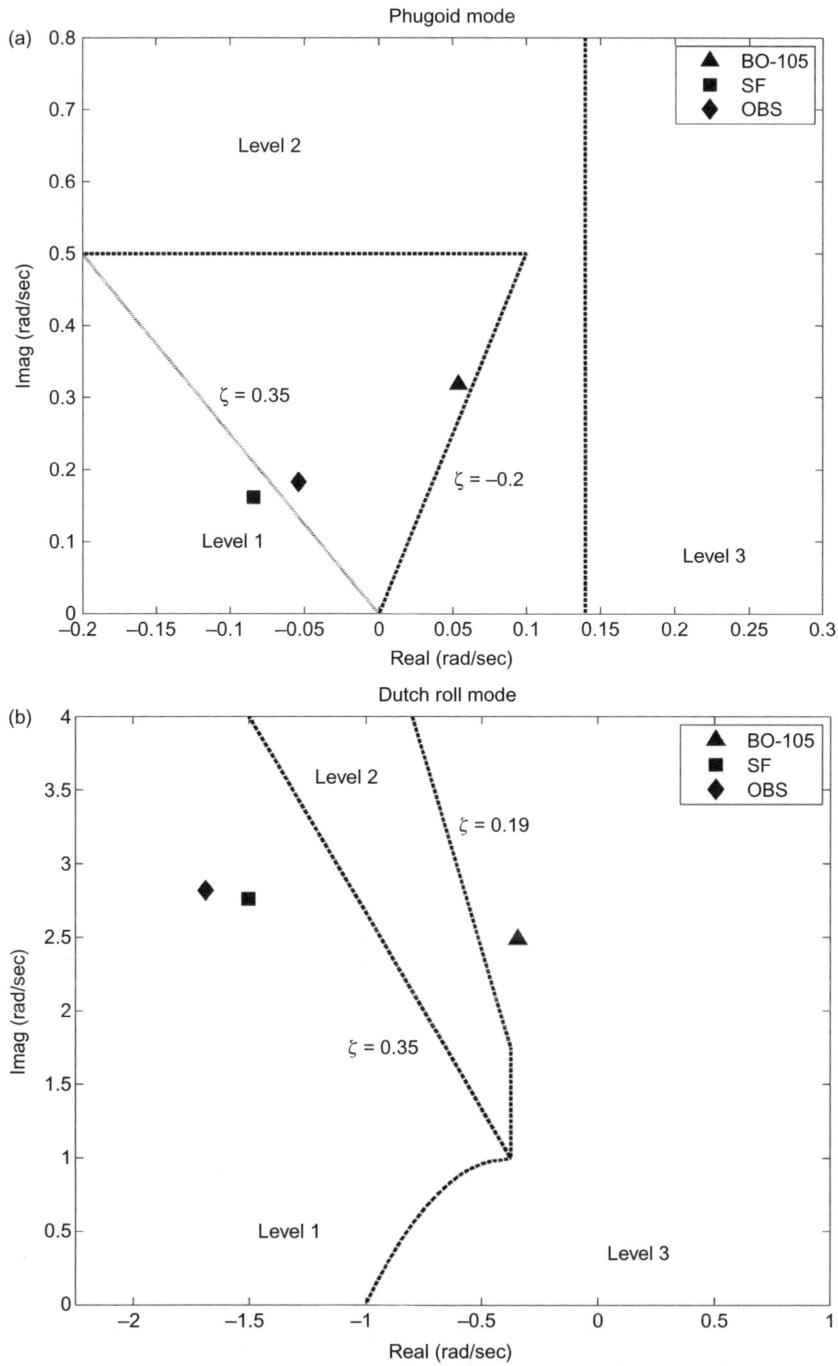

Figure 11.4 Dynamic stability characteristics (a) Pitch dynamic stability (ADS-33E-PRF); (b) Roll dynamic stability (ADS-33E-PRF)

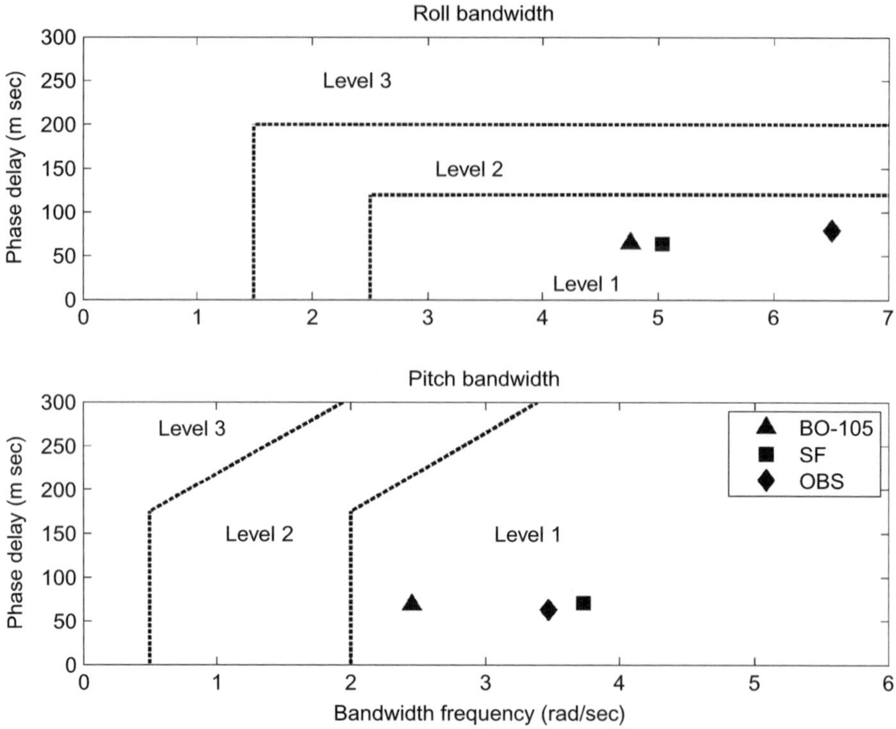

Figure 11.5 Pitch and roll axes bandwidth (ADS-33E-PRF) (small-amplitude target acquisition and tracking)

11.5.3 Pitch–roll–yaw inter-axis coupling

Figures 11.6–11.8 show the q/p, p/q and r/q frequency response ratio over the rigid body frequency range. The BO-105 does exhibit substantial off-axis response in pitch, roll and yaw channels, especially near the Dutch roll frequency. Feedback optimisation, using the objective function (11.2), has resulted in the reduction of the off-axis response throughout the rotor coupled rigid body model frequency range for both SF and OBS designs. The ADS-required frequency range is labelled as BW in the figures. The ADS-defined average inter-axis coupling cross plot is shown in Figure 11.9. Both feedback designs achieve level 1 performance.

11.5.4 Stability margins

The loop transfer functions for the OBS control law are shown in Figure 11.10. The δ_A channel is close to the exclusion boundary near the phugoid frequency. Improvement of margin by increasing the phugoid natural frequency results in higher contamination of the phugoid mode in the pitch rate response. A compromise solution is sought. The δ_P channel has excessive stability margin since in

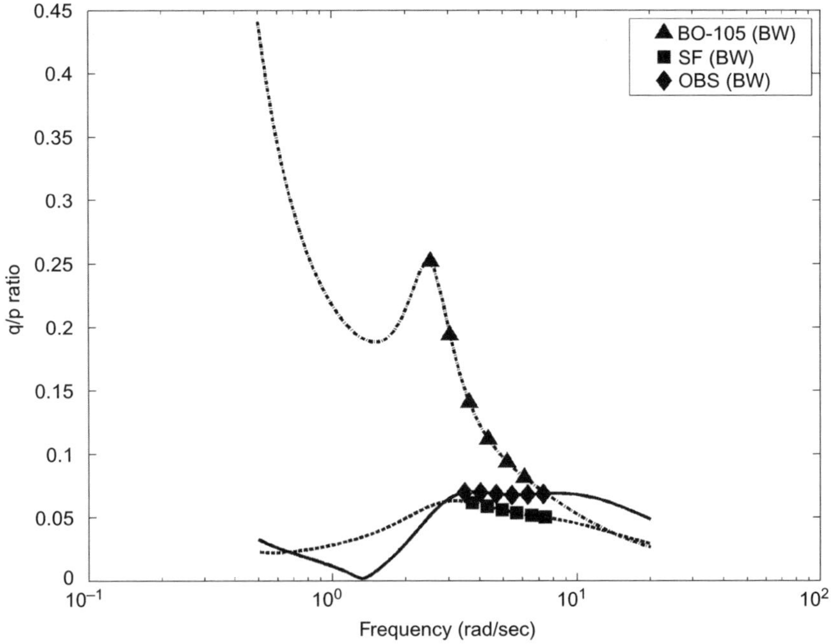

Figure 11.6 Pitch–roll (q/p) coupling due to roll input (δ_E)

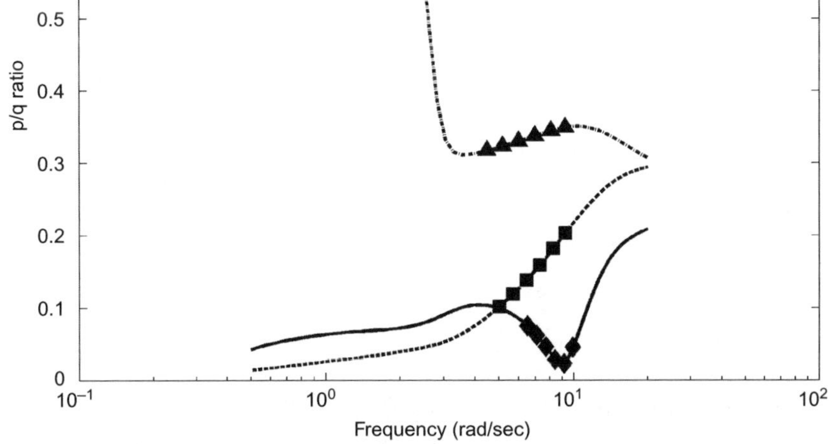

Figure 11.7 Roll–pitch (p/q) coupling due to pitch input (δ_A)

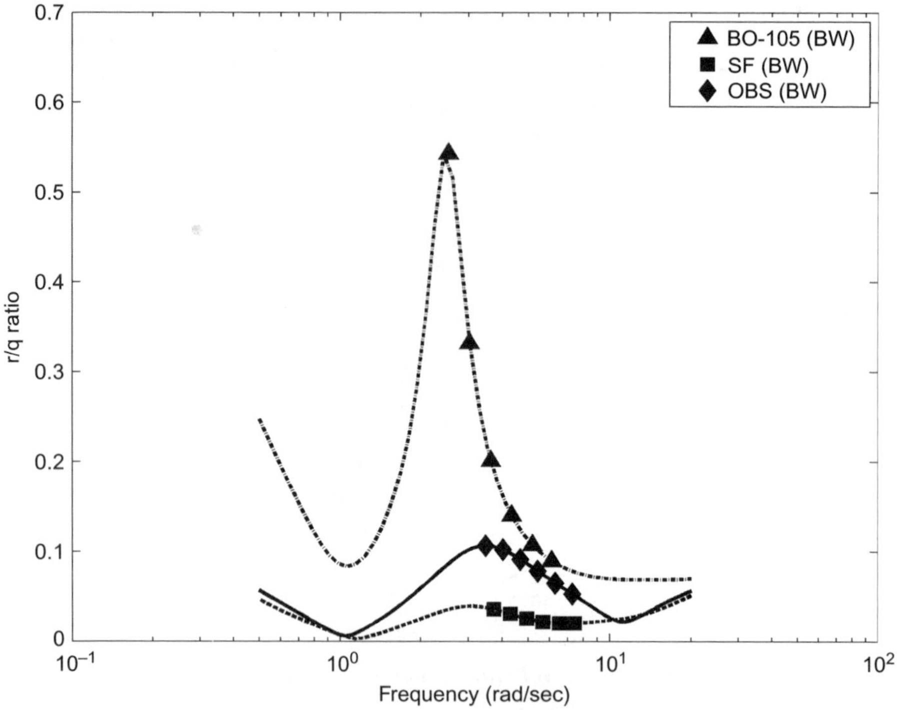

Figure 11.8 Yaw–pitch (r/q) coupling due to pitch input (δ_A)

this channel, the phase roll-off due to the flapping modes is absent. The observer also acts as a lead filter in this control channel. Design modification is possible by forcing the loop transfer function to be closer to the exclusion boundary by adding the objective function defined in Chapter 8 (section 8.6.1 (8.11)) to the performance index. System disturbance rejection properties improve with the loop transfer function closer to the exclusion boundary (higher feedback gains). It should be remembered however that control actuator rate demand also increases with increase in feedback gains.

The addition of computational delay and sensor filter lag in each control channel ($T_2(s)$, Appendix C) does not affect the stability margin of δ_E and δ_P channels since adequate margins exist. However, for the δ_E channel, there is a reduction in gain margin and the loop transfer function crosses the exclusion boundary. A lead filter $T_p(s) = \dfrac{2.5(s+9)}{s+30}$ in the δ_E channel restores the gain margin. The filter adds a high frequency gain of +8 dB. Figure 11.11 illustrates the compensation scheme.

The single-loop gain and phase margins are summarised for the control laws in Table 11.6. It is seen that the margins near the phugoid mode are low in all cases. The margins near the crossover frequency are adequate.

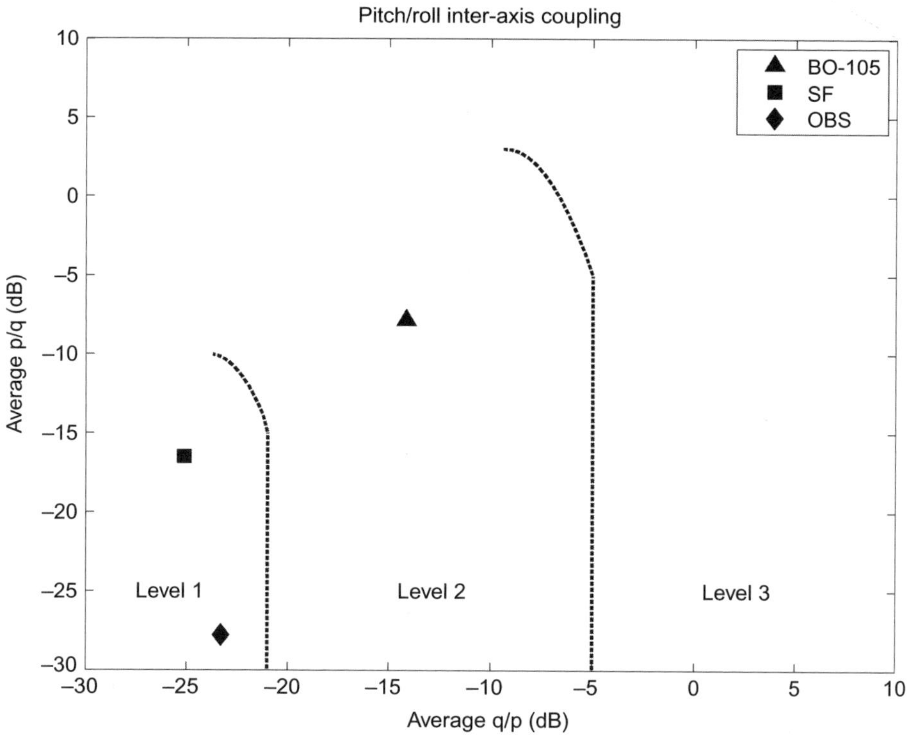

Figure 11.9 Average pitch/roll inter-axis coupling (ADS-33E-PRF)

Table 11.6 Single-loop stability margins

Control law	δ_A		δ_E		δ_P	
	GM	**Frequency***	**GM**	**Frequency**	**GM**	**Frequency**
(a) Gain margin (GM in decibels)						
SF	6.44	0.0	−12.32	0.0	−10.15	0.0
OBS	−7.55	0.28	7.31	10.95	−9.51	0.0

*Frequency in radians per second.

Control law	δ_A		δ_E		δ_P	
	PM	**Frequency**	**PM**	**Frequency**	**PM**	**Frequency**
(b) Phase margin (PM in degrees)						
SF	−28.27	0.15	85.65	0.07	93.35	0.09
OBS	−27	0.17	86.12	0.06	92.5	0.10

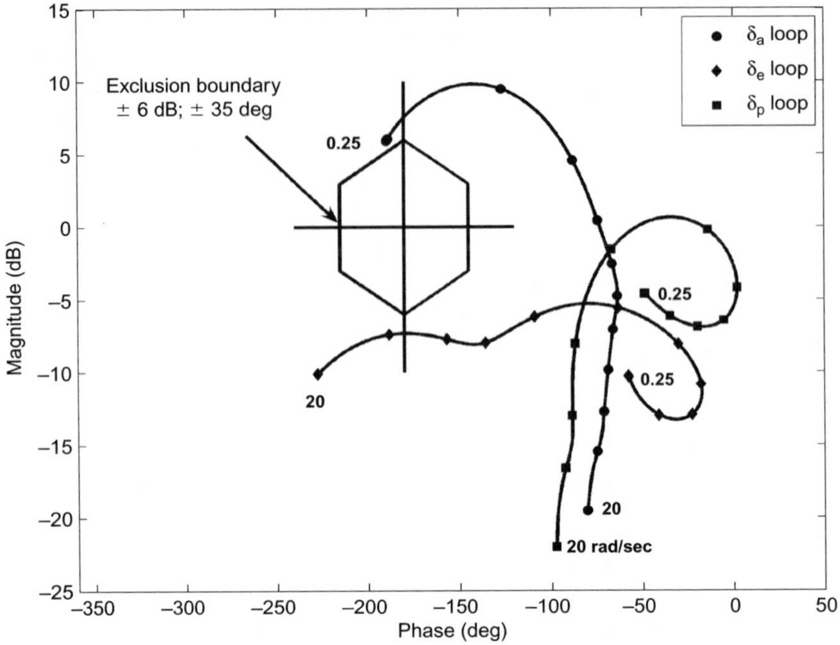

Figure 11.10 Loop transfer functions for stability margin estimates (control law: OBS)

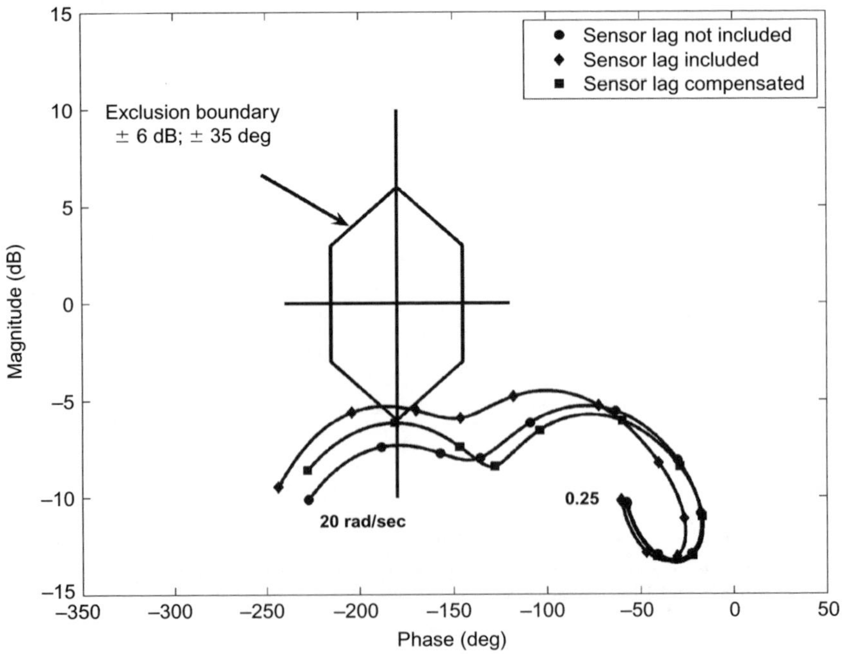

Figure 11.11 Compensation of sensor filter lag in δ_E control channel (control law: OBS)

11.5.5 Multivariable gain locus

The multivariable root locus using scalar gain and phase offsets is useful in assessing the stability margins as discussed in Chapter 8 (section 8.6.2 (8.13)). Figure 11.12a and b shows a multivariable gain locus as a scalar gain (ρ) is varied from 0 to 1. The OBS control law is chosen to demonstrate how to locate the observer roots so that they properly coalesce with the other helicopter modes. The longitudinal flap mode and pitch #1 modes merge to form a conjugate pair. The expanded root locus (Figure 11.12b) shows the migration of the lower frequency roots. As the gain ρ is increased, observer modes #1 and #2 initially form a conjugate pair and eventually split into two real roots at $\rho = 1$. Observer mode #3 and pitch #2 merge to form a conjugate pair. Observer mode #4, Dutch roll mode, phugoid mode and spiral mode root locus branches remain independent.

11.5.6 Time response characteristics

Figure 11.13a demonstrates the pitch–roll decoupled response achieved by the control laws for δ_A and δ_E doublet inputs. The feedback control substantially reduces yaw rate and sideslip response to δ_A input (Figure 11.13b). The control deflections are given in Figure 11.13c. The control excursions are within acceptable limits.

11.6 Command path controller

The command path controller is required to provide response shaping to pilot control inputs in varied UCE and MTE. The requirements are specified in terms of *response types* [10]. A brief definition of some of the *response types* is given in Chapter 8. The feedback controller discussed in section 11.4 falls in the category of *rate response type* as pilot input to δ_A, δ_E and δ_P commands pitch, roll and yaw rates, respectively. In this section, a decoupled attitude command (AC) system design will be illustrated. The tunable CGT control algorithm discussed in Chapter 7 will be used for the controller design. The problem formulation of Chapter 7 is briefly summarised for completeness.

Consider a *strictly proper* system (plant)

$$\begin{aligned} \dot{x} &= A_p x + Bu \\ y &= C_p x \end{aligned} \tag{11.14}$$

and a *strictly proper* command model

$$\begin{aligned} \dot{x}_m &= A_m x_m + B_m u_m \\ y_m &= C_m x_m \end{aligned} \tag{11.15}$$

where $x \in R^n$, $x_m \in R^p$, u, y, $y_m \in R^m$ (the sizes of the plant inputs, plant outputs and model outputs are equal) and $u_m \in R^q$ with $q \le m$. The matrices A, B, C, A_m, B_m and C_m are of compatible dimensions. Then an EMF control law of the form

$$u = Q_c x_m + R_c u_m \tag{11.16}$$

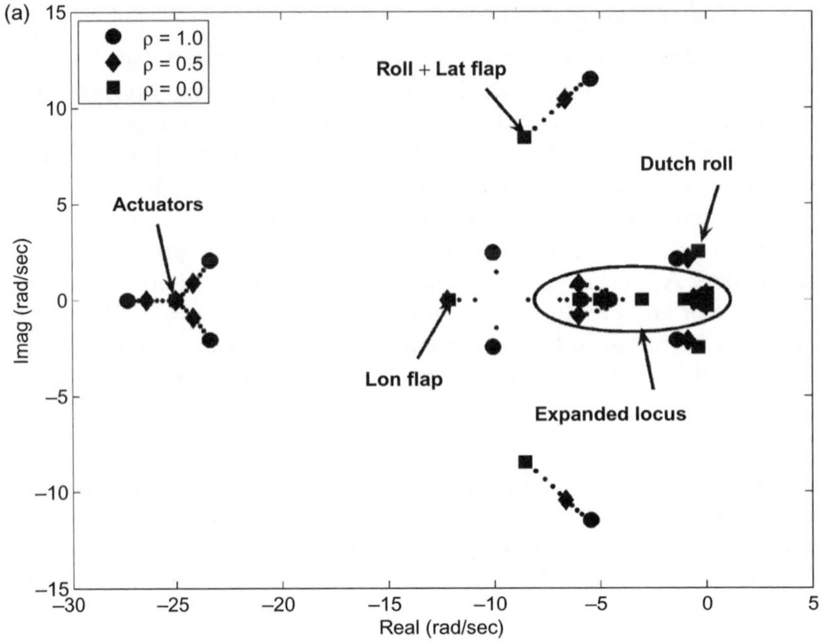

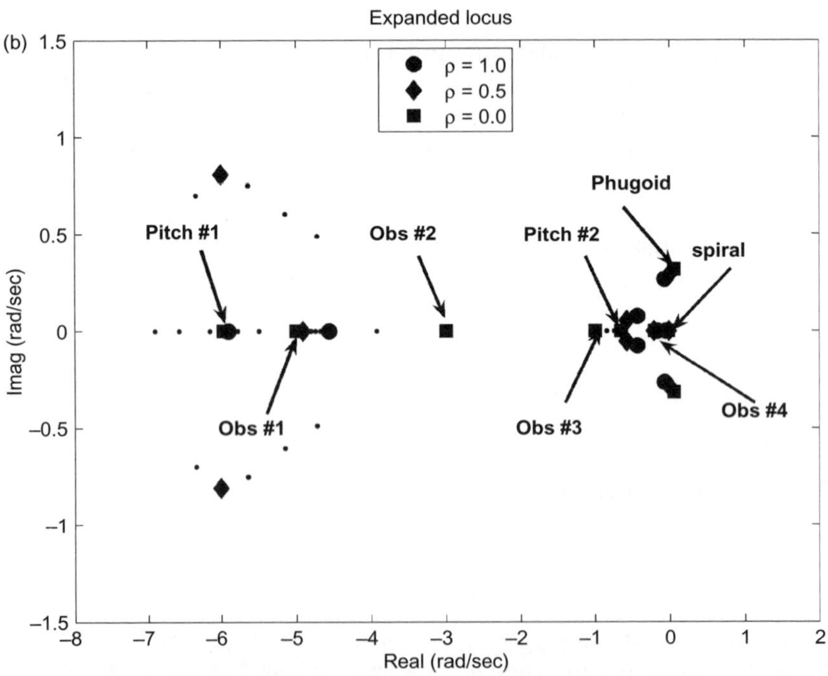

Figure 11.12 Multivariable gain root locus (control law: OBS) (a) Root locus of helicopter and observer roots; (b) Expanded root locus showing observer roots' interaction

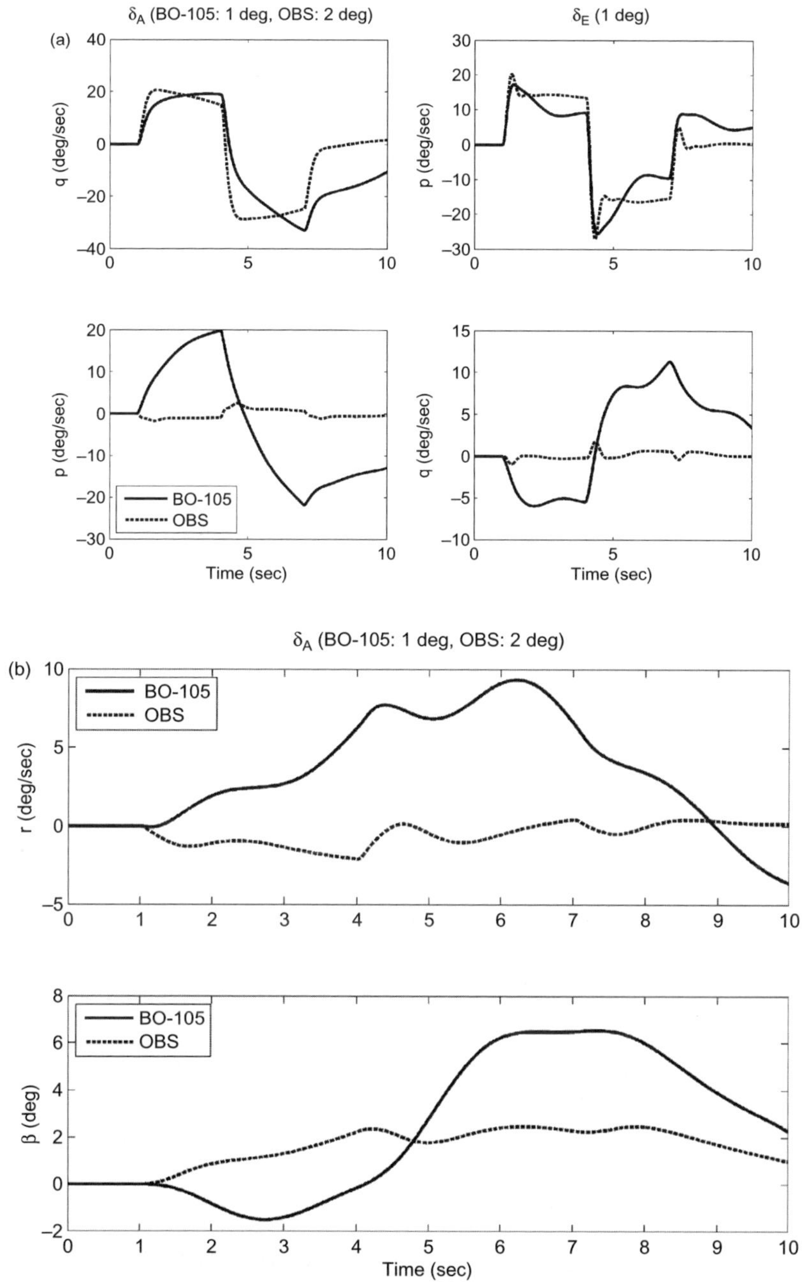

Figure 11.13 Pitch/roll decoupled response to doublet command inputs (control law: OBS) (a) Pitch rate and roll rate response; (b) Yaw rate and sideslip response; (c) Control actuator response

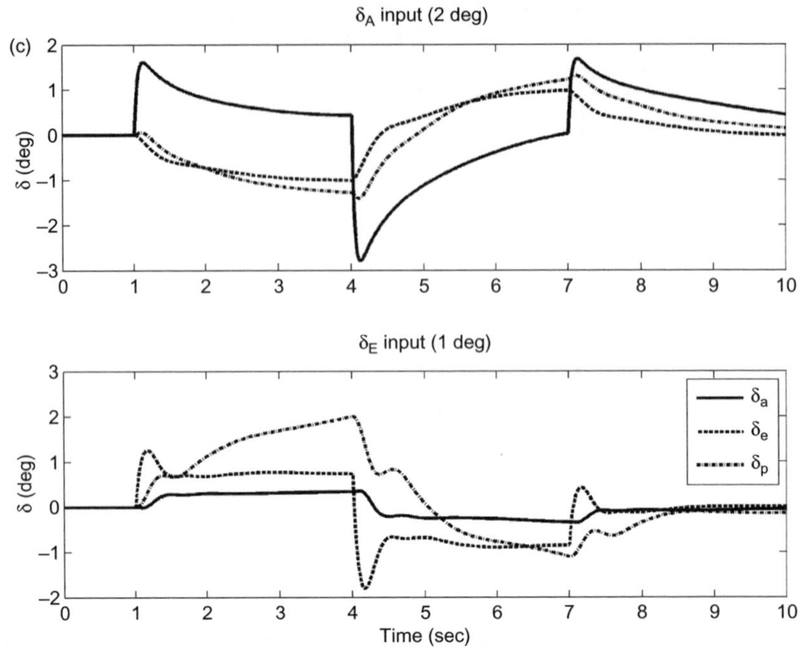

Figure 11.13 Concluded

is chosen to minimise the transient and steady-state errors between the plant and model outputs for commanded inputs u_m. Figure 11.14 gives a schematic of the command path and observer-based feedback controller.

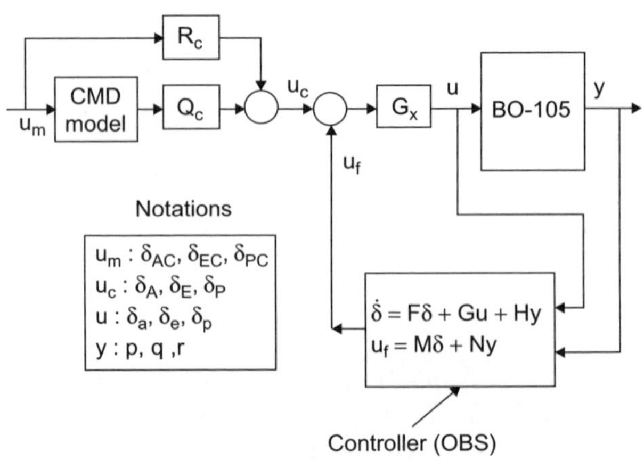

Figure 11.14 Schematic of rotorcraft command path and feedback controller

For the EMF design, the respective plant matrices of (11.14) are given by

1. SF

$$A_p = A + BK, \qquad B_p = B \tag{11.17a}$$

2. OBS (from (11.7) and (11.10))

$$A_p = \begin{bmatrix} A + BNC & BM \\ HC + GNC & F + GM \end{bmatrix}, \qquad B_p = \begin{bmatrix} B \\ G \end{bmatrix} \tag{11.17b}$$

The commanded responses are (i) pitch attitude (θ) using δ_{AC}, (ii) roll attitude (φ) using δ_{EC} and (iii) sideslip angle (β) using δ_{PC}.

A standard second-order unity gain model of the form

$$T_c(s) = \frac{\omega_n^2}{s^2 + 2\varsigma\omega_n s + \omega_n^2} \tag{11.18}$$

is used for each of the three command channels (q = 3, m = 3). The natural frequency and damping are listed in Table 11.7.

Table 11.7 Command model parameters

	Pitch	**Roll**	**Sideslip**
Damping ratio, ς	0.7	0.7	0.7
ω_n (rad/sec)	2.0	4.0	3.0

The composite three-axis command model (sixth-order decoupled state variable model using the second-order dynamics of (11.18) for each channel) is used to minimise the following objective function:

$$J_3 = \min_{C_m} \left\{ \mathrm{rms}\left(\sum_{i=1}^{q} \sum_{j=1}^{m} e_{ij} \right) \right\} \tag{11.19}$$

where e_{ij} is the error between the plant and model outputs for a pulse input to the respective command channels.

Figure 11.15a–c shows the command following characteristics of the EMF controller. The decoupling of the uncommanded response variables is acceptable for all command channels. However, the pitch attitude response does not follow the commanded signal accurately primarily due to the phugoid mode contamination of the pitch rate response. As discussed earlier, lowering the phugoid natural frequency while reducing the phugoid mode corruption in pitch rate response also reduces the low-frequency stability margin.

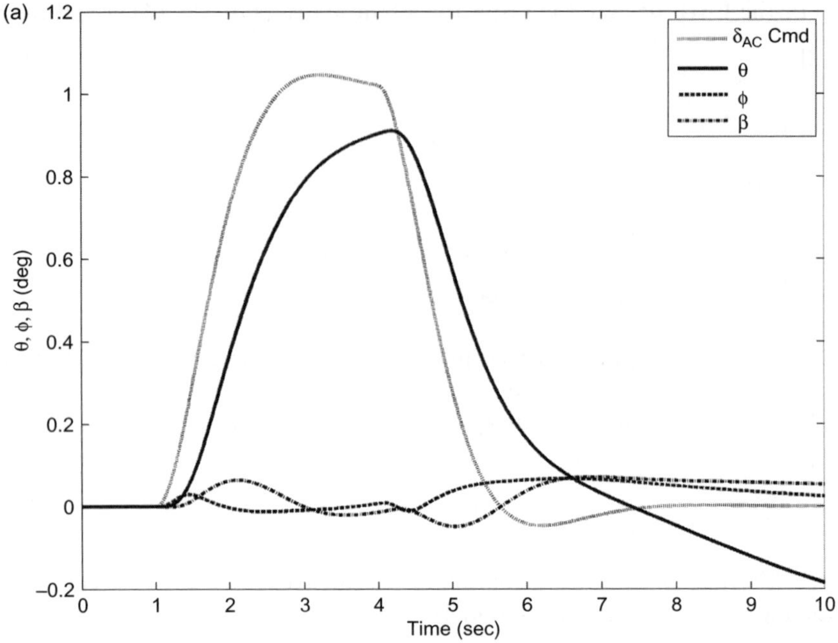

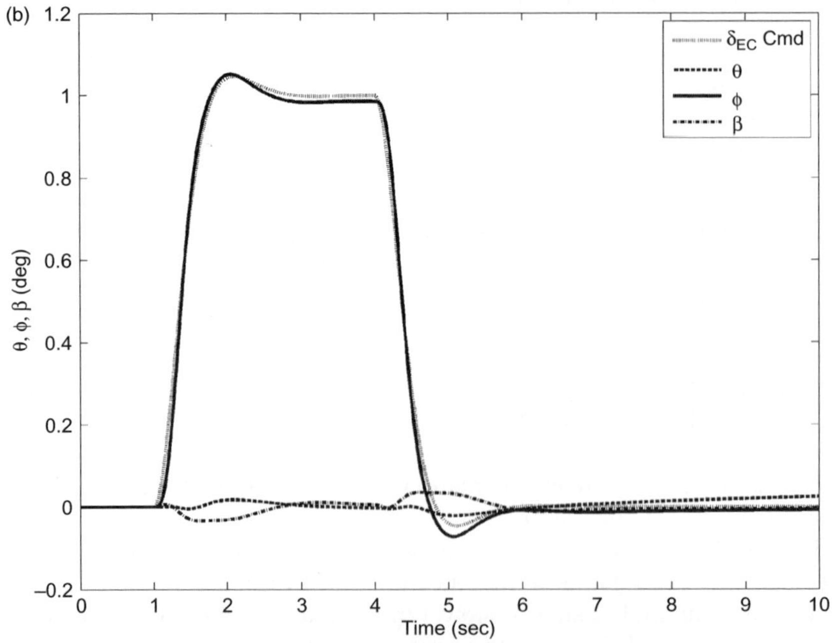

Figure 11.15 Model following decoupled response to command inputs (control law: OBS) (a) Pitch attitude command input; (b) Roll attitude command input; (c) Sideslip command input

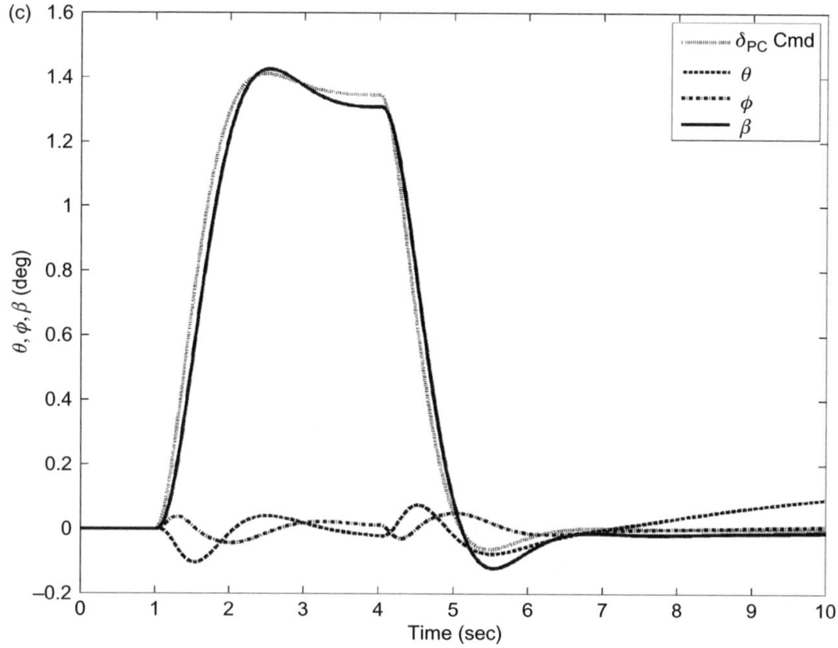

Figure 11.15 Concluded

Table 11.8 lists the EMF gain matrices Q_c and R_c (Figure 11.14) for the observer-based (OBS) controller.

Table 11.8 EMF gain matrices (control law: OBS)

	Q_c-Matrix (deg/deg)					
	θ_m	$\dot{\theta}_m$	φ_m	$\dot{\varphi}_m$	β_m	$\dot{\beta}_m$
δ_A	0.0189	0.0624	−0.0185	−0.0003	0.1076	0.0561
δ_E	−0.0056	−0.0137	−0.0771	0.0242	−0.0919	−0.0122
δ_P	0.0349	−0.0228	0.3741	0.0507	−0.1278	0.0379

	R_c-Matrix (deg/deg)		
	δ_{AC}	δ_{EC}	δ_{PC}
δ_A	−0.0012	0.0184	−0.0863
δ_E	0.0110	0.0775	−0.0207
δ_P	−0.0149	−0.3902	−0.5916

Table 11.8 (Continues)

Optimum model output matrix C_m
$$C_{m_{opt}} = \begin{bmatrix} 0.1865 & -0.3093 & 0.0623 & -0.0661 & 0.1498 & -0.1504 \\ 0.0992 & 0.1274 & 0.9216 & 0.0162 & 0.0386 & -0.0404 \\ 0.0603 & 0.1594 & 0.0008 & -0.0490 & 0.9784 & 0.0956 \end{bmatrix}$$

11.7 Summary

In this chapter the design of a control system to improve the flying characteristics of a BO-105 helicopter model is addressed. The study is aimed at meeting the level 1 handling qualities requirements specified in ADS-33E-PRF for the pitch, roll and yaw axes. The control system consists of two parts, namely: (i) a feedback controller to reduce the severe inter-axis coupling and (ii) a command path controller that shapes the response to pilot control inputs to reduce pilot workload when the helicopter is operated in a degraded UCE.

The feedback controller is designed using eigenstructure assignment techniques coupled with minimal order functional observer design. A state variable feedback is first designed, as a benchmark reference, using longitudinal cyclic, lateral cyclic and pedal controls. The SF design is replaced by designing a fourth-order, functional observer based, dynamic feedback controller using inertial rate sensors as feedback variables. It is shown that this low-order dynamic controller closely tracks the SF performance. The command path controller is designed using an EMF control formulation. The controller is an AC system for pitch and roll axes and a sideslip command system for the yaw axis.

The study demonstrates that a *viable practical feedback control scheme*, using only inertial rate feedback sensors coupled with a fourth-order compensator, can be designed to meet the level 1 handling qualities requirements in the pitch and roll axes. The observer-based controller exhibits good stability and performance robustness. The very strong longitudinal to lateral coupling of the BO-105 helicopter makes it difficult to meet level-1 pitch to roll cross axis coupling requirements, using eigenstructure modification alone that can only reduce the coupling due to off-axis stability derivatives. Thus, control interconnects between the longitudinal and lateral cyclic channels become essential to reduce the coupling due to off-axis control derivatives. Similarly control interconnects from the pedal to the cyclic controls are needed to reduce the pitch to yaw cross coupling. The AC controller exhibits excellent decoupled off-axis response characteristics.

References

1. MORALEZ, III, E., HINDSON, W.S., FROST, C.R., TUCKER, G.E., ARTERBURN, D.R., KALINOWSKI, K.F., *et al.*: 'Flight Research Qualification of the Army/NASA RASCAL Variable-Stability Helicopter',

Proceedings of the AHS International, 58th Annual Forum, June 2002, Montreal, Canada, pp. 499–514

2. KALETKA, J., KURSCHEID, H., and BUTTER, U.: 'FHS, the new research helicopter: ready for service', *Aerospace Sci. Technol.*, 2005, **9**, (5), pp. 456–467
3. POSTLETHWAITE, I., PREMPAIN, E., TURKOGLU, E., TURNER, M.C., ELLIS, K., and GUBBELS, A.W.: 'Design and flight testing of various H_∞ controllers for the Bell 205 helicopter', *Control Eng. Pract.*, 2005, **13**, pp. 383–398
4. KOTHMANN, B.D., and ARMBRUST, J.: 'RAH-66 Comanche Core AFCS Control Law Development: DEMVAL to EMD', *Proceedings of the AHS International, 58th Annual Forum*, June 2002, Montreal, Canada
5. ENNS, D., BUGAJSKI, D., HENDRICK, R., and STEIN, G.: 'Dynamic inversion: an evolving methodology for flight control design', *Int. J. Control*, 1994, **59**, (1), pp. 71–91
6. POSTLETHWAITE, I., SMERLAS, A., WALKER, D.J., GUBBELS, A.W., BAILLIE, S.W., STRANGE, M.E., *et al.*: 'H_∞ control of the NRC Bell 205 fly-by-wire helicopter', *J. Am. Helicopter Soc.*, 1999, **44**, pp. 276–284
7. MANNES, M.A., and MURRAY-SMITH, D.J.: 'Aspects of multivariable flight control law design for helicopters using eigenstructure assignment', *J. Am. Helicopter Soc.*, 1992, **37**, (3), pp. 18–32
8. LOW, E., and GARRARD, W.L.: 'Design of flight control systems to meet rotorcraft handling qualities specifications', *J. Guidance Control Dyn.*, 1993, **16**, (1), pp. 69–78
9. SRINATHKUMAR, S.: 'Design of Helicopter Stability Augmentation System using Eigenstructure Assignment and Minimal Order Functional Observers', *Proceedings of the AHS International, 61st Annual Forum*, June 2005, Grapevine, Texas
10. ANONYMOUS: 'Aeronautical Design Standard Performance Specification – Handling Qualities Requirements for Military Rotorcraft', 2000, ADS-33E-PRF

[illegible]

[illegible] 2002, [illegible]

[illegible] and [illegible]

[illegible]

Chapter 12
Aircraft flutter control system design

12.1 Introduction

Modern aircraft designs tend to maximise aerodynamic efficiency with minimum structural weight. The resulting airframe flexibility often leads to aeroelastic instabilities, such as flutter. The operational envelope of an aircraft in such cases may be limited by the flutter boundary. With the advances in the capability and reliability of high-response flight control sensors and actuators, it is not unrealistic to expect that active suppression of flutter modes using feedback control will be attempted to extend the flutter boundary of an aircraft. The first successful flight test demonstration of using active control to extend the flutter boundary of a B-52 airplane is reported in Reference 1. A flight research program, using a remotely piloted research vehicle, to demonstrate active flutter suppression techniques has also been reported [2]. In light of the high risk involved in flight testing, wind tunnel test techniques have been successfully used for preliminary validation of aeroelastic modelling and flutter control concepts. The Active Flexible Wing (AFW) wind tunnel test project [3] is an example of such a research program. The present chapter is primarily based on eigenstructure assignment and classical control techniques used in the AFW program [4–9], with additional extensions to improve upon the results presented therein.

12.2 AFW flutter suppression problem

12.2.1 Wind tunnel model

The AFW wind tunnel model is an aeroelastically scaled model equipped with accelerometer sensors and hydraulic servo-controlled control surfaces for active control of the aeroelastic modes. Four control surfaces are located on each wing semispan: leading edge outboard (LEO), leading edge inboard (LEI), trailing edge outboard (TEO) and trailing edge inboard (TEI). Similarly the model has four accelerometers on each wing semispan at LEO, TEO, TEI and wing tip (TIP). Additional references on the model details are found in Reference 5. For the flutter suppression control laws discussed in this chapter, the trailing edge control surfaces (δ_{teo}, δ_{tei}) along with the tip and trailing edge accelerometers (Z_{teo}, Z_{tip}) are used as shown in Figure 12.1.

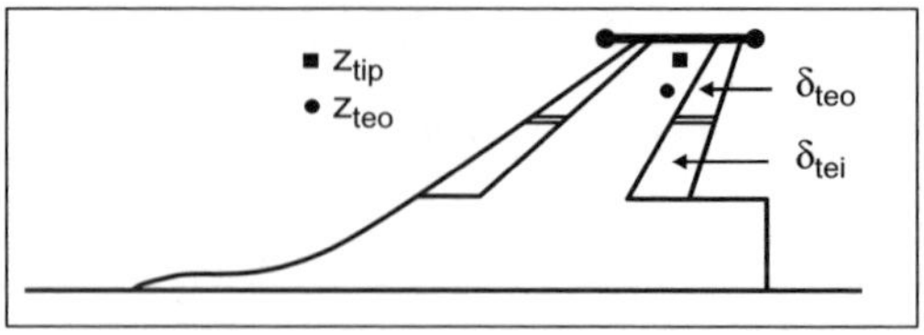

Figure 12.1 Schematic of AFW wind tunnel model

12.2.2 Mathematical model

The aeroelastic mathematical model consists of representing the structural and aerodynamic mode coupling using appropriate finite element and unsteady aerodynamic model formulations. These high-order models are suitably approximated to derive a reduced-order linear, time-invariant, state variable design model, which is used for control analysis and synthesis.

The design model consists of eight elastic modes, eight unsteady aerodynamic modes, TEO and TEI actuator dynamics (third order each) and a Dryden turbulence model (second order). This results in a state variable model with 32 states, 2 control inputs, 1 disturbance input and 2 outputs (TIP and TEO accelerations). These linear models are generated over a range of dynamic pressures.

The AFW model exhibits two distinct flutter modes (symmetric and antisymmetric). The model flutter characteristics differ depending on the density of the test medium. The symmetric flutter model, at M = 0.9 (Mach number), with a heavy gas test medium (Freon), will be used for the present design study. The actual wind tunnel tests were, however, conducted with air as the test medium (M = 0.5). These wind tunnel test results [3,9] will also be briefly reviewed in light of the present design study.

The dynamic characteristic of the aeroelastic model over a range of dynamic pressures ($\bar{q}$) in lbf/ft^2 (psf) is depicted in Figure 12.2. All the elastic modes from mode 1 (lowest frequency) to mode 8 (highest frequency), except for mode 2, remain stable as the dynamic pressure is increased. Mode 1 (*sting* mode), associated with the model support system, along with modes 4 and 7 does not appreciably change with dynamic pressure.

Figure 12.3 shows the reduction in the flutter mode damping as a function of dynamic pressure (labelled AFW). The *critical flutter* dynamic pressure is 232 psf. The flutter frequency is 10.2 Hz (64 rad/sec). The figure also shows the violent nature of flutter mode divergence. In Figure 12.3, the damping characteristics labelled as eigenstructure assignment controller (EAC) and flutter filter controller (FFC) correspond to the damping augmentation by use of feedback control. These controller characteristics will be discussed in section 12.4.

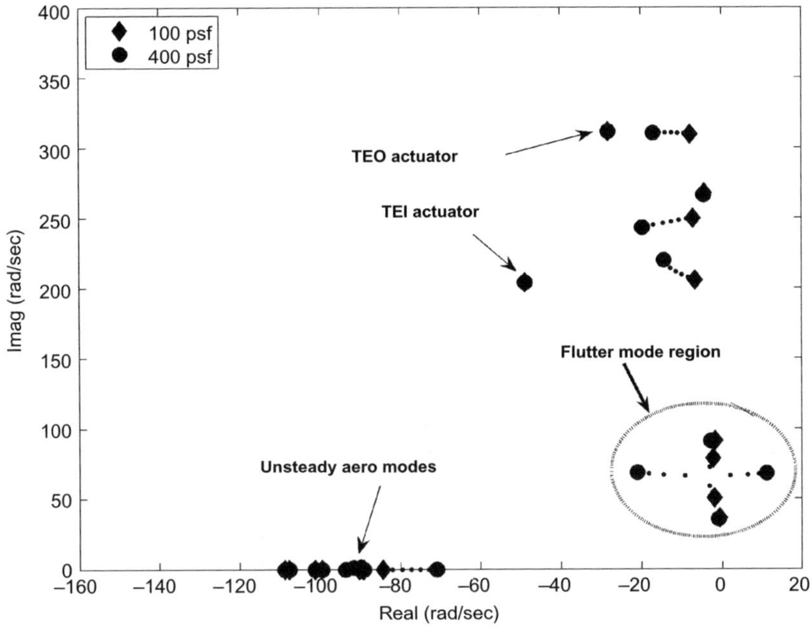

Figure 12.2 Dynamic pressure locus – all modes ($\bar{q} = 100–400\,\text{psf}$)

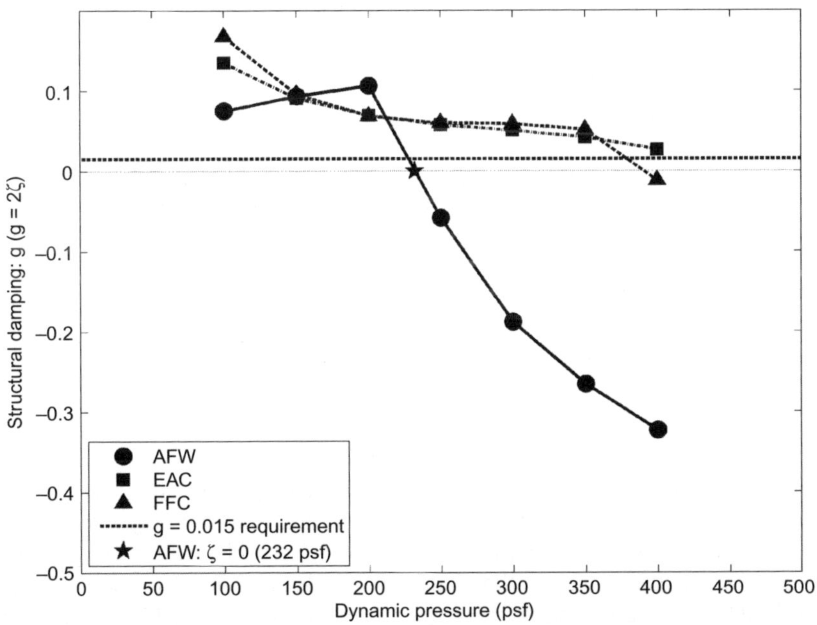

Figure 12.3 Flutter mode damping characteristics

12.3 Flutter mechanism

The aeroelastic dynamics at a fixed dynamic pressure ($\bar{q}$) can be represented by a finite-dimensional, constant coefficient differential equation in state variable form as

$$\begin{bmatrix} \dot{x} \\ \ddot{x} \end{bmatrix} = \begin{bmatrix} 0 & I_n \\ -M^{-1}K & -M^{-1}D \end{bmatrix} \begin{bmatrix} x \\ \dot{x} \end{bmatrix} + \begin{bmatrix} 0 \\ -M^{-1}B \end{bmatrix} u \qquad (12.1)$$
$$y = C\ddot{x}$$

where x is a set of generalised co-ordinates associated with the flexible modes. M, K and D are the generalised mass, stiffness and damping matrices, respectively. B and C are the control and output distribution matrices, respectively, u is the control input vector and y is the acceleration output vector at physical locations on the wing.

In order to understand the *principal* characteristics of flutter, from a feedback control perspective, a single-input/single-output control problem will be examined using only flexible mode dynamics. Throughout the text and figures, the following shorthand notation will be used to represent second-order transfer functions:

$$s^2 + 2\varsigma\omega s + \omega^2 = (\omega, \varsigma) \qquad (12.2)$$

A scalar transfer function representation of (12.1) can be written as

$$\frac{y(s)}{u(s)} = \frac{k_0 s^2}{(\omega_b, \varsigma_b)} \prod_{i=1}^{n-1} \frac{(\omega_{z_i}, \varsigma_{z_i})}{(\omega_{p_i}, \varsigma_{p_i})} \qquad (12.3)$$

where Π indicates continued product, zeros at the origin are due to the output being acceleration and n is the number of flexible modes. The lightly damped zero/pole pairs, $(\omega_{z_i}, \varsigma_{z_i})$ and $(\omega_{p_i}, \varsigma_{p_i})$, are typically in close proximity. Initially ignoring the high-frequency modes, a dominant two-mode approximation to (12.3) for the AFW model is of the form

$$\frac{Z_{tip}(s)}{\delta_{teo}(s)} = \frac{k_0 s^2}{(\omega_b, \zeta_b)} \frac{(\omega_c, \zeta_c)}{(\omega_t, \zeta_t)} \qquad (12.4)$$

where Z_{tip} is the wing tip acceleration and δ_{teo} is the TEO control surface (Figure 12.1). The flutter dynamics can be classically represented as the strong interaction between a bending mode $p_b(\omega_b, \zeta_b)$ and a torsion mode $p_t(\omega_t, \zeta_t)$. Associated with these modes is a *critical* zero $z_c(\omega_c, \zeta_c)$ whose relative location in the complex plane with respect to the dominant modes p_b and p_t plays a key role in the feedback stabilisation of the unstable flutter mode. Figure 12.4 shows a locus of p_b, p_t and z_c as a function of dynamic pressure. Also shown in the figure is the equivalent multivariable zero (invariant zero) for the two-input/two-output pair of Figure 12.1. In the multivariable design, the invariant zero plays a similar role as z_c.

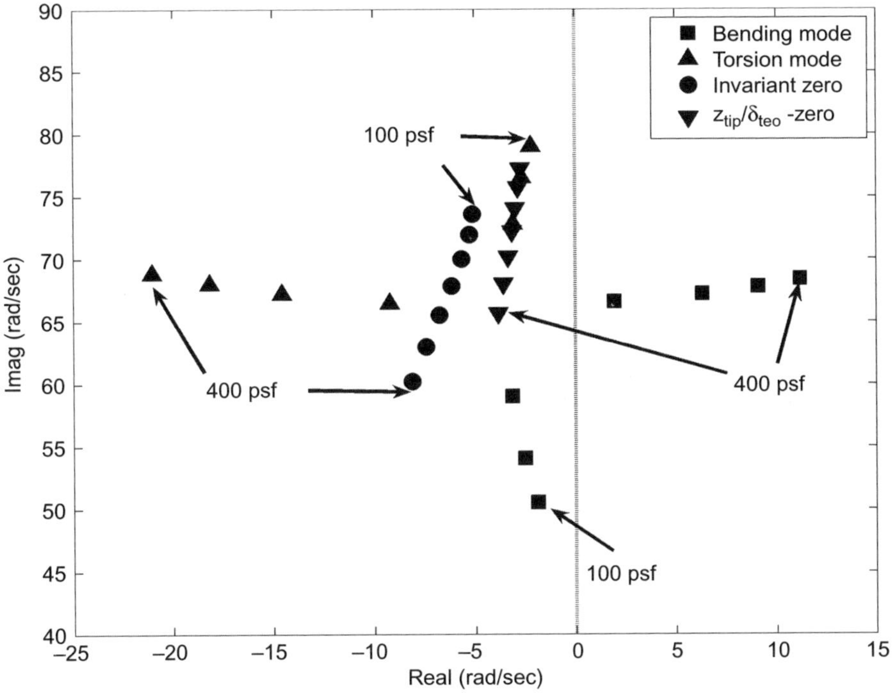

Figure 12.4 Dynamic pressure locus – dominant modes ($\bar{q}$ = 100–400 psf in 50-psf increments)

At low dynamic pressures, the bending mode and torsion mode are distinct with associated mode shapes. As the dynamic pressure is increased, these modes tend to coalesce exhibiting both bending and torsion mode shapes. Eventually one of the modes becomes unstable while the other mode becomes stable. Furthermore, at low dynamic pressures the torsion mode and the zero (z_c) are close to each other. At higher dynamic pressures, the separation between the pole and zero increases. Finally the triple (p_b, p_t and z_c) fully characterise the flutter mechanism.

12.4 Flutter control problem formulation

12.4.1 Design objectives and specifications

The primary objective of control design is to extend the stable flutter boundary of the model. An additional objective is to perform rapid rolling manoeuvres while simultaneously guaranteeing flutter stability in the presence of wind tunnel turbulence. This implies ensuring no actuator rate saturation occurs while meeting the actuator demands of both flutter and roll manoeuvre controllers (as would be the case in an actual aircraft). The controller has to meet the following performance criteria:

1. gain and phase margins as defined in section 8.6.1;
2. no occurrence of RMS control rate saturation for 1 ft/sec RMS wind tunnel turbulence;
3. stability robustness throughout the operating dynamic pressure range;
4. preferable to have a fixed compensator for the entire dynamic pressure range (no scheduling of parameters based on $\overline{q}$).

Remarks:

1. The TEO surface actuator has a peak no load rate of 225 deg/sec. This translates to an RMS rate of 75 deg/sec (section 8.3, point 9) for a random turbulence excitation. Allowing for roll manoeuvre controller requirements, the net RMS rate available for the flutter controller is fixed at 60 deg/sec. The TEI surface actuator has a peak no load rate of 470 deg/sec.
2. For the above actuator rate constraint, if the actuator is dominantly excited at the flutter frequency, the corresponding RMS surface deflection is approximately 1 deg (assuming sinusoidal excitation). The TEO and TEI control surfaces have maximum displacement of ±5 deg.

12.4.2 Feedback controller evolution

The primary goal in design is to devise a controller of minimum complexity to meet the objectives stated earlier. Towards this end only two of the four available control surfaces are selected as control variables. The trailing edge surfaces are preferred as they are predicted to effectively control the first bending and torsion modes. While the classical colocated sensor/actuator pair philosophy would result in the selection of TEO and TEI accelerometer sensors, it is found that the TIP accelerometer, in addition to having a high flutter mode content, has very good roll-off characteristics at higher frequencies. This alleviates the use of additional filters to get the desired roll-off. Figure 12.5a and b shows typical frequency response characteristics. Thus, TEO and TIP sensors are selected as feedback sensors.

This results in a two-input/two-output system for designing the multivariable controller EAC. A single-input/output controller FFC using the TIP sensor/TEO actuator pair will also be designed to assess its relative performance. The FFC controller was successfully tested in the wind tunnel experiments [9].

As a first step in understanding the feedback problem, consider a gain feedback from Z_{tip} to δ_{teo}. As the gain is increased, two types of root locus branches are possible as shown in Figure 12.6. The loci shown in dotted, wherein the bending mode approaches the roots at the origin, are non-stabilising. Thus, it is imperative that the controller produces a stabilising locus. In the stabilising loci case, the bending mode approaches the critical zero (z_c).

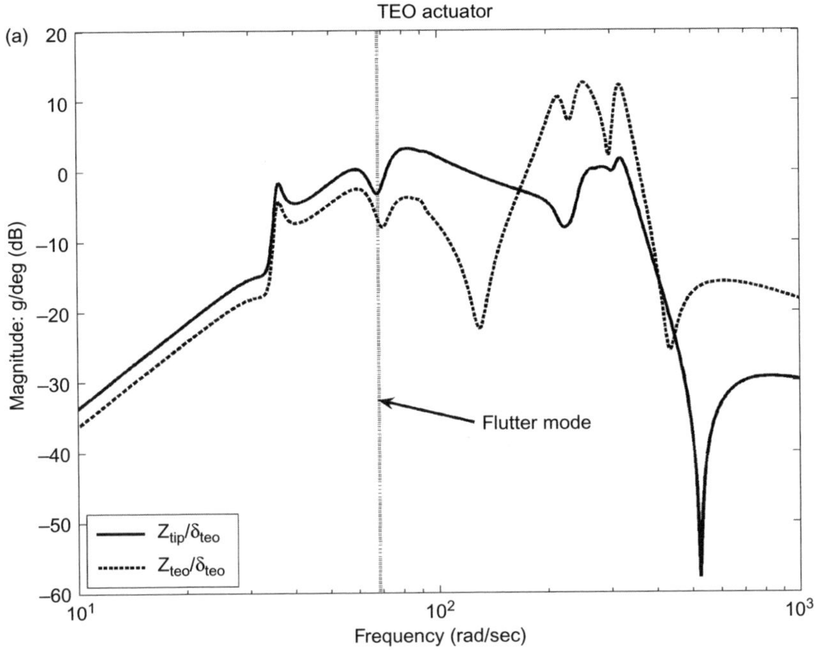

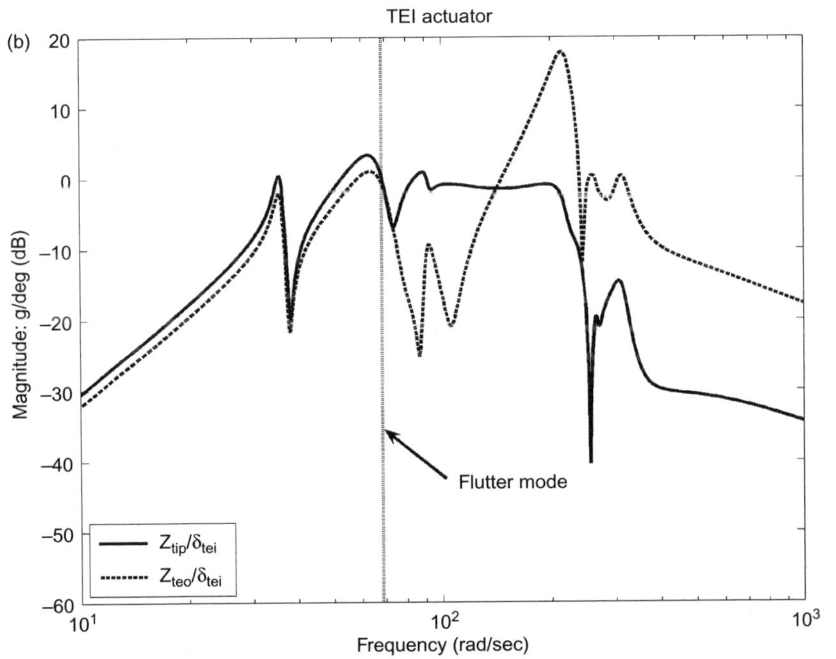

Figure 12.5 Sensor frequency response ($\bar{q} = 350$ psf)
(a) TEO channel; (b) TEI channel

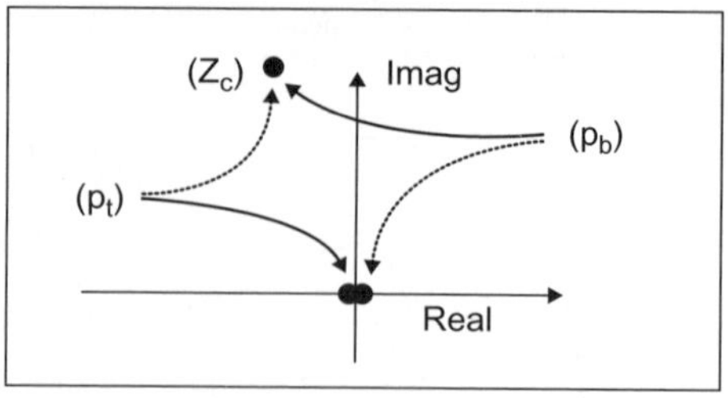

Figure 12.6 Generic gain root locus

The incremental gain required to improve the flutter mode damping progressively increases as p_b approaches z_c. In the limit as k tends to ∞ the flutter mode p_b coalesces with z_c. Thus, the location of z_c strongly influences the gain required to stabilise the flutter mode. Figure 12.7 shows the root locus plot for stabilising the flutter mode using a gain feedback from Z_{tip} to δ_{teo}. As the unstable flutter mode migrates towards the stable region, mode 8 and *sting* mode become unstable. The TEO actuator roots also tend towards instability. The high feedback gain also implies higher control activity with the likelihood of actuator rate saturation in the presence of wind tunnel turbulence. Thus, static gain feedback solution is not feasible, and dynamic compensation is required. One simple solution is to close the feedback loop through a *flutter filter* of the form

$$H(s) = \frac{(\omega_z, \varsigma_z)}{(\omega_f, \varsigma_f)} \tag{12.5}$$

and a gain constant k_f. Choosing the flutter filter pole to (i) cancel the critical zero (pole–zero cancellation!) and (ii) having the flutter filter zero farther into the left half plane will facilitate the uninhibited migration of the flutter mode into the stable region as gain k_f is increased. Since there will be uncertainty in estimating the zero, the solution scheme should be robust to the variations in the critical zero location. A sensitivity study of the effect of the filter pole location with respect to the critical zero is shown in Figure 12.8. Table 12.1 lists the flutter filter constants for the cases studied.

Figure 12.8 shows a root locus of the flutter modes for feedback gain $k = \rho^* k_f$ variation. The gain scale factor ρ is varied from 0 to 1. The gain $k_f = -0.3$ deg/g. Case (a) represents pure gain feedback (no flutter filter). In this case flutter mode cannot be stabilised with the feedback gain k_f. Case (b) illustrates the design with the flutter filter pole cancelling the critical zero. The bending mode is stabilised and migrates towards the origin, while the torsion mode migrates towards the flutter filter zero. Cases (c) and (d) illustrate the designs with the flutter filter pole in the proximity of the critical zero. The stabilisation characteristics are retained except that either the torsion mode

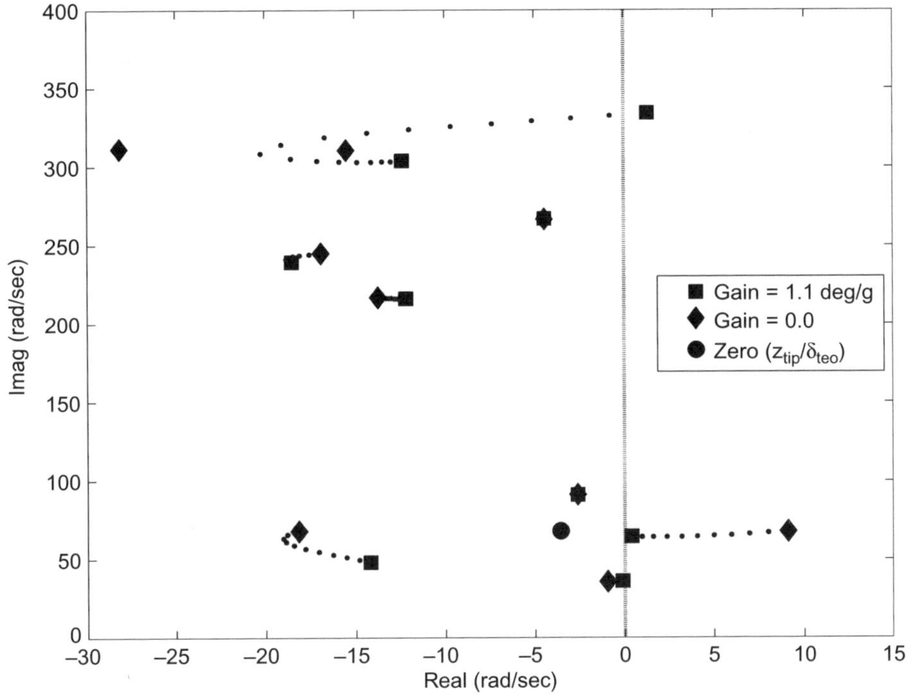

Figure 12.7 Scalar gain feedback (Z_{tip} to δ_{teo}) root locus ($\bar{q}$ = 350 psf)

Table 12.1 Flutter filter constants for sensitivity study (Figure 12.8)

Case no.	Flutter filter		Remarks
	Pole (p)	Zero (z)	
a	–	–	Gain feedback
b	–3.3 ± j70	–40 ± j75	Critical zero (z_c) cancellation
c	–5.0 ± j75		Real (p) < real (z_c)
d	–2.0 ± j75		Real (p) > real (z_c)

(case (c)) or bending mode (case (d)) migrates towards the critical zero. The sensitivity study thus reveals that the compensation scheme proposed is robust to variation of flutter filter pole location in relation to the critical zero. Further, using the flutter filter, the bending mode has been stabilised with low gains in all cases.

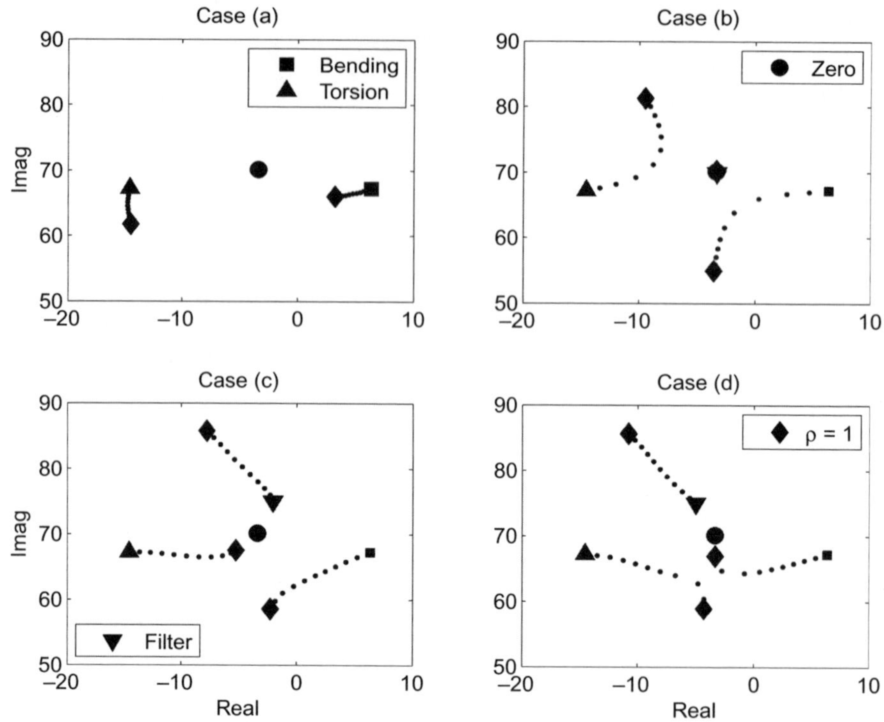

Figure 12.8 Effect of flutter filter location on gain root locus ($\bar{q}$ = 300 psf, $k_f = -0.3$ deg/g)

12.4.3 Controller structure

A schematic of the AFW flutter control system is given in Figure 12.9. It depicts a two-channel controller used by the EAC. Each control channel has an anti alias filter (a = 150 rad/sec) and a washout filter (b = 5 rad/sec). The washout filter attenuates accelerometer low-frequency noise. These filters do not contribute to the basic flutter mode stabilisation. $H_1(s)$ and $H_2(s)$ are second-order filters of the form in (12.5). A gain matrix with four gain elements completes the feedback loop. The single-input/single-output FFC uses only the Z_{tip} to δ_{teo} feedback channel with compensator $H_1(s)$ and scalar gain k_1.

12.4.4 Flutter filter controller

The single-input/single-output feedback controller concept developed in the previous section is used to synthesise a system to meet the design objectives stated earlier. The design parameters available for optimisation are the flutter filter constants of $H_1(s)$. The key parameters affecting the performance are the filter pole location $\lambda_f(\omega_f, \zeta_f)$ and scalar gain k_1. The flutter filter zero $z_f(\omega_z, \zeta_z)$ has a secondary influence.

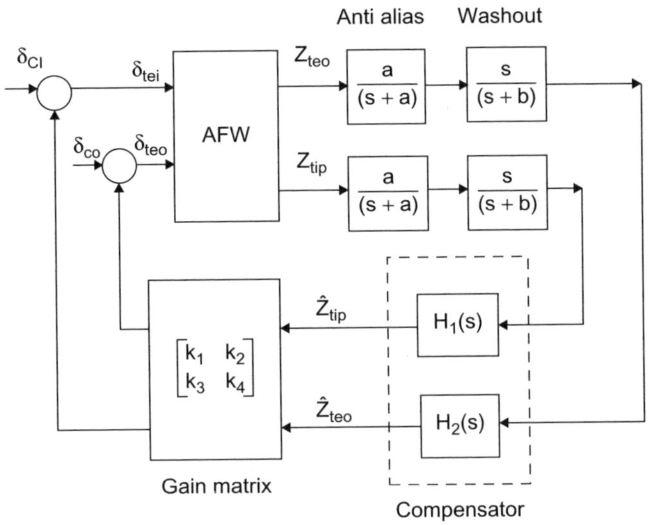

Figure 12.9 Schematic of the AFW flutter control system

The following design procedure is followed for performance optimisation. The reference design model corresponds to $\overline{q} = 300\,\text{psf}$.

For a set of λ_f in the neighbourhood of z_c,

1. Using graphical root locus, determine feedback gain k_1 to get an acceptable flutter mode damping.
2. For the full dynamic pressure range, compute the performance metrics: gain margin, phase margin and TEO actuator RMS rate for turbulence input.
3. Select the best acceptable design from the set of solutions that do not violate the limits set on the performance metrics.

Remarks: An iterative optimisation procedure could well have been used to derive a satisfactory solution. However, since the number of design parameters is small (only three), the design process described above is deemed adequate. Further the method yielded good insight in determining the effect of change in a parameter on the performance metrics. For example, the following properties could be identified:

(i) Cancelling the critical zero with the filter pole results in excellent gain margin but very poor phase margin.
(ii) Increasing the filter pole damping has the effect of reduction in gain margin and increase in phase margin. Increased filter pole damping also requires higher gains for flutter mode stabilisation.
(iii) Feedback gain directly controls the closed-loop flutter mode damping, and higher gains imply higher turbulence response.
(iv) Marginal improvement in both gain and phase margins is possible by varying the damping of the flutter filter zero.

The controller parameters are listed in Table 12.2. Figure 12.10 shows the location of the closed-loop modes in the flutter region as a function of dynamic pressure. Notice that at sub-critical dynamic pressures the damping of the flutter mode is better than the critical zero because of the attraction of the filter root to the zero. The FFC design does not stabilise the flutter mode at $\bar{q} = 400\,\text{psf}$.

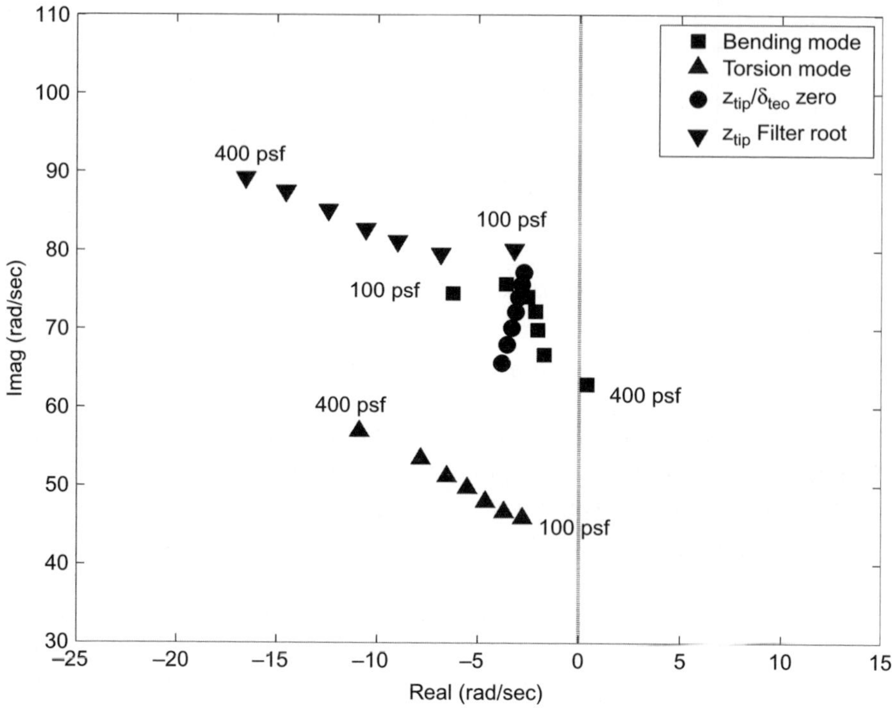

Figure 12.10 Dynamic pressure locus of FFC closed-loop modes

12.4.5 Eigenstructure controller

The schematic of Figure 12.9 shows a dynamic output feedback controller to stabilise the flutter mode. Output feedback algorithms (Chapter 4) will be used to design the system. The AFW system has 38 (n) states (including compensator states), 2 (m) inputs and 2 (r) outputs. From Theorem 4.1 four design parameters, namely two eigenvalues and two eigenvector elements, can be chosen arbitrarily to optimise the performance of the system. Additional eight parameters of the second order compensator filters ($H_1(s)$ and $H_2(s)$) are also available for optimisation.

In the FFC design only the TEO surface is used for feedback and its rate capability limits the highest dynamic pressure that can be stabilised. In Reference 7, blending of both TEI and TEO surfaces was used to alleviate this problem and successfully tested in the wind tunnel experiments [9]. However, the EAC method brings in additional design freedom to facilitate systematic optimisation of the control activity of both TEO and TEI actuators. Unlike the FFC design, a formal optimisation procedure is used to select the best solution.

The design is initiated with the filters set to fixed values. This is to start searching for an acceptable solution with minimum number of design parameters. Compensator $H_1(s)$ is chosen as that used in the FFC design. The filter constants are shown in Table 12.2. This leaves only the eigensystem parameters for optimisation. The eigenvalue freedom is used to assign the flutter mode at $-2 \pm j70$. The real part of the eigenvalue is fixed taking into consideration the presence of an invariant zero (Figure 12.4). The invariant zero plays the same role as that of the critical zero in the single-input/single-output case. Attempt to increase the flutter mode damping results in excessive gains. The eigenvector parameters are arbitrary. The following objective function for wind tunnel turbulence disturbance input is optimised:

$$J = \min\{\mathrm{RMS}(\dot{\delta}_{tei} + w\dot{\delta}_{teo})\} \tag{12.6}$$

where w (w > 1) is a weighting factor used to penalise the rate response of TEO actuator, which has a low rate capability.

The iterative design procedure, similar to the FFC design, is followed for performance optimisation. For different control rate weighting factors w, the performance of the resulting solution is assessed over the range of dynamic pressures. The weight factor is adjusted based on the outcome of the design.

An acceptable design resulted with gain feedback

$$\begin{bmatrix} \delta_{tei} \\ \delta_{teo} \end{bmatrix} = \begin{bmatrix} -0.3051 & -0.0212 \\ -0.3623 & -0.0081 \end{bmatrix} \begin{bmatrix} \hat{Z}_{tip} \\ \hat{Z}_{teo} \end{bmatrix} \tag{12.7}$$

The gain units are in deg/g. The gains in (12.7) indicate that only the tip accelerometer output is dominantly used to drive both the control surfaces. The feedback gains from $\hat{Z}_{teo}$ sensor are very small. This is to be expected since $\hat{Z}_{teo}$ sensor's higher response at higher frequencies tends to increase the actuator rates that the objective function J in (12.6) penalises. Setting the feedback gains from the TEO accelerometer ($\hat{Z}_{teo}$) to zero marginally changed the performance of the system. This results in a controller with only one filter corresponding to the FFC design and two gains to drive the TEI and TEO surfaces. The controller constants for the EAC design are given in Table 12.2. The filter $H_2(s)$ in the controller structure of EAC design is *not required for this particular solution*. However, $H_2(s)$ has an important role to play, in some situations, as will be discussed in section 12.7. Figure 12.11 shows the multivariable root locus at the highest dynamic pressure. It is seen that the high-frequency elastic modes undergo minimal change. The TEI actuator root and one unsteady aero root undergo some change. Most of the control activity is confined only to the flutter region (Figure 12.12). Figure 12.13 is a summary of the closed-loop roots for the full dynamic pressure range. It is seen that unlike in the FFC design, the invariant zero acts as a strong barrier to improve the flutter mode damping even in sub-critical dynamic pressures.

12.5 Controller performance assessment

The performance of both EAC and FFC controllers is evaluated over the complete dynamic pressure range. Table 12.2 summarises the controller parameters.

Table 12.2 EAC and FFC controller parameters (Figure 12.9)

Filter	Pole			Zero		
	λ	ω	ζ	λ	ω	ζ
(a) Filter constants						
$H_1(s)$	$-7 \pm j70$	70.35	0.10	$40 \pm j75$	85.0	0.47
$H_2(s)$	$-10 \pm j100$	100.50	0.10	$-50 \pm j100$	111.8	0.45

Controller	Gain matrix elements (deg/g)			
	k_1	k_2	k_3	k_4
(b) Gain matrix				
FFC	–0.4000	0	0	0
EAC	–0.3051	0	–0.3623	0

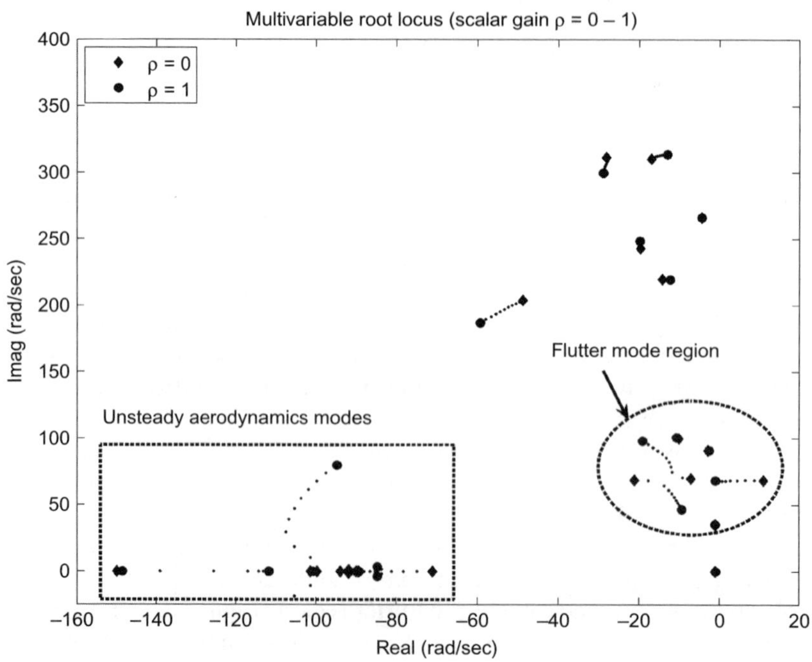

Figure 12.11 Multivariable root locus ($\bar{q} = 400$ psf) (controller: EAC)

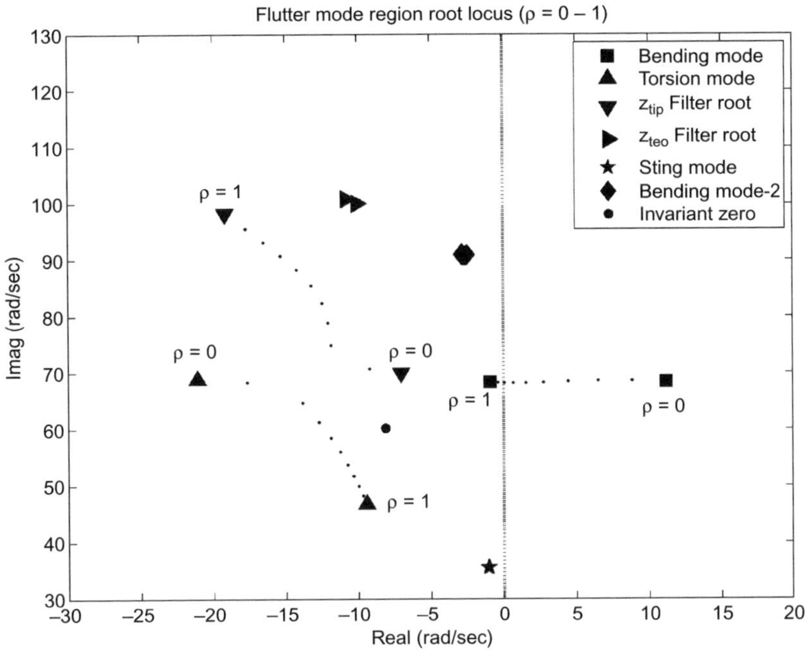

Figure 12.12 *Expanded multivariable root locus ($\bar{q} = 400$ psf) (flutter mode region; controller: EAC)*

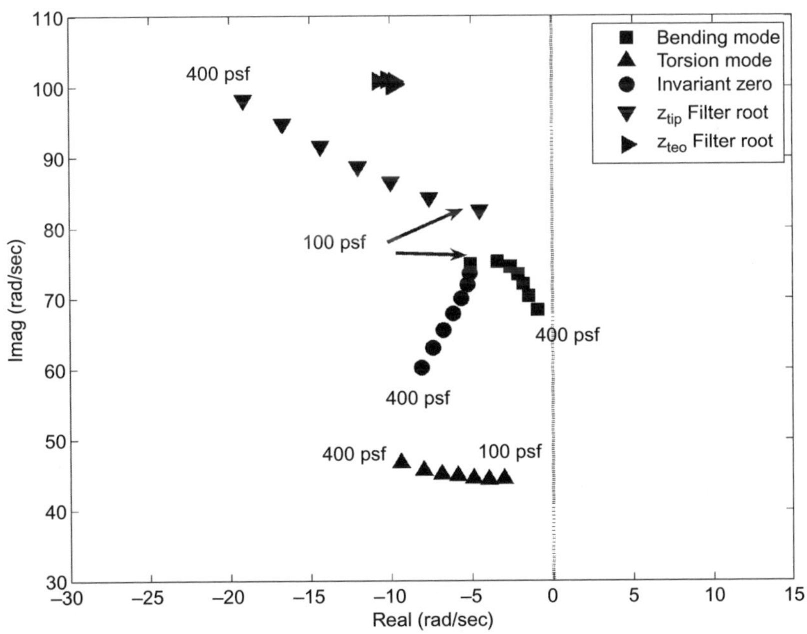

Figure 12.13 *Locus of EAC closed-loop modes ($\bar{q} = 100$–400 psf)*

12.5.1 System robustness

System robustness is evaluated using the robustness criteria defined in section 8.6.1. Using the exclusion boundary criterion, Figure 12.14 shows the loop transfer function for the entire dynamic pressure range for the FFC design. Figures 12.15 and 12.16 show the corresponding EAC design loop transfer functions for the TEI and TEO channels, respectively. Based on the requirement of *simultaneously* meeting both gain and phase margins as defined by the exclusion boundary, the FFC controller increases the stable flutter margin to 300 psf. The corresponding EAC margin is 350 psf for both the TEI and TEO channels. System robustness based on conventional independent gain and phase margin criteria shown in Figures 12.17 and 12.18, respectively, also indicates similar margins. The EAC–TEI channel has infinite phase margin up to 150 psf. The PPF, EAC–TEO and EAC–TEI channels have negative infinite gain margins up to 200, 250 and 350 psf, respectively.

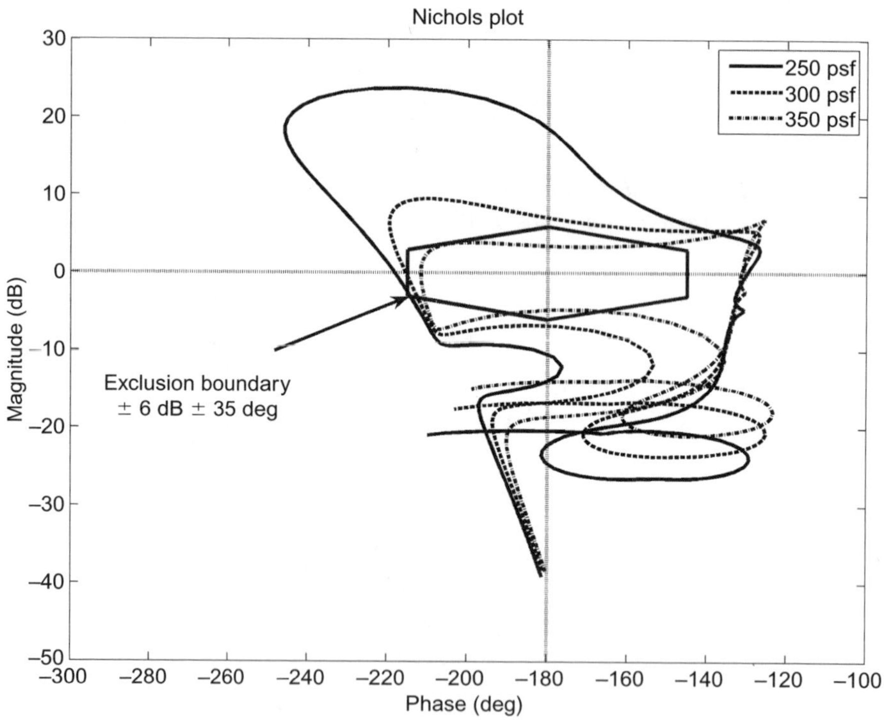

Figure 12.14 FFC controller loop transfer function (TEO channel)

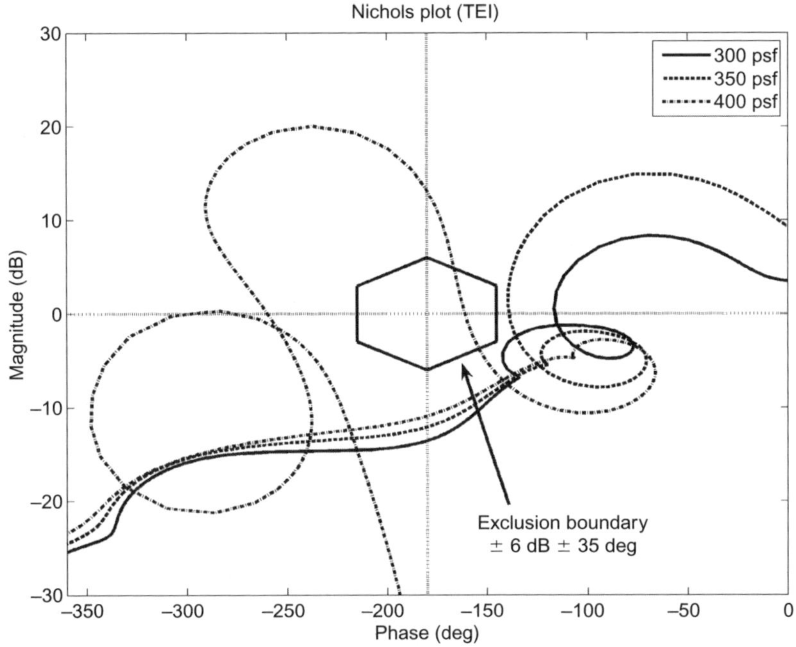

Figure 12.15 EAC controller loop transfer function (TEI channel)

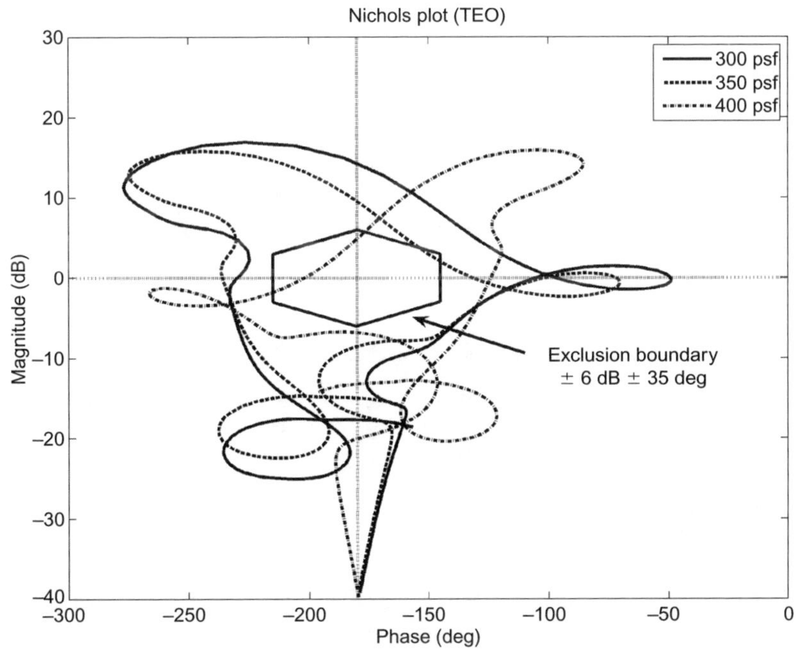

Figure 12.16 EAC controller loop transfer function (TEO channel)

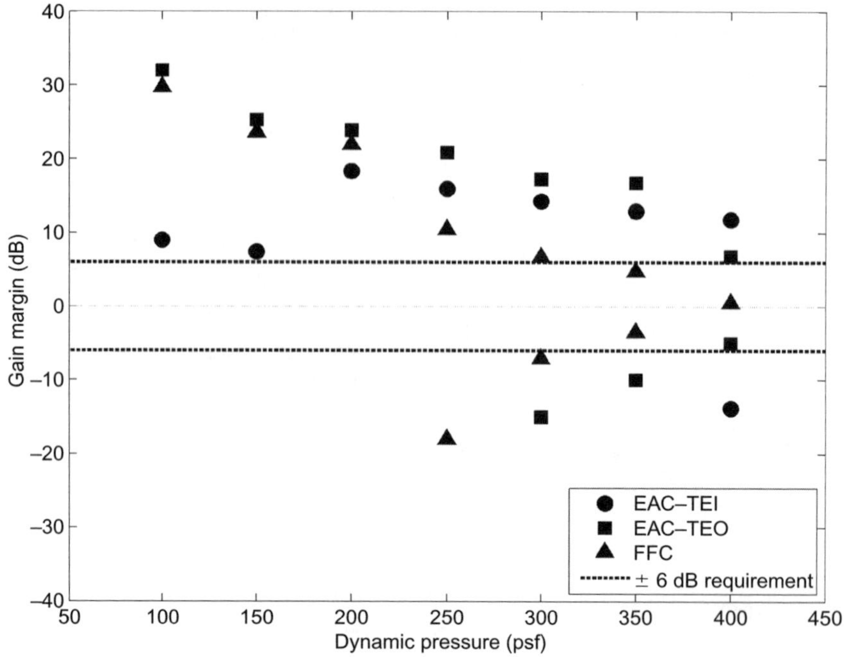

Figure 12.17 Classical gain margin

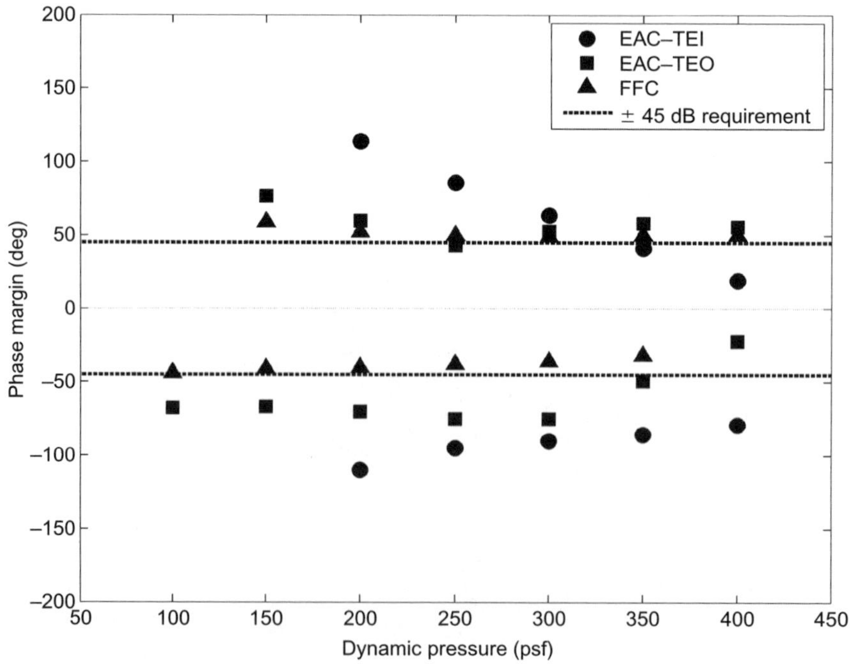

Figure 12.18 Classical phase margin

12.5.2 Wind tunnel turbulence response

The actuator displacement and rate response to wind tunnel turbulence disturbance is computed using the method described in Appendix A ((A.17) and (A.18)). Figure 12.19 summarises the results. It is seen that FFC controller demands higher TEO actuator rate response at higher dynamic pressures. The TEO actuator rate limit is reached at $\bar{q} = 320\,\mathrm{psf}$. The EAC design succeeds in controlling the rate response of both actuators within limits even till $\bar{q} = 400\,\mathrm{psf}$.

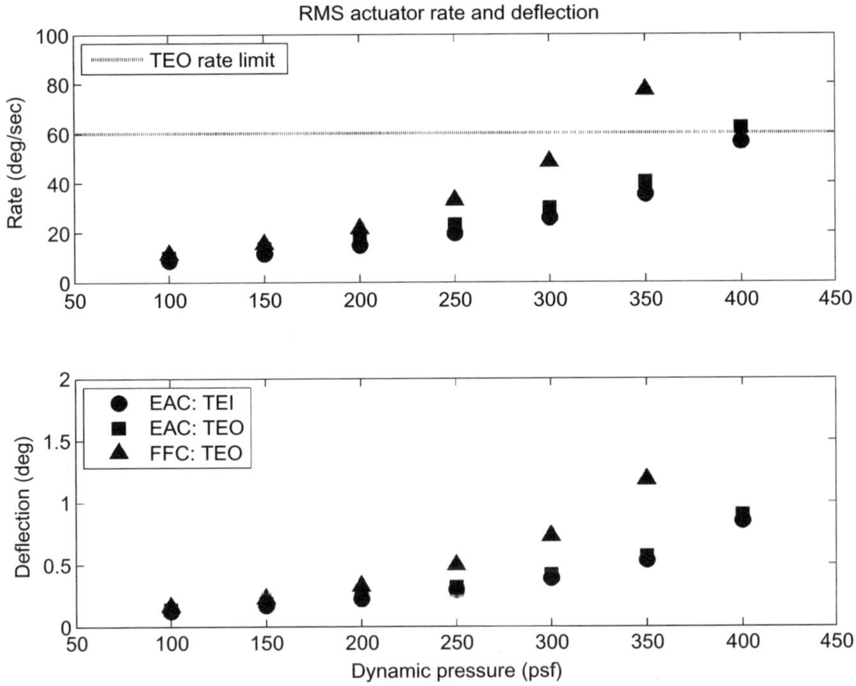

Figure 12.19 Wind tunnel turbulence response (1 ft/sec)

In summary the closed-loop system analysis has revealed that the FFC design can increase the stable flutter boundary by 29% by using the TIP sensor/TEO control surface pair. The EAC design is able to optimise the actuator activity of both the TEI and TEO surfaces while achieving a 50% increase in the stable flutter boundary. The EAC optimisation process naturally selects the TIP acceleration sensor with its high flutter mode content and high-frequency roll-off as the ideal feedback sensor. The low feedback gains from the TEO sensor indicate this preference. A clear understanding of the flutter phenomenon as it impacts feedback stabilisation has helped in designing a simple, low-complexity controller (EAC) consisting of a second-order filter and two gains.

The objective of avoiding scheduling of parameters with dynamic pressure has been achieved. The EAC controller robustly stabilises the system up to a dynamic pressure of 350 psf compared with the FFC design robust boundary of 300 psf.

12.6 Wind tunnel experiment

In this section a brief review of the results of the AFW wind tunnel tests [3,6–8] will be presented to illustrate the experimental verification of the design concepts discussed in this chapter.

12.6.1 Test objectives

The design and simulation results presented so far are based on the AFW symmetric flutter model using a heavy gas medium (Freon). The actual wind tunnel tests are conducted with air as the test medium. The symmetric flutter characteristics showed two major differences from the heavy gas case: (i) the torsion mode becomes unstable instead of the bending mode and (ii) modes 5 and 6 coalesce to form a secondary flutter mode. Figure 12.20 illustrates these characteristics. The arrows in the figure indicate increase in dynamic pressure.

With the wind tunnel model fixed-in-roll, the model is predicted to have (i) primary symmetric flutter mode at 11.2 Hz ($\bar{q} = 248\,\text{psf}$), (ii) a secondary symmetric flutter mode at 35 Hz ($\bar{q} = 350\,\text{psf}$) and (iii) an antisymmetric flutter mode at 10.9 Hz ($\bar{q} = 233\,\text{psf}$). With the wind tunnel model free-to-roll, only the symmetric flutter is predicted to occur in the operating range of the wind tunnel.

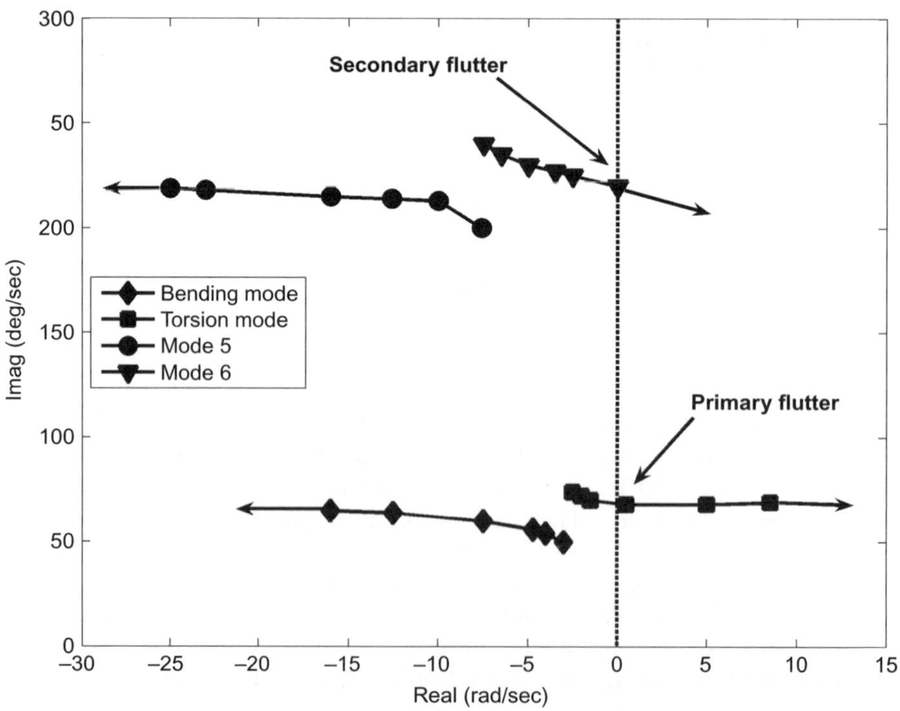

Figure 12.20 Dynamic pressure locus of AFW wind tunnel model (test medium: air, $\bar{q} = 100 - 350$ psf)

The wind tunnel multiple function test objectives [3] are:

- *Free-to-roll configuration*:
  1. Symmetric flutter suppression with no roll manoeuvre
  2. Symmetric flutter suppression in the presence of rapid roll manoeuvres
- *Fixed-in-roll configuration*:
  1. Simultaneous symmetric and antisymmetric flutter suppression

12.6.2 Controller configuration

The FFC configuration is used to suppress the symmetric flutter mode. The TEO and TEI control-gearing concept primarily used to reduce the TEO actuator demands [7] is used to suppress the antisymmetric flutter mode with better stability margins. Since the secondary flutter is predicted to occur beyond the wind tunnel operational limits, no attempt is made to modify the controller to improve the damping of this mode.

12.6.3 Wind tunnel test results

- *Free-to-roll configuration*:
  1. Symmetric flutter suppression with no roll manoeuvre: The flutter suppression was effective till the wind tunnel dynamic pressure limit was reached. This corresponded to a 23% increase in flutter boundary. Extrapolation of the experimental results indicated that the controller would have increased the flutter boundary by 40%.
  2. Symmetric flutter suppression in the presence of rapid roll manoeuvres: The rapid roll manoeuvres were tested to 275 psf (17% above open loop flutter) with the flutter suppression controller operational. The experimental data indicated that the flutter controller was operating well within its budgeted actuator rates. This successfully demonstrated the multifunction active control features of the wind tunnel test program.
- *Fixed-in-roll configuration*:
  1. Simultaneous symmetric and antisymmetric flutter suppression: The flutter dynamic pressure boundary was increased by over 24%. The testing was terminated due to preset model load limits being reached. The experimental data suggested that the controller was providing sufficient damping to the flutter modes. Extrapolation of the experimental data suggested that the controller would have increased the flutter boundary by 37%.

12.7 Multiple flutter mode suppression

In this section, design of controllers to simultaneously improve the damping of two flutter modes, as it occurred in the wind tunnel test model, is examined. For illustration purposes, the heavy gas symmetric flutter model at a dynamic pressure of 350 psf is modified to produce a secondary flutter. The modification involves shifting the mode 6 from $-16.89 \pm j244.87$ to $+0.5 \pm j244.87$ while retaining the original mode shape (eigenvectors). Examination of Figure 12.20 reveals that the FFC concept is applicable to both the primary and secondary flutter modes. Thus, an additional second-order filter suitably located near the secondary flutter mode is adequate to stabilise that mode. Indeed the controller structure of the EAC in Figure 12.9 provides for this simultaneous stabilisation of two modes. It was also noted in the design of the single-mode flutter suppression (section 12.5.2) that the role of $H_2(s)$ filter (Figure 12.9) was only secondary. The final design even eliminated this filter. However, in the two-mode flutter suppression controller (TMC), this filter will contribute to secondary flutter mode stabilisation. In the present study, the filters are fixed as in Table 12.3. The primary flutter mode filter $H_1(s)$ is retained as that in the single-mode design. Output feedback algorithm (Chapter 4) is used to assign the secondary flutter mode at $\lambda = -2.5 \pm j245$. The performance index (12.6) is optimised using the eigenvector freedom. The resulting gains are shown in Table 12.3.

Table 12.3 TMC controller parameters (Figure 12.9)

Filter	**Pole**			**Zero**		
	λ	ω	ζ	λ	ω	ζ
(a) Filter constants						
$H_1(s)$	$-7 \pm j70$	70.35	0.100	$40 \pm j75$	85.0	0.47
$H_2(s)$	$-5 \pm j220$	220.06	0.023	$-50 \pm j220$	225.6	0.22

Controller	**Gain matrix elements (deg/g)**			
	k_1	k_2	k_3	k_4
(b) Gain matrix				
TMC	−0.2448	0.02	−0.4349	0.013

The multivariable root locus of the resulting solution is shown in Figures 12.21 and 12.22. Both the flutter modes have been stabilised with minimum control effort. The absence of a minimum phase multivariable zero in the secondary flutter mode zone (the zeros are non-minimum phase in this region) facilitates assigning a higher damping to this mode. Figure 12.23 illustrates the robustness of the solution in terms of the TEI and TEO single-loop stability boundaries based on the loop transfer functions. Other performance results are presented in Table 12.4. The results indicate that robust stabilisation of two flutter modes with acceptable RMS actuator rate response to wind tunnel disturbance is achievable.

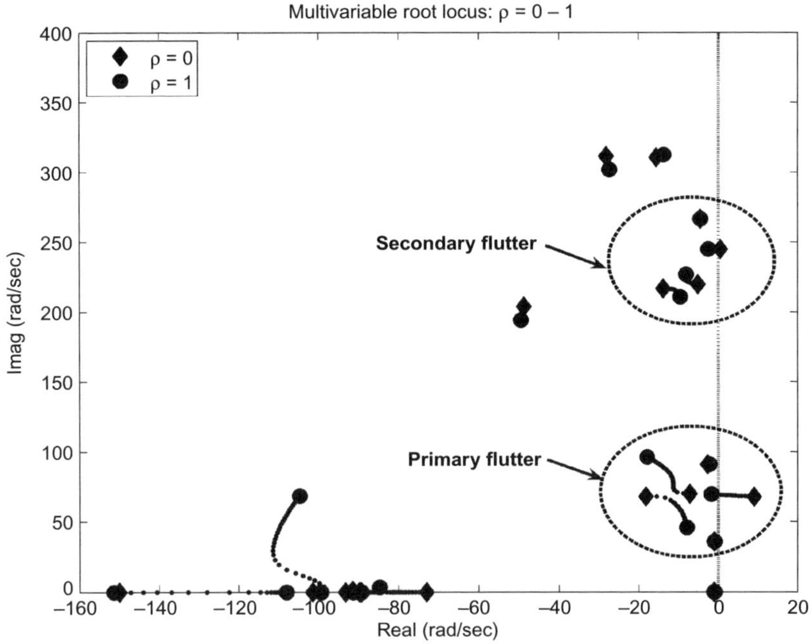

Figure 12.21 Multivariable root locus of two flutter modes stabilisation (controller: TMC)

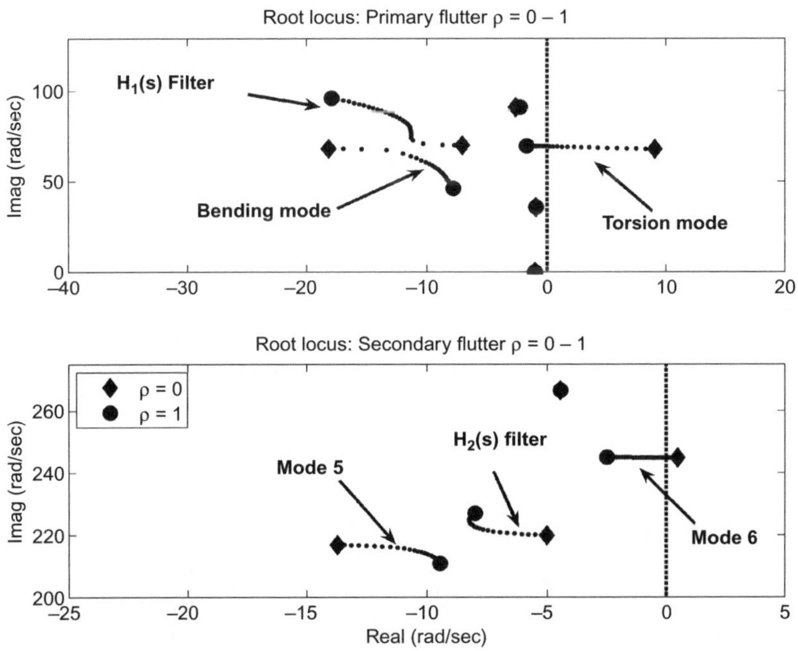

Figure 12.22 Expanded root locus of flutter mode zones

Table 12.4 Performance results

Channel	Gain margin (dB)		Phase margin (deg)		RMS actuator response*	
	Margin	Frequency (rad/sec)	Margin	Frequency (rad/sec)	δ (deg)	$\dot{\delta}$ (deg/sec)
TEI	13.66	151.8	51.66	69.39	0.444	47.51
TEO	6.19	35.9	52.05	100.67	0.679	56.14

*1 ft/sec wind tunnel turbulence.

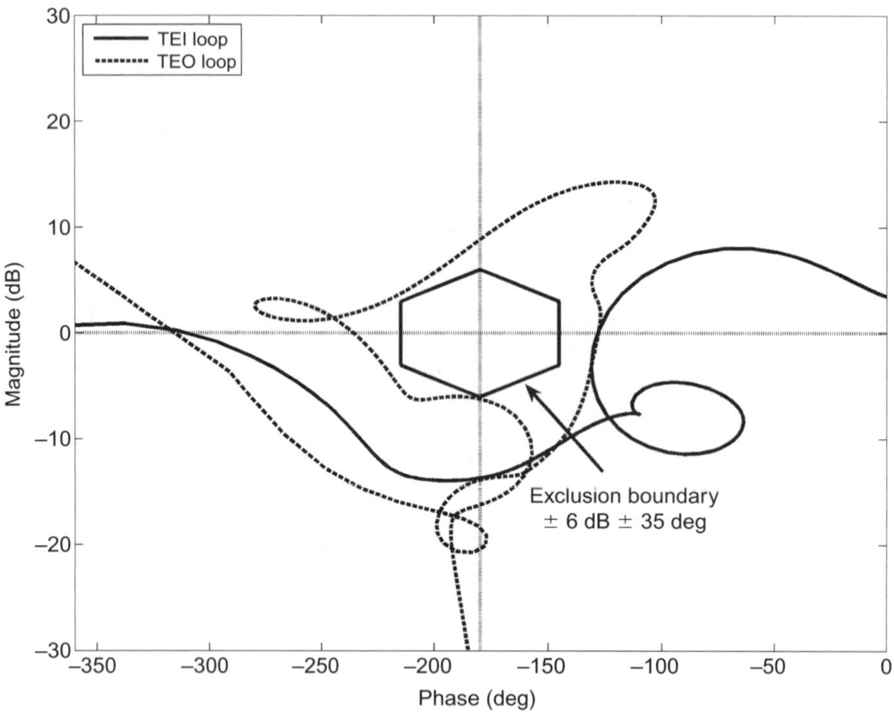

Figure 12.23 TMC controller loop transfer function (TEO and TEI channels)

12.8 Summary

In this chapter, problem of stabilising aeroelastic modes has been examined. The study of the flutter mechanism, from a feedback control point of view, reveals that associated with each pair of interacting elastic modes (poles) participating in flutter, there exists a zero called the *critical zero*. This zero determines the achievable improvement in the damping of the unstable flutter mode. For the AFW model, the critical zero is located close to the imaginary axis. Thus, using only gain feedback from the acceleration sensors to the control surfaces, it is impossible to stabilise the flutter mode with adequate damping. A second-order filter, judiciously located near the critical zero, alleviates this problem. This results in stabilising the flutter mode with low feedback gains. This design concept provides a *fundamental solution* to the flutter stabilisation problem. The study reveals that *no more than a second-order compensator* is needed to stabilise *each flutter mode*.

The novel flutter suppression control concept has been validated in a wind tunnel experiment designed to demonstrate active flutter suppression control strategies. The simple compensator-based control scheme has been successfully tested in the wind tunnel experiments using a single acceleration sensor/actuator pair. The test results demonstrated that the controller increased in excess of 24% the wind tunnel model's flutter boundary, when the testing was terminated at the wind tunnel operational boundary.

In the present chapter, the above design concept has also been extended to stabilise multiple flutter modes. The role of EAC to optimally deploy multiple control effectors for this control problem is also highlighted.

References

1. ROGER, K.L., and HODGES, G.E.: 'Active flutter suppression – a flight test demonstration', *J. Aircraft*, 1975, **12**, (6), pp. 551–556
2. EDWARDS, J.W.: 'Flight test results of an active flutter suppression system', *J. Aircraft*, 1983, **20**, (3), pp. 267–274
3. PERRY, III, B., COLE, S.R., and MILLER, G.D.: 'Special section on active flexible wing: summary of an active flexible wing program', *J. Aircraft*, 1995, **32**, (1), pp. 10–15
4. SRINATHKUMAR, S., and ADAMS, W.M.: 'Active flutter suppression using invariant zeros/eigensystem assignment', *Proceedings of the AIAA Guidance, Navigation and Control Conference*, 1989, Boston, MA, AIAA-1989-3610
5. WASZAK, M.R., and SRINATHKUMAR, S.: 'Active flutter suppression: control system design and experimental validation', *Proceedings of the AIAA Guidance, Navigation and Control Conference*, August 12–14, 1991, New Orleans, LA, AIAA-1991-2629
6. WASZAK, M.R., and SRINATHKUMAR, S.: 'Flutter suppression for the active flexible wing: control system design and experimental validation', *AIAA Dynamic Specialists Conference*, 16–17 April 1992, Dallas, TX, AIAA-1992-2097
7. WASZAK, M.R., and SRINATHKUMAR, S.: 'Design and Experimental Validation of a Flutter Suppression Controller for the Active Flexible Wing', August 1992, NASA TM 4381
8. SRINATHKUMAR, S., and WASZAK, M.R.: 'Active Flutter Suppression using "DIPOLE" filters', September 1992, NASA TM 107594
9. WASZAK, M.R., and SRINATHKUMAR, S.: 'Flutter suppression for the active flexible wing: a classical design', *J. Aircraft*, 1995, **32**, (1), pp. 61–67

Epilogue

An astute engineer,
Irrespective of the design methodology,
Finds good solutions.

Eigenstructure control, as a multivariable synthesis tool, has fascinated many researchers, including this author, for many years. Application studies, mostly in flight control, have revealed the many facets of this interesting theory. The computational simplicity and the direct relation of the synthesis parameters to the dynamic response of a system has been the driving force for the method to become attractive for feedback design. The application of the method for the design of the A-320 lateral–directional autopilot in the late 1980s and the NASA F/A-18 HARV lateral–directional control laws during mid-1990s are two notable applications culminating in successful flight tests.

However, there has been still some reluctance in the aircraft industry to infuse multivariable control techniques into their design practices. The NATO report [1] is an excellent reference document that details the complexities of flight critical control law design and examines the role multivariable control design methodologies need to play in the design of future-generation combat aircraft. In general, some of the apprehensions, perhaps justifiable, cited against the use of multivariable control methods are

1. Multivariable design methods, using generalised controller structures, do not provide good insight into the design process. Specific controller structures built on intimate knowledge of flight mechanics are essential for evolving successful flight control designs.
2. The multivariable controller structure invariably has too many design parameters, and with the design process being iterative in nature, tuning of these parameters becomes quite cumbersome and time-consuming and also results in loss of physical insight.
3. Aircraft design techniques, primarily based on classical control concepts, have been well established. What additional advantages do the 'modern' control design techniques really offer to justify their use?

The approach taken in this book to address these issues has been:

(a) Construct simple-to-use, synthesis algorithms that retain the transparency between design parameter change and consequent response deviations. The eigenstructure synthesis formulation, as developed in this book, does result in minimal set of tunable parameters required for iterative design refinement process.

(b) Build the controller structure from the simplest possible, based on flight mechanics analysis, and progressively increase the complexity as the demand for better system performance is deemed essential.

(c) Formulate suitable non-linear constrained optimisation problems to fine-tune those design parameters that cannot be easily determined by direct synthesis. This aspect of using eigenstructure synthesis algorithms, as a core part of an overall optimisation problem, has not been well emphasised in the literature, thus leading to some of the apprehensions cited earlier.

(d) Based on the above design process, demonstrate the utility of the approach by detailed design studies of aircraft and rotorcraft flight control laws to meet stringent dynamic performance requirements as defined in the appropriate handling qualities specification documents (aircraft: MIL-HDBK-1797; rotorcraft: ADS-33E-PRF).

The principal findings of the studies in this book can be summarised as follows:

1. The algorithm developments in Chapters 2–4 highlight the concept of direct eigenvector element assignment that preserves the transparency of relation between design parameters and system response, an essential characteristic alluded to earlier. In Chapter 5, the numerical ill-conditioning of the matrix of eigenvectors is shown to be an indicator of robustness of design. This leads to the definition and properties of modal robustness metrics. The optimisation of these robustness metrics forms the basis for the design of robust feedback systems. The importance of combined optimisation of both eigenvalues and available eigenvector freedom (not usually considered in application papers) to improve modal robustness is highlighted.
2. The formulation of modal canonical observer design as an eigenstructure assignment problem in Chapter 6 leads to the determination of minimal dynamic order 'functional' observers that estimate a state variable feedback control law. This results in a low-order 'dynamic compensator'–based design that replicates a benchmark state feedback solution. The functional observers also find applications in sensor fault detection and isolation schemes wherein the analytically derived sensor response from the observer is used to identify a faulty hardware sensor in a dual redundant hardware sensor set.
3. The two-degree-of-freedom controller structure consisting of forward path and feedback elements is a generic structure used in multivariable control. The novel concept of *tunable* command generator tracker proposed in Chapter 7 plays an important role in the forward path controller design.
4. Aircraft lateral–directional control law design example in Chapter 9 introduces the concept of building the controller complexity starting from the traditional roll/yaw damper structure to a more complex dynamic compensator design based on eigenstructure synthesis. This study reveals the additional

benefits that accrue as the controller design parameter set is progressively increased. The analysis of eigenvector structure properties of the Dutch roll mode enables non-interacting aileron and rudder loop response optimisation design possible. The property of the invariance of the eigenstructure assignment to the aileron to rudder interconnect gain again leads to another non-interacting design optimisation.

5. In Chapter 10, the aircraft pitch axis control problem is addressed. With only a single input available for control, it is shown that the traditional controller structures such as pitch rate command/attitude hold can be formulated as an eigenvalue assignment problem. If the aircraft is equipped with multiple control surfaces, the pitch axis control can be cast as a model following control problem. The study reveals the benefits of using multiple inputs in (i) improving the pitch axis handling qualities using an implicit model following design and (ii) design of advance control modes such as 'pitch pointing' using the explicit model following controller structure. The role of the tunable command generator tracker design in arriving at optimal design is again highlighted.
6. The rotorcraft control law design study in Chapter 11 exemplifies the need for using a generalised controller structure to meet the exacting handling qualities specifications. The rotorcraft exhibits significant inter-axis coupling, and identification of a specific controller structure, as was possible in aircraft examples of Chapters 9 and 10, especially to reduce the inter-axis cross coupling, becomes difficult. This is especially true since use of acceleration sensors as surrogate signals for flow angles, as in case of aircraft, is not feasible. This is due to the corruption of rigid body accelerations with rotorcraft vibration modes. Thus, use of the generic two-degree-of-freedom multivariable controller structure becomes inevitable. The design study in Chapter 11 reveals that a compensator-based controller using only inertial rate sensors as feedback sensors can meet the handling qualities specification and in particular the reduction of the pitch/roll cross axis coupling. The dynamic compensator consists of a fourth-order functional observer that estimates a state variable feedback design tuned to meet handling qualities design objectives. A forward path controller using the tunable command generator tracker concept is used to design a multi-axis decoupled attitude command system.
7. The flutter control problem discussed in Chapter 12 reveals that a fundamental understanding of the flutter phenomenon from a control point of view suggests that no more than a second-order compensator is needed to stabilise each flutter mode. This concept has been experimentally demonstrated in a wind tunnel test of an aero-elastic wing model. The problem of stabilising/improving the damping of multiple aero-elastic modes using eigenstructure control concepts is also highlighted.

In conclusion, in this book, an attempt has been made to present a unified suite of algorithms, based on eigenstructure control theory, for flight control law design. The use of this *design tool set* to evolve practical control laws for aircraft and rotorcraft to meet the handling qualities specifications has been demonstrated. The presentation of design results has been intentionally made extensive. The purpose behind this approach has been that a serious reader will be able to verify the intermediate design steps. Towards this end, the entire state variable model and other hardware filter assumption details, which are required to reconstruct the results, have been included in the appendices.

Reference

1. ANONYMOUS: 'Flight Control Design – Best Practices', December 2000, NATO RTO Technical Report 29

Appendix A

Relevant flight mechanics models

A.1 Aircraft lateral–directional state variable model

The dynamics of the lateral–directional of an aircraft, in literal form, and polar co-ordinates are given in the standard state variable form

$$\begin{aligned} \dot{\mathbf{x}} &= \mathbf{A}\mathbf{x} + \mathbf{B}\mathbf{u} \\ \mathbf{y} &= \mathbf{C}\mathbf{x} + \mathbf{D}\mathbf{u} \end{aligned} \tag{A.1}$$

where

$$A = \begin{bmatrix} L_p & L_r & L_\beta & 0 \\ N_p & N_r & N_\beta & 0 \\ Y_p & Y_r & Y_\beta & \dfrac{g\cos\theta_0}{V_0} \\ 1 & \tan\theta_0 & 0 & 0 \end{bmatrix}, \qquad B = \begin{bmatrix} L_{\delta_a} & L_{\delta_r} \\ N_{\delta_a} & N_{\delta_r} \\ Y_{\delta_a} & Y_{\delta_r} \\ 0 & 0 \end{bmatrix}$$

$$\mathbf{x} = [p \quad r \quad \beta \quad \varphi], \qquad \mathbf{u} = [\delta_a \quad \delta_r], \qquad \mathbf{y} = n_y$$

The state variables are p – body axis roll rate, r – body axis yaw rate, β – sideslip angle and φ – bank angle. The control variables are δ_a – aileron deflection and δ_r – rudder deflection. V_0 – aircraft trim velocity (ft/sec), θ_0 – pitch trim attitude, g – acceleration due to gravity (ft/sec^2) and n_y – lateral acceleration at centre of gravity in g-units. All angular rotation and rates are in radian units. The entries in the A- and B-matrices are the dimensional stability and control derivatives, respectively. The derivative definition includes the effect of coupling inertia I_{xz} (usually represented as primed derivatives). The literal form of the derivatives will be used while discussing their importance in shaping the dynamic response of the augmented aircraft.

To assess the aircraft departure resistance characteristics, stability metrics based on non-dimensional derivatives are used (8.2). The dimensional form, defined in (A.1), is converted to the non-dimensional form using the relations

$$L_i = \frac{\bar{q}Sb}{I_z} c_{l_i}, \quad N_i = \frac{\bar{q}Sb}{I_x} c_{n_i}, \quad i = \beta, \delta_a \tag{A.2}$$

where $\bar{q}$ is the dynamic pressure (lb/ft^2), S is the wing area (ft^2), b is the wing span (ft), I_x is the roll inertia (slug-ft^2) and I_z is the yaw inertia (slug-ft^2).

A.2 Aileron to rudder interconnect characteristics

A.2.1 Dutch roll mode contamination in roll rate response

The aileron to rudder interconnect (ARI) plays an important role in decoupling the Dutch roll mode from the roll rate response to aileron input. Using a simplified model of the aircraft model described in (A.1), this aspect is easily established. Consider a third-order simplified state variable model with states p, r and β and aileron control input (δ):

$$A = \begin{bmatrix} L_p & L_r & L_\beta \\ N_p & N_r & N_\beta \\ Y_p & -1 & Y_\beta \end{bmatrix}, \qquad B = \begin{bmatrix} L_\delta \\ N_\delta \\ 0 \end{bmatrix} \tag{A.3}$$

Assume $N_p = 0$. The derivative $Y_p = \sin \alpha$. For low angle of attack assume $Y_p = 0$. Then the approximate roll rate to aileron transfer function can be written as

$$\frac{p(s)}{\delta(s)} = \frac{L_\delta(s^2 + 2\zeta_\varphi \omega_\varphi s + \omega_\varphi^2)}{\Delta(s)} \tag{A.4}$$

where

$$\begin{aligned} &\Delta(s) = (s - L_p)(s^2 + 2\varsigma_d \omega_d s + \omega_d^2) \\ &2\varsigma_d \omega_d = -(N_r + Y_\beta), \quad \omega_d^2 = (N_r Y_\beta + N_\beta), \quad k_\delta = \left(\frac{N_\delta}{L_\delta}\right) \\ &2\varsigma_\varphi \omega_\varphi = 2\varsigma_\delta \omega_\delta + k_\delta L_r, \quad \omega_\varphi^2 = \omega_d^2 - k_\delta (L_r Y_\beta + L_\beta) \end{aligned} \tag{A.5}$$

If $\omega_\varphi = \omega_d$ and $\zeta_\varphi = \zeta_d$ in (A.4), the Dutch roll mode is fully decoupled from roll rate response. The adverse yaw ratio k_δ can be used to modify the natural frequency and damping of the zero in (A.4). This can be achieved by using the ARI in (A.1) as

$$\hat{B} = B \begin{bmatrix} 1 & 0 \\ g_x & 1 \end{bmatrix} \begin{bmatrix} \delta_a \\ \delta_r \end{bmatrix} \tag{A.6}$$

resulting in the augmented moment ratio

$$\hat{k}_{\delta_a} = \left(\frac{N_{\delta_a} + g_x N_{\delta_r}}{L_{\delta_a} + g_x L_{\delta_r}} \right) \tag{A.7}$$

where g_x is the interconnect gain. Thus, the interconnect gain can be effectively used in reducing the contamination of the Dutch roll mode in the roll rate response.

A.2.2 Reduction of sideslip response to aileron input

The adverse yaw ratio k_δ also affects the ratio of sideslip/roll rate response for a given aileron input. This is easily shown by deriving the approximate transfer function ratio of sideslip/roll rate to aileron input, assuming $N_p \neq 0$ as

$$\frac{\beta(s)}{p(s)} = \frac{-k_\delta(s+[(N_p/k_\delta)-L_p])}{s^2+2\varsigma_\varphi\omega_\varphi s+\omega_\varphi^2} \tag{A.8}$$

Thus, a judicious choice of the interconnect gain g_x can be made to reduce both Dutch roll mode contamination in roll rate response and sideslip response to aileron input.

A.2.3 Improving aircraft departure resistance

The aircraft departure resistance metrics $c_{n_\beta}\text{dyn}$ and LCDP are defined in (8.2). The LCDP parameter is dependent on the non-dimensional equivalent of the adverse yaw ratio k_δ. Using ARI (A.7), k_δ can be suitably augmented to improve the departure resistance of the aircraft.

A.3 Aircraft response to turbulence disturbance

The Dryden model of power spectral density function (PSD) for lateral atmospheric turbulence is represented as [1]

$$\varphi(\omega) = \left(\frac{\sigma^2 L}{\pi V}\right)\frac{1+3(L\omega/V)^2}{[1+(L\omega/V)^2]^2} \tag{A.9}$$

where the spatial frequency Ω (rad/ft) has been replaced by the observed frequency ω (rad/sec) using the relation $\omega = \Omega V$, V is the aircraft velocity (ft/sec), L is the turbulence scale length (ft) and σ is the RMS lateral translational velocity turbulence intensity (ft/sec).

The PSD function of (A.9) has an equivalent frequency response function $G(j\omega)$, given by

$$\varphi(\omega) = |G(s)|^2_{s=j\omega} \tag{A.10}$$

The resulting linear filter G(s) when driven by a unity PSD white noise source generates the PSD of (A.9). The filter takes the form

$$G(s) = \sigma\sqrt{\frac{L}{\pi V}}\frac{1+\sqrt{3}(L/V)s}{[1+(L/V)s]^2} \tag{A.11}$$

To construct a state variable model, (A.11) can be written in an equivalent form

$$G(s) = \sqrt{k}\,\frac{s+z}{(s+\lambda)^2} \tag{A.12}$$

where $\lambda = \frac{V}{L}$, $z = \frac{\lambda}{\sqrt{3}}$ and $k = \frac{3\sigma^2\lambda}{\pi}$. The resulting Dryden state variable model is defined as

$$\begin{aligned} \dot{x}_g &= A_g x_g + B_g \eta \\ v_g &= C_{gv} x_g \end{aligned} \tag{A.13}$$

where η is white noise input, v_g is the RMS aircraft lateral velocity gust disturbance and the matrices are given by

$$A_g = \begin{bmatrix} 0 & 1 \\ -\lambda^2 & -2\lambda \end{bmatrix}, \quad B_g = \begin{bmatrix} 0 \\ 1 \end{bmatrix}, \quad C_{gv} = \sqrt{k}\begin{bmatrix} \frac{\lambda}{\sqrt{3}} & 1 \end{bmatrix} \tag{A.14}$$

For the aircraft model in polar co-ordinates (A.1), the equivalent sideslip gust disturbance β_g is given by $\beta_g = \frac{-v_g}{V}$ [2] and the output matrix in (A.14) is modified as

$$C_g = \frac{-\sqrt{k}}{V}\begin{bmatrix} \frac{\lambda}{\sqrt{3}} & 1 \end{bmatrix} \tag{A.15}$$

Combining (A.1), (A.14) and (A.15), the composite state variable model for computing the turbulence response to a sideslip gust (β_g) is given by

$$\begin{aligned} \dot{x}_a &= F x_a + G\eta \\ y &= H x_a \end{aligned} \tag{A.16}$$

where $x_a = [x \quad x_g]^T$ is the augmented state vector and

$$F = \begin{bmatrix} A & EC_g \\ 0 & A_g \end{bmatrix}, \; G = \begin{bmatrix} 0 \\ B_g \end{bmatrix}, \; H = [C \quad 0], \; E = [L_\beta \quad N_\beta \quad Y_\beta \quad 0]^T$$

The isotropic turbulence model for the turbulence intensity (σ) as a function of altitude is given in Table A.1. The turbulence intensities at the flight conditions used in Chapter 9 are listed in Table A.2.

The RMS turbulence response for the sideslip gust β_g is easily computed for the system in (A.16), by solving for Q using the Lyapunov's equation [2]

$$FQ + QF^T = GG^T \tag{A.17}$$

The RMS responses of the state and output variables are, respectively, given by

$$x_{RMS} = \sqrt{\mathrm{Diag}(Q)}, \qquad y_{RMS} = \sqrt{\mathrm{Diag}(HQH^T)} \tag{A.18}$$

Table A.1 Isotropic turbulence model (L = 1,750 for altitude > 1,750 ft)*

Classification	Altitude range, h (kft)	RMS turbulence intensity, σ (ft/sec)
Light	2–9	σ = 5
	9–30	σ = –0.2381*h + 7.1429
Moderate	2–11	σ = 10
	11–59	σ = –0.2083*h + 12.2917
Severe	2–5	σ = 2.5*h + 10
	5–20	σ = 22.5
	20–79	σ = –0.3347*h + 29.1949
	>79	σ = 2.75

*Reference 1, Figure 7.

Table A.2 RMS turbulence intensities (altitude = 20,000 ft)

Variable	V (ft/sec)	Intensity		
		Light	Moderate	Severe
σ (ft/sec)	–	2.38	8.13	22.5
βg (deg)	695	0.111	0.378	1.047
βg (deg)	623	0.124	0.422	1.168

References

1. ANONYMOUS: 'Military Specifications – Flying Qualities of Piloted Aircraft', November 1980, MIL-8785C
2. MCLEAN, D.: *Automatic Flight Control Systems* (Prentice-Hall International, London, 1990)

Appendix B
F-8C aircraft state variable models

B.1 F-8C aircraft rigid body models

The F-8C aircraft state variable models are used for numerical examples used in the book. The linear perturbation rigid body models of the aircraft in Reference 1 are given in Cartesian (rectangular) co-ordinates. These are converted to spherical (polar) co-ordinates (convenient for stability and control analysis) as follows.

Let the translational dynamics in rectangular co-ordinates be given by

$$\dot{x}_c = A_c x_c + B_c u, \quad x_c = [u \quad v \quad w]^T \tag{B.1}$$

The transformation from rectangular to polar co-ordinates is given by [2]

$$x_c = T_p x_p, \quad x_p = [V \quad \beta \quad \alpha]^T \tag{B.2}$$

where

$$T_p = T_r E$$

$$T_r = \begin{bmatrix} \cos\alpha_0 & 0 & -\sin\alpha_0 \\ 0 & 1 & 0 \\ \sin\alpha_0 & 0 & \cos\alpha_0 \end{bmatrix} \begin{bmatrix} \cos\beta_0 & -\sin\beta_0 & 0 \\ \sin\beta_0 & \cos\beta_0 & 0 \\ 0 & 0 & 1 \end{bmatrix}, \quad E = \begin{bmatrix} 1 & 0 & 0 \\ 0 & V_0 & 0 \\ 0 & 0 & V_0 \end{bmatrix} \tag{B.3}$$

$$V = \sqrt{u^2 + v^2 + w^2}, \quad w = V\alpha, v = V\beta$$

V_0, α_0 and β_0 in (B.3) are the trim velocity, angle of attack and sideslip, respectively.

The translational dynamics in polar co-ordinates are given by

$$\dot{x}_p = T_p^{-1} A_c T_p x_p + T_p^{-1} B_c u \tag{B.4}$$

The state variables for the combined longitudinal and lateral dynamics of the aircraft in the polar co-ordinates are p – body axis roll rate, q – body axis pitch rate, r – body axis yaw rate, V – aircraft velocity, α – angle of attack, β – sideslip angle, θ – pitch angle and φ – bank angle. The control variables are δ_e – elevator deflection, δ_f – symmetric aileron deflection, δ_a – differential aileron deflection and δ_r – rudder deflection. The aircraft accelerations are a_z – vertical acceleration and n_y – lateral acceleration. The normal acceleration is defined as $n_z = -a_z$. All angular rotation and rates are in radian units, accelerations in g-units and velocity V is in ft/sec.

The general aircraft model parameters and selected flight conditions [1] are given in Table B.1. Three reference flight conditions are chosen for study in the book. Flight condition FC1 corresponds to a nominal cruise condition. FC2 represents a high angle of condition. FC3 corresponds to landing approach condition representing a low control effectiveness flight condition. The state variable models, for the longitudinal and lateral dynamics, in polar co-ordinates, for the three reference flight conditions, are listed in Tables B.2 and B.3, respectively.

Table B.1 F-8C aircraft parameters

(a) General parameters

Flight condition	Mass (slug)	I_x (slug-ft^2)	I_y (slug-ft^2)	I_z (slug-ft^2)	I_{xz} (slug-ft^2)
FC1 and FC2	639.8	0.92e + 4	0.866e + 5	0.91e + 5	0.295e + 4
FC3	590.0	0.90e + 4	0.85e + 5	0.897e + 5	0.257e + 4

(b) Flight conditions

Flight condition	Mach no.	Altitude (ft)	Dynamic pressure (lb/ft^2)	Trim velocity (ft/sec)	Angle of attack (deg)
FC1 (cruise)	0.67	20,000	305	695	3.45
FC2 (3 g-climb)	0.60	20,000	245	623	15.45
FC3 (power approach)	0.219	0.0	71.0	211	7.48

Wing area = 376 ft^2; span = 35.67 ft.

B.2 Lateral–directional design model

The dynamic feedback (DF) design in Chapter 9 uses the following feedback variables:

1. Sideslip rate estimate,

$$\dot{\beta}_e = \left(\frac{s}{s + p_w} \right) (\alpha p - r) \tag{B.5}$$

2. Sideslip estimate,

$$\beta_e = \frac{p_L}{s+p_L} n_y \tag{B.6}$$

The *design model* is constructed by augmenting the dynamic filters in (B.5) and (B.6) to the rigid body dynamics (Table B.3) as follows:

$$A_D = \left[\begin{array}{c|c|c} A & 0 & 0 \\ \hline C & -p_L & 0 \\ \hline [\alpha \ \ -1 \ \ 0 \ \ 0] & 0 & -p_W \end{array}\right], \qquad B_D = \left[\begin{array}{c} B \\ \hline D \\ \hline 0 \end{array}\right] \tag{B.7}$$

with state variables $[p \ \ r \ \ \beta \ \ \varphi \ \ x_L \ x_W]^T$ and control variables $[\delta_a \ \ \delta_r]^T$, where x_L and x_W are the lag and washout filter states with $\beta_e = p_L x_L$ and $\dot{\beta}_e = \alpha p - r - p_w x_w$.

B.3 Truth model for lateral–directional control design

The flight control system (FCS) hardware elements add phase lags associated with (i) sensors, (ii) actuators, (iii) filters to suppress structural mode response and (iv) digital computer delays. These effects have to be accounted in the assessment of the feedback system performance. Table B.4 lists these filters associated with aircraft and FCS hardware. These lag elements are augmented to the *design model* to construct the *truth model* for each control axis.

Table B.2 F-8C aircraft rigid body longitudinal state variable models

$$\dot{x} = Ax + Bu$$
$$y = Cx + Du$$
$$x = [\alpha \ \ q \ \ \theta \ \ V]^T, \qquad u = [\delta_e \ \ \delta_f]^T, \qquad y = a_z$$

(a) Flight condition: FC1 (α = 3.45 deg)

$$A = \begin{bmatrix} -1.0485 & 1 & 0 & -0.0002 \\ -6.3697 & -0.616 & 0 & 0.0001 \\ 0 & 1 & 0 & 0 \\ 14.3353 & 0.0389 & -32.1586 & -0.0119 \end{bmatrix}, \qquad B = \begin{bmatrix} -0.1565 & -0.1107 \\ -13.8 & -2.27 \\ 0 & 0 \\ -6.0183 & -1.154 \end{bmatrix}$$

$$C = [-22.6694 \ \ -0.0013 \ \ -0.0001 \ \ -0.0034], \qquad D = [-3.3844 \ \ -2.3935]$$

(b) Flight condition: FC2 (α = 15.45 deg)

$$A = \begin{bmatrix} -0.6932 & 0.9997 & 0 & -0.0006 \\ -9.4893 & -0.54 & 0 & -0.0015 \\ 0 & 1 & 0 & 0 \\ -128.0 & -0.1629 & -32.1655 & -0.1376 \end{bmatrix}, \quad B = \begin{bmatrix} -0.151 & -0.099 \\ -11.0 & -1.82 \\ 0 & 0 \\ -7.731 & 21.427 \end{bmatrix}$$

$$C = [-12.961 \quad -0.0048 \quad -0.0003 \quad -0.0115], \quad D = [-2.8237 \quad -1.8538]$$

(c) Flight condition: FC3 (α = 7.48 deg)

$$A = \begin{bmatrix} -0.4824 & 1.0 & 0 & -0.0014 \\ -0.6239 & -0.354 & 0 & -0.0003 \\ 0 & 1 & 0 & 0 \\ 21.197 & 0.0126 & -32.2075 & -0.0592 \end{bmatrix}, \quad B = \begin{bmatrix} -0.0823 & -0.0687 \\ -1.86 & -0.397 \\ 0 & 0 \\ 1.3688 & 1.8481 \end{bmatrix}$$

$$C = [-3.1406 \quad 0.0005 \quad 0.0001 \quad -0.0091], \quad D = [-0.5355 \quad -0.4472]$$

Table B.3 F-8C aircraft rigid body lateral–directional state variable models

$$\dot{x} = Ax + Bu$$
$$y = Cx + Du$$
$$x = [p \quad r \quad \beta \quad \varphi]^T, \quad u = [\delta_a \quad \delta_r]^T, \quad y = n_y$$

(a) Flight condition: FC1 (α = 3.45 deg)

$$A = \begin{bmatrix} -3.78 & 0.0406 & -51.9389 & 0 \\ -0.134 & -0.359 & 4.2344 & 0 \\ 0.0604 & -0.995 & -0.272 & 0.0462 \\ 1.0 & 0.0603 & 0 & 0 \end{bmatrix}, \quad B = \begin{bmatrix} 24.9 & 9.82 \\ 1.3 & -4.19 \\ 0.005 & 0.0502 \\ 0 & 0 \end{bmatrix}$$

$$C = [0.0049 \quad 0.0635 \quad -5.8916 \quad 0], \quad D = [0.1084 \quad 1.0872]$$

(b) Flight condition: FC2 (α = 15.45 deg)

$$A = \begin{bmatrix} -2.24 & 0.76 & -36.3034 & 0 \\ -0.0699 & -0.406 & -0.0581 & 0 \\ 0.2666 & -0.9603 & -0.283 & 0.0498 \\ 1.0 & 0.276 & 0 & 0 \end{bmatrix}, \quad B = \begin{bmatrix} 15.20 & -0.865 \\ 1.04 & -3.71 \\ 0.0002 & 0.0475 \\ 0 & 0 \end{bmatrix}$$

$$C = [0.0036 \quad 0.0685 \quad -5.4898 \quad -0.0343], \quad D = [0.0036 \quad 0.9221]$$

(c) Flight condition: FC3 ($\alpha = 7.48$ deg)

$$A = \begin{bmatrix} -1.97 & 0.959 & -16.247 & 0 \\ -0.0779 & -0.179 & 0.6794 & 0 \\ 0.1431 & -0.9858 & -0.198 & 0.1512 \\ 1.0 & 0.139 & 0 & 0 \end{bmatrix}, \quad B = \begin{bmatrix} 2.91 & 1.74 \\ 0.0263 & -0.912 \\ 0.0034 & 0.0374 \\ 0 & 0 \end{bmatrix}$$

$$C = [0.0339 \quad 0.0306 \quad -1.3015 \quad -0.0062], \quad D = [0.0226 \quad 0.246]$$

Table B.4 Filters associated with truth model

S. no.	Definition	Remarks
(a) Aircraft FCS hardware assumptions		
1	$T_1(s) = \dfrac{25}{s+25}$	Aileron primary servo (rate limit: 100 deg/sec) Deflection: TEU = 25 deg, TED = 45 deg
2	$T_2(s) = \dfrac{30}{s+30}$	Rudder primary servo (rate limit: 80 deg/sec) Deflection: ± 25 deg
3	$T_3(s) = \dfrac{\omega_n^2}{s^2 + 2\varsigma\omega_n + \omega_n^2}$ $\omega_n = 62.8$ rad/s $\varsigma = 0.7$	Secondary servo (for both channels)
(b) FCS filter assumption		
1	$T_4(s) = \dfrac{26}{s+26}$	Lumped effect of structural filters and computation delay (80-Hz sampling) Phase lag at 1 Hz = 13.6° (in each control channel)
(c) Control law filter (Chapter 9)		
1	$T_5(s) = \dfrac{2.0(s+9.5)}{s+19}$	Phase advance filter (high frequency gain = +6 dB) (in aileron control channel)

References

1. HARTMANN, G.L., HAUGE, J.A., and HENDRICK, R.C.: 'F-8C Digital CCV Flight Control Laws', February 1976, NASA CR-2629
2. MAINE, R.E., and ILIFF, K.W.: 'Application of Parameter Estimation to Aircraft Stability and Control', June 1986, NASA Reference Publication 1168

Appendix C
BO-105 helicopter state variable models

C.1 Design model

The BO-105 helicopter has been chosen for the study since its dynamics have been well characterised from flight data [1]. For a forward speed of 80 knots (41.12 m/sec), flight data have been generated exclusively for system identification purposes. Six different working groups have used these flight data to estimate the stability and control derivatives of the helicopter. Each model is different due to (i) different model structure employed, (ii) number of parameters identified and (iii) parameter estimation methodologies employed. For the design studies in Chapter 11, two candidate models identified by Aero Flight Dynamics Directorate (AFD, United States) and the German Aerospace Research Establishment (DLR, Germany) are used.

Assuming small perturbations, the translational and angular accelerations are given by [1]

$$
\begin{aligned}
\dot{u} &= \Delta a_x - g\theta - w_0 q + v_0 r \\
\dot{v} &= \Delta a_y + g\varphi - u_0 r + w_0 p \\
\dot{w} &= \Delta a_z - v_0 p + u_0 q \\
\dot{p} &= \Delta L \\
\dot{q} &= \Delta M \\
\dot{r} &= \Delta N
\end{aligned}
\tag{C.1}
$$

where p, q and r are the roll, pitch and yaw rates. θ and φ are the pitch and bank angles. Δa_x, Δa_y and Δa_z are the specific aerodynamic forces and ΔL, ΔM and ΔN are the specific aerodynamic moments. u_0, v_0 and w_0 are the longitudinal, lateral and vertical trim velocities. The aerodynamic forces and moments are expanded in terms of translational velocities (u, v and w), angular rates (p, q and r) and control deflections δ_c (collective), δ_a (longitudinal cyclic), δ_e (lateral cyclic) and δ_p (pedal). This results in an eighth-order rigid body state variable 'design' model.

C.2 Rotor model

The design model, postulated above, accounts only for the rigid body dynamics of the helicopter. However, for a helicopter, the rotor/fuselage coupling is significant and thus cannot be neglected especially for high-gain feedback systems. These high-frequency modes originate from the coupling of the body – rotor flapping dynamics and a regressive lead – lag dynamics coupling into the roll response [2]. For the BO-105 helicopter, rotor mode coupling models have been identified using flight data [2,3] to generate high-bandwidth models needed for control system design. However, for the studies in Chapter 11, a rotor coupling dynamics using the model structure proposed in Reference 4 is used. For the lateral flapping motion, the first-order approximation is

$$\begin{bmatrix} \dot{p} \\ \dot{\beta}_e \end{bmatrix} = \begin{bmatrix} 0 & \frac{-1}{\tau^* \tau_p} \\ 1 & \frac{-1}{\tau} \end{bmatrix} \begin{bmatrix} p \\ \beta_e \end{bmatrix} + \begin{bmatrix} 0 \\ \frac{1}{\tau} \end{bmatrix} \delta_e \tag{C.2}$$

$$\tau = \frac{[\gamma^2/16] + 16}{\gamma \Omega}, \qquad \tau_p = -\frac{1}{L_p}$$

where p is roll rate, δ_e is lateral cyclic control, β_e is the lateral flapping angle, γ is the lock number and Ω is main rotor angular velocity. For BO-105, $\gamma = 7.6$ and $\Omega =$ rad/sec [3]. L_p is the roll damping derivative. For the longitudinal flapping motion, the rotor model is exactly as in (C.2) with q (pitch rate), β_a (longitudinal flapping angle) and δ_a (longitudinal cyclic control) replacing p, β_e and δ_e in

Table C.1 Rotor model time constants

Model	τ^*	τ_p	τ_q
AFD	0.0581	0.1139	0.2226
DLR	0.0581	0.1176	0.2860

*Units – seconds.

(C.2), respectively. Further $\tau_q = -\frac{1}{M_q}$, where M_q is the pitch damping derivative. With the addition of the above rotor coupling models, a tenth-order state variable model results. Table C.1 lists the rotor time constants for AFD and DLR models.

C.3 Scaling of state and control variables

The aerodynamic data in Reference 1 are specified in radian units for angular rates and percent travel for control deflections. A uniform scaling of degree units for all angular deflections is employed for generating the state variable models used in this book. Thus, the control deflections have to be converted to degree units. For example, to determine the percent to radian scale factor for δ_c (collective control), the steady-state pitch rate response of a first-order pitch dynamics of the form

$$\dot{q} = M_q q + M_\delta \delta_c, \qquad q_{ss} = -M_q^{-1} M_\delta \tag{C.3}$$

is matched between the AFD model and the BO-105 model (given in radian units) in Reference 4. Similarly the steady-state yaw rate response model is used to derive the δ_p (pedal) scale factor. The scale factor for δ_a (δ_e) is derived by matching the steady-state response q_{ss} (p_{ss}) of a first-order pitch (roll) dynamics with the corresponding second-order rotor model given in (C.2). This results in a full-scale (100%) deflection of 20 deg for δ_c, 7 deg for δ_a, 7 deg for δ_e and 20 deg for δ_p.

Table C.2 FCS hardware assumption (in each control channel)

S. no.	Definition	Remarks
1	$T_1(s) = \dfrac{25}{s+25}$	Primary servo
2	$T_2(s) = \dfrac{26}{s+26}$	Lumped effect of structural filters and computation delay (80-Hz sampling) Phase lag at 1 Hz = 13.6 deg

C.4 Truth model

The flight control system (FCS) hardware elements add phase lags associated with (i) sensors, (ii) actuators, (iii) filters to suppress structural mode response and (iv) digital computer delays. These effects have to be accounted in the design of the feedback system. Table C.2 lists the filters associated with FCS hardware.

The FCS dynamics of Table C.2 is augmented in each control channel of the rotor model to derive an 18th-order truth model. The truth model is used in assessment of stability and performance of the control system design discussed in Chapter 11.

Table C.3 AFD model

A: Matrix

−8.7790	3.1820	0.9910	9.6257	−3.4950	−11.8602	0	0
−0.9980	−4.4930	−0.4380	5.5004	0	2.8648	0	0
−0.4660	5.4320	−1.0700	−6.8182	0	4.6983	0	0
−0.0066	0.8205	0	−1.1870	0.2457	0	0	0
0.0132	0.1317	0.0524	−0.0610	−0.0385	0	−0.1712	0
−0.1227	0.0842	−0.6816	−0.0830	0	−0.2210	0	0.1712
0	1.0000	0	0	0	0	0	0
1.0000	0	0	0	0	0	0	0

B: Matrix

16.6158	59.7513	146.5135	–7.7349
20.9130	80.2141	0	3.7242
–14.6104	–61.3883	27.0109	16.3293
–1.9400	–1.4714	0	0
–0.2300	–1.0286	0	0
–0.1600	0	0.9429	0.0750
0	0	0	0
0	0	0	0

Table C.4 DLR model

A: Matrix

−8.5010	3.0370	0.4100	6.6463	−4.6410	−15.5272	0	0
−0.4190	−3.4960	−0.1170	3.0367	1.6616	2.7502	0	0
−1.0570	0.8090	−0.8580	1.9481	0	6.7036	0	0
−0.0524	0.7856	0	−0.9980	0.0140	0	0	0
0	0.0873	0.0524	0.0360	−0.0590	0	−0.1712	0
−0.0873	0	−0.6749	0	0	−0.1700	0	0.1712
0	1.0000	0	0	0	0	0	0
1.0000	0	0	0	0	0	0	0

B: Matrix			
9.1673	19.6443	151.4246	−8.0214
16.3293	76.1215	−7.3666	0
0	0	21.2813	14.0375
−1.7450	−4.3286	0	0
0	−0.4000	0	0
0	0	0.0429	−0.0550
0	0	0	0
0	0	0	0

The rigid body state variable models are defined with the following state variable and control variable ordering:

1. State variable ordering: $x = [p, q, r, w, u, v, \theta, \varphi]$.
2. Control variable ordering: $u = [\delta_c, \delta_a, \delta_e, \delta_p]$.

The rigid body models are listed in Tables C.3 and C.4 for AFD and DLR models, respectively.

References

1. ANONYMOUS: 'Rotorcraft System Identification', September 1991, AGARD Advisory Report 280
2. TISCHLER, M.B., and CAUFFMAN, M.G.: 'Frequency response method for rotorcraft system identification with applications to the BO-105 Helicopter', *J. Am. Helicopter Soc.*, 1992, **37**, (3), pp. 99–136
3. KALETKA, J.: 'Identification of mathematical derivative models for the design of a model following control system', *Vertica*, 1989, **13**, (3), pp. 213–228
4. PADFIELD, G.D.: *Helicopter Flight Dynamics: The Theory and Application of Flying Qualities and Simulation Modeling*. AIAA Education Series (Blackwell, Washington, 1996)

Appendix D
Properties of singular matrix pencils

D.1 Introduction

Consider a linear time-invariant state space system

$$\begin{aligned} \dot{\mathbf{x}} &= A\mathbf{x} + B\mathbf{u} \\ \mathbf{y} &= C\mathbf{x} + D\mathbf{u} \end{aligned} \tag{D.1}$$

where $\mathbf{x} \in R^n$, $\mathbf{u} \in R^m$ and $\mathbf{y} \in R^r$. The matrix pencil of the system is given by $(S - \lambda E)$, where

$$S = \left[\begin{array}{c:c} A & B \\ \hdashline C & D \end{array}\right], \quad E = \left[\begin{array}{c:c} \lambda I_n & 0 \\ \hdashline 0 & 0 \end{array}\right] \tag{D.2}$$

The system matrix pencil plays an important role in characterising many control theoretical properties of multivariable state space system [1,2]. In particular, the characterisation of multivariable zeros is of interest in the design of observers (Chapter 6). A brief definition of multivariable zeros [2] is included here for reference.

The *smith zeros* are commonly called the *invariant zeros* of the system. In (D.2) when $r = 0$, they are called the *input decoupling zeros* of the system. When $m = 0$, they are called the *output decoupling zeros* of the system. When the system is minimal, the *smith zeros* are called the *transmission zeros*. The *smith zeros* are points in the complex plane where the rank of the pencil $S(\lambda)$ drops below its normal rank $(n + \min(m, r))$. The Kronecker canonical structure, to be discussed in this appendix, reveals the zeros of the system as the finite eigenvalues of the matrix pencil.

The system pencil in (D.2) is a special form of the general matrix pencil $A - \lambda B$, where the matrices are $m \times n$. When $m = n$, the matrix pencil is called *regular*. The eigenstructure of the matrix pencil is given by

$$(A - \lambda B)v = 0 \tag{D.3}$$

where the scalar λ is called the generalised eigenvalue and v is called the generalised eigenvector of the matrix pencil. A regular matrix pencil can have both finite and infinite generalised eigenvalues. When A and B are rectangular or when $\det(A - \lambda B) \equiv 0$ (for all λ), the matrix pencil is called *singular*.

D.2 Kronecker canonical form of a matrix pencil (KCF)

Kagstrom [3] provides a comprehensive review of properties of matrix pencils. Just as the Jordan canonical form describes the eigenvalues and invariant subspaces of a square matrix A, there is a Kronecker canonical form (KCF) that describes the generalised eigenvalues and generalised eigenspaces associated with a regular matrix pencil $A - \lambda B$. If the matrix pencil is singular, the KCF has four types of diagonal structural blocks. These blocks have the generic form [4] as shown below:

$$J_j(\alpha) = \begin{bmatrix} \alpha & 1 & & \\ & \ddots & \ddots & \\ & & \ddots & 1 \\ & & & \alpha \end{bmatrix} - \lambda \begin{bmatrix} 1 & 0 & & \\ & \ddots & \ddots & \\ & & \ddots & 0 \\ & & & 1 \end{bmatrix}, \quad N_j = \begin{bmatrix} 1 & 0 & & \\ & \ddots & \ddots & \\ & & \ddots & 0 \\ & & & 1 \end{bmatrix} - \lambda \begin{bmatrix} 0 & 1 & & \\ & \ddots & \ddots & \\ & & \ddots & 1 \\ & & & 0 \end{bmatrix}$$

$$L_j = \begin{bmatrix} 0 & 1 & & \\ & \ddots & \ddots & \\ & & 0 & 1 \end{bmatrix} - \lambda \begin{bmatrix} 1 & 0 & & \\ & \ddots & \ddots & \\ & & 1 & 0 \end{bmatrix}, \quad L_j^T = \begin{bmatrix} 0 & & \\ 1 & \ddots & \\ & \ddots & 0 \\ & & 1 \end{bmatrix} - \lambda \begin{bmatrix} 1 & & \\ 0 & \ddots & \\ & \ddots & 1 \\ & & 0 \end{bmatrix} \tag{D.4}$$

where $J_j(\alpha)$ is a $j \times j$ Jordan block corresponding to finite eigenvalue α, N_j is a $j \times j$ block corresponding to infinite eigenvalues, L_j is a $j \times (j + 1)$ singular block of right (or column) minimal indices and L_j^T is a $(j + 1) \times j$ singular block of left (or row) minimal indices.

An $m \times n$ singular rectangular matrix pencil is called *generic* when for $m < n$ it has only L blocks and when $m > n$ it has only L^T blocks in its KCF. Thus, *generic* matrix pencils have no regular part. A matrix pencil is called *degenerate* if it has other blocks as well. Construction of the KCF is recognised to be a numerically ill-conditioned problem [1,3]. Hence, only a staircase canonical form (SCF) that reveals all the structural properties of KCF are computed. Many numerically stable algorithms have been proposed to compute the SCF [1,3,4]. The algorithm described by Van Dooren [1] is used to compute the SCF for the examples in Chapter 6. A brief outline of this algorithm is given in the next section.

D.3 Staircase canonical form of a matrix pencil (SCF)

The Van Dooren algorithm [1] consists of the following steps:

Step 1 (Algorithm 4.1 [1]). Extract the structural elements: (i) Kronecker column indices and (ii) infinite elementary divisors. This reduces the matrix pencil $\lambda B - A$ to the following form:

$$\tilde{P}(\lambda B - A)\tilde{Q} = \left[\begin{array}{c|c} \lambda\bar{B} - \bar{A} & 0 \\ \hline X & \lambda\hat{B} - \hat{A} \end{array}\right] \tag{D.5}$$

where $\tilde{P}$ and $\tilde{Q}$ are unitary matrices and X is a non-zero matrix pencil.

Note that the block $\lambda\bar{B} - \bar{A}$ in (D.5) contains Kronecker row indices and finite elementary divisors, and the block $\lambda\hat{B} - \hat{A}$in (D.5) contains Kronecker column indices and infinite elementary divisors.

Step 2 (Algorithm 4.5 [1]). If Algorithm 4.5 is applied to the pencil $\lambda\bar{B} - \bar{A}$ in (D.5), it separates the structural elements: (i) Kronecker row indices and (ii) finite elementary divisors. This reduces the matrix pencil $\lambda\bar{B} - \bar{A}$ to the following form:

$$\bar{P}(\lambda\bar{B} - \bar{A})\bar{Q} = \left[\begin{array}{c|c} \lambda B_r - A_r & 0 \\ \hline X & \lambda B_f - A_f \end{array}\right] \tag{D.6}$$

where $\bar{P}$ and $\bar{Q}$ are unitary matrices and X is a non-zero matrix pencil. The matrix pencil $\lambda B_r - A_r$ contains the Kronecker row indices (L^T blocks), and the matrix pencil $(\lambda B_f - A_f)$ contains the finite elementary divisors. The eigenvalues of $(\lambda B_f - A_f)$ are the finite generalised eigenvalues of the original matrix pencil $(\lambda B - A)$. The resulting SCF of the original matrix pencil using (D.5) and (D.6) has the form

$$P(\lambda B - A)Q = \left[\begin{array}{c|c|c} \lambda B_r - A_r & 0 & 0 \\ \hline X_r & \lambda B_f - A_f & 0 \\ \hline X_1 & X_2 & \lambda\hat{B} - \hat{A} \end{array}\right] \tag{D.7}$$

The matrix pencils marked as X_r, X_1 and X_2 in (D.7) are non-zero matrix pencils. Finally it should be noted that the finite and infinite elementary divisor blocks constitute the *regular* part of the matrix pencil. The presence of L^T and J blocks of the matrix pencil plays a key role in the existence of solution of functional observers discussed in Chapter 6.

D.4 Generic matrix pencil

As indicated earlier, for a generic matrix pencil the Kronecker indices are solely dependent on the size of the pencil ($m \times n$). The pencil has only L ($m < n$) or L^T ($m > n$) blocks. For the case $m < n$, the structure of the L blocks has the form [3]

$$\mathrm{Diag}(L_e,\ldots,L_e,L_{e+1},\ldots,L_{e+1}) \tag{D.8}$$

For the case $m > n$, a structure similar to (D.8) with L^T blocks is constructed. It should be noted that there are only two structural indices (L_e and L_{e+1}). In Reference 3 an algorithm is given to compute the structure (D.8) for a given matrix pencil size. As an example a (3×5) matrix pencil has an index $e = 1$ with blocks $L_1 \oplus L_2$. The symbol $\oplus$ indicates the block structure summation notation [4].

While using the SCF algorithm discussed in section D.3, care should be taken to determine if the pencil is generic while using the information provided by the structural indices. In Reference 1, an analysis of the special nature of the staircase form for a generic matrix pencil is given. The following example illustrates the subtlety that is involved.

Consider a (3×5) random matrix pencil. Applying Algorithm 4.1 [1] to this pencil results in the following structural indices:

$$\begin{bmatrix} s \\ r \end{bmatrix} = \left[\begin{array}{c:c} 2 & 2 \\ 2 & 1 \end{array}\right], \quad e = \left[\begin{array}{c:c} L_0 & L_1 \\ \hline 0 & 1 \end{array}\right], \quad d = \left[\begin{array}{c:c} N_1 & N_2 \\ \hline 0 & 1 \end{array}\right] \tag{D.9}$$

From (D.9), one can wrongly conclude that the KCF structure is $L_1 \oplus N_2$. A recheck on the size of the matrix pencil, based on this KCF structure, yields the size of the matrix pencil as 3×4, which is incorrect. However, applying the generic matrix pencil algorithm [3] based on the pencil size results in a KCF structure as $L_1 \oplus L_2$, yielding the correct matrix pencil size as 3×5. Thus, re-deriving the size of the matrix pencil from the KCF structural indices computed by the Van Dooren SCF algorithm provides a simple method to determine if the given matrix pencil is generic. However, if it is determined that the matrix pencil is indeed generic, algorithm based on the size of the matrix pencil [3] is a simple way to determine the generic matrix pencil structure given by (D.8).

D.5 Example

The example in this section illustrates the construction of the SCF using the Van Dooren algorithm [1].

Example D.1: This example is taken from example 11 of Reference 2. The state space system matrices (D.1) are given by

$$A = \begin{bmatrix} -2 & -6 & 3 & -7 & 6 \\ 0 & -5 & 4 & -4 & 8 \\ 0 & 2 & 0 & 2 & -2 \\ 0 & 6 & -3 & 5 & -6 \\ 0 & -2 & 2 & -2 & 5 \end{bmatrix}, \quad B = \begin{bmatrix} -2 & 7 \\ -8 & -5 \\ -3 & 0 \\ 1 & 5 \\ -8 & 0 \end{bmatrix}$$
$$C = \begin{bmatrix} 0 & -1 & 2 & -1 & -1 \\ 1 & 1 & 1 & 0 & -1 \\ 0 & 3 & -2 & 3 & -1 \end{bmatrix}, \quad D = \begin{bmatrix} 0 & 0 \\ 0 & 0 \\ 0 & 0 \end{bmatrix} \tag{D.10}$$

The SCF is computed for the 8 × 7 matrix pencil $S(\lambda)$ given by (D.2).

Step 1. Applying Algorithm 4.1 [1] results in the following structural indices:

$$\begin{bmatrix} s \\ r \end{bmatrix} = \begin{bmatrix} 2 & 2 \\ 2 & 2 \end{bmatrix}, \quad e = \begin{bmatrix} L_0 & L_1 \\ \hline 0 & 0 \end{bmatrix}, \quad d = \begin{bmatrix} N_1 & N_2 \\ \hline 0 & 2 \end{bmatrix} \tag{D.11}$$

Step 2. Applying Algorithm 4.5 [1] to the matrix pencil $(\lambda\bar{B} - \bar{A})$ of (D.5) that has L^T and J blocks results in the following structural indices:

$$\begin{bmatrix} \hat{s} \\ \hat{r} \end{bmatrix} = \begin{bmatrix} 1 & 1 \\ 1 & 0 \end{bmatrix}, \quad \hat{e} = \begin{bmatrix} L_0^T & L_1^T \\ \hline 0 & 1 \end{bmatrix}, \quad \hat{d} = \begin{bmatrix} N_1 & N_2 \\ \hline 0 & 0 \end{bmatrix} \tag{D.12}$$

This results in the block L_1^T. The residual matrix pencil $\lambda B_f - A_f$ representing the J block is given by

$$A_f = \begin{bmatrix} -0.3239 & 0.6583 \\ 3.1160 & -0.5411 \end{bmatrix}, \quad B_f = \begin{bmatrix} 0.7080 & -0.2068 \\ 0.0594 & 0.2035 \end{bmatrix} \tag{D.13}$$

The generalised eigenvalues of the finite block in (D.13) are $\lambda_1 = -3$ and $\lambda_2 = 4$. Thus, the original matrix pencil has Kronecker indices $L_1^T \oplus J_1(-3) \oplus J_1(4) \oplus 2N_2$. The SCF of the matrix pencil, on the application of Algorithms 4.1 and 4.5 [1] as described above, has the form (D.7)

$$
PAQ = \left[\begin{array}{c|cc|cccc}
2.8716 & 0 & 0 & 0 & 0 & 0 & 0 \\
2.0135 & 0 & 0 & 0 & 0 & 0 & 0 \\
\hline
x & -0.3239 & 0.6583 & 0 & 0 & 0 & 0 \\
x & 3.1160 & -0.5411 & 0 & 0 & 0 & 0 \\
\hline
x & x & x & -0.5207 & 1.7852 & 0 & 0 \\
x & x & x & 0.6585 & 0.1921 & 0 & 0 \\
x & x & x & 2.7236 & -3.6756 & 11.1445 & 5.8332 \\
x & x & x & 1.2622 & 1.3793 & -4.2191 & 8.0606
\end{array}\right]
\begin{array}{c}
L_1^T \\ \\ \hline J(-3,4) \\ \\ \hline \\ 2N_2 \\ \\ \\ \hline
\end{array}
$$

$$
PBQ = \left[\begin{array}{c|cc|cccc}
0 & 0 & 0 & 0 & 0 & 0 & 0 \\
0.9991 & 0 & 0 & 0 & 0 & 0 & 0 \\
\hline
x & 0.7080 & -0.2068 & 0 & 0 & 0 & 0 \\
x & 0.0594 & 0.2035 & 0 & 0 & 0 & 0 \\
\hline
x & x & x & 0 & 0 & 0 & 0 \\
x & x & x & 0 & 0. & 0 & 0 \\
0 & 0 & 0 & -0.1479 & -0.9890 & 0 & 0 \\
0 & 0 & 0 & -0.9890 & 0.1479 & 0 & 0
\end{array}\right]
\begin{array}{c}
L_1^T \\ \\ \hline J(-3,4) \\ \\ \hline \\ 2N_2 \\ \\ \\ \hline
\end{array}
\qquad (D.14)
$$

The elements in (D.14) marked as 'x' are non-zero entries. The 4×4 matrix pencil, $2N_2$ given in (D.14), yields four infinite eigenvalues of the pencil.

References

1. VAN DOOREN, P.: 'The computation of Kronecker's canonical form of a singular pencil', *Lin. Algebra Appl.*, 1979, **27**, pp. 103–141
2. EMAMI-NAEINI, A., and VAN DOOREN, P.: 'Computation of zeros of linear multivariable systems', *Automatica*, 1982, **18**, (4), pp. 415–430
3. KAGSTROM, B.: 'Singular Matrix Pencils (Section 8.7)' In BAI, Z., DEMMEL, J., DONGARRA, J., RUHE, A., and VAN DER VORST, H. (eds.). *Templates for the Solution of Algebraic Eigenvalue Problems: A Practical Guide* (SIAM, Philadelphia, 2000)
4. JOHANSSON, S.: 'Stratification of matrix pencils in systems and control: theory and algorithms', Licentiate Thesis, Department of Computer Science, Umea University, Sweden, May 2005

Appendix E

Conversion of FPS units to SI units

The aircraft parameters and flight condition data used in the application chapters 9–12 have been given in FPS units. This has been done in order to preserve the original authenticated data of the referenced sources. In order to facilitate easy conversion of these data to SI units, the appropriate conversion factors from FPS to SI units are given in Table E.1.

Table E.1 Constants for converting FPS units to SI units

Variable	FPS units	SI units	Multiplying factor of unit FPS to SI
Altitude	ft	m	0.30480
Area	ft^2	m^2	0.09290
Force	lbf	N (Newton)	4.44822
Mass	slug	kg	14.59428
Inertia	slug-ft^2	kg-m^2	1.35582
Dynamic pressure	lbf/ft^2 (psf)	N/m^2 (Pascal)	47.88026

Bibliography

1. ETKIN, B.: *Dynamics of Atmospheric Flight* (John Wiley & Sons, New York, 1972)
2. MCRUER, D.T., ASHKENAS, I.L., and GRAHAM, D.C.: *Aircraft Dynamics and Automatic Control* (Princeton University Press, Princeton, NJ, 1975)
3. KAILATH, T.: *Linear Systems* (Prentice-Hall, Englewood Cliffs, NJ, 1980)
4. MCLEAN, D.: *Automatic Flight Control Systems* (Prentice-Hall International, London, 1990)
5. PADFIELD, G.D.: *Helicopter Fight Dynamics: The Theory and Application of Flying Qualities and Simulation Modeling*. AIAA Education Series (Blackwell Washington, 1996)
6. HODGKINSON, J.: *Aircraft Handling Qualities*. AIAA Education Series (Blackwell, Washington, 1999)
7. STEVENS, B.L., and LEWIS, F.L.: *Aircraft Control and Simulation*, 2nd edn (John Wiley & Sons, New Jersey, 2003)

Index

Page numbers followed by "*f*" indicate figure; and those followed by "*t*" indicate table.